Innovative Processing Technologies for Foods with Bioactive Compounds

Contemporary Food Engineering

Series Editor

Professor Da-Wen Sun, Director
Food Refrigeration & Computerized Food Technology
National University of Ireland, Dublin
(University College Dublin)
Dublin, Ireland
http://www.ucd.ie/sun/

Advances in Heat Transfer Unit Operations: Baking and Freezing in Bread Making, *edited by Georgina Calderon-Dominguez, Gustavo F. Gutierrez-Lopez, and Keshavan Niranjan* (2016)

Innovative Processing Technologies for Foods with Bioactive Compounds, *edited by Jorge J. Moreno* (2016)

Edible Food Packaging: Materials and Processing Technologies, *edited by Miquel Angelo Parente Ribeiro Cerqueira, Ricardo Nuno Correia Pereira, Oscar Leandro da Silva Ramos, Jose Antonio Couto Teixeira, and Antonio Augusto Vicente* (2016)

Handbook of Food Processing: Food Preservation, *edited by Theodoros Varzakas and Constantina Tzia* (2015)

Handbook of Food Processing: Food Safety, Quality, and Manufacturing Processes, *edited by Theodoros Varzakas and Constantina Tzia* (2015)

Edible Food Packaging: Materials and Processing Technologies, *edited by Miquel Angelo Parente Ribeiro Cerqueira, Ricardo Nuno Correia Pereira, Oscar Leandro da Silva Ramos, Jose Antonio Couto Teixeira, and Antonio Augusto Vicente* (2015)

Advances in Postharvest Fruit and Vegetable Technology, *edited by Ron B.H. Wills and John Golding* (2015)

Engineering Aspects of Food Emulsification and Homogenization, *edited by Marilyn Rayner and Petr Dejmek* (2015)

Handbook of Food Processing and Engineering, Volume II: Food Process Engineering, *edited by Theodoros Varzakas and Constantina Tzia* (2014)

Handbook of Food Processing and Engineering, Volume I: Food Engineering Fundamentals, *edited by Theodoros Varzakas and Constantina Tzia* (2014)

Juice Processing: Quality, Safety and Value-Added Opportunities, *edited by Víctor Falguera and Albert Ibarz* (2014)

Engineering Aspects of Food Biotechnology, *edited by José A. Teixeira and António A. Vicente* (2013)

Engineering Aspects of Cereal and Cereal-Based Products, *edited by Raquel de Pinho Ferreira Guiné and Paula Maria dos Reis Correia* (2013)

Fermentation Processes Engineering in the Food Industry, *edited by Carlos Ricardo Soccol, Ashok Pandey, and Christian Larroche* (2013)

Modified Atmosphere and Active Packaging Technologies, *edited by Ioannis Arvanitoyannis* (2012)

Advances in Fruit Processing Technologies, *edited by Sueli Rodrigues and Fabiano Andre Narciso Fernandes* (2012)

Biopolymer Engineering in Food Processing, *edited by Vânia Regina Nicoletti Telis* (2012)

Operations in Food Refrigeration, *edited by Rodolfo H. Mascheroni* (2012)

Thermal Food Processing: New Technologies and Quality Issues, Second Edition, *edited by Da-Wen Sun* (2012)

Physical Properties of Foods: Novel Measurement Techniques and Applications, *edited by Ignacio Arana* (2012)

Handbook of Frozen Food Processing and Packaging, Second Edition, *edited by Da-Wen Sun* (2011)

Advances in Food Extrusion Technology, *edited by Medeni Maskan and Aylin Altan* (2011)

Enhancing Extraction Processes in the Food Industry, *edited by Nikolai Lebovka, Eugene Vorobiev, and Farid Chemat* (2011)

Emerging Technologies for Food Quality and Food Safety Evaluation, *edited by Yong-Jin Cho and Sukwon Kang* (2011)

Food Process Engineering Operations, *edited by George D. Saravacos and Zacharias B. Maroulis* (2011)

Biosensors in Food Processing, Safety, and Quality Control, *edited by Mehmet Mutlu* (2011)

Physicochemical Aspects of Food Engineering and Processing, *edited by Sakamon Devahastin* (2010)

Infrared Heating for Food and Agricultural Processing, *edited by Zhongli Pan and Griffiths Gregory Atungulu* (2010)

Mathematical Modeling of Food Processing, *edited by Mohammed M. Farid* (2009)

Engineering Aspects of Milk and Dairy Products, *edited by Jane Sélia dos Reis Coimbra and José A. Teixeira* (2009)

Innovation in Food Engineering: New Techniques and Products, *edited by Maria Laura Passos and Claudio P. Ribeiro* (2009)

Processing Effects on Safety and Quality of Foods, *edited by Enrique Ortega-Rivas* (2009)

Engineering Aspects of Thermal Food Processing, *edited by Ricardo Simpson* (2009)

Ultraviolet Light in Food Technology: Principles and Applications, *Tatiana N. Koutchma, Larry J. Forney, and Carmen I. Moraru* (2009)

Advances in Deep-Fat Frying of Foods, *edited by Serpil Sahin and Servet Gülüm Sumnu* (2009)

Extracting Bioactive Compounds for Food Products: Theory and Applications, *edited by M. Angela A. Meireles* (2009)

Advances in Food Dehydration, *edited by Cristina Ratti* (2009)

Optimization in Food Engineering, *edited by Ferruh Erdoğdu* (2009)

Optical Monitoring of Fresh and Processed Agricultural Crops, *edited by Manuela Zude* (2009)

Food Engineering Aspects of Baking Sweet Goods, *edited by Servet Gülüm Sumnu and Serpil Sahin* (2008)

Computational Fluid Dynamics in Food Processing, *edited by Da-Wen Sun* (2007)

Innovative Processing Technologies for Foods with Bioactive Compounds

EDITED BY
Jorge J. Moreno

CRC Press is an imprint of the
Taylor & Francis Group, an **informa** business

CRC Press
Taylor & Francis Group
6000 Broken Sound Parkway NW, Suite 300
Boca Raton, FL 33487-2742

© 2017 by Taylor & Francis Group, LLC
CRC Press is an imprint of Taylor & Francis Group, an Informa business

No claim to original U.S. Government works

Printed on acid-free paper
Version Date: 20160617

International Standard Book Number-13: 978-1-4987-1484-6 (Hardback)

This book contains information obtained from authentic and highly regarded sources. Reasonable efforts have been made to publish reliable data and information, but the author and publisher cannot assume responsibility for the validity of all materials or the consequences of their use. The authors and publishers have attempted to trace the copyright holders of all material reproduced in this publication and apologize to copyright holders if permission to publish in this form has not been obtained. If any copyright material has not been acknowledged please write and let us know so we may rectify in any future reprint.

Except as permitted under U.S. Copyright Law, no part of this book may be reprinted, reproduced, transmitted, or utilized in any form by any electronic, mechanical, or other means, now known or hereafter invented, including photocopying, microfilming, and recording, or in any information storage or retrieval system, without written permission from the publishers.

For permission to photocopy or use material electronically from this work, please access www.copyright.com (http://www.copyright.com/) or contact the Copyright Clearance Center, Inc. (CCC), 222 Rosewood Drive, Danvers, MA 01923, 978-750-8400. CCC is a not-for-profit organization that provides licenses and registration for a variety of users. For organizations that have been granted a photocopy license by the CCC, a separate system of payment has been arranged.

Trademark Notice: Product or corporate names may be trademarks or registered trademarks, and are used only for identification and explanation without intent to infringe.

Visit the Taylor & Francis Web site at
http://www.taylorandfrancis.com

and the CRC Press Web site at
http://www.crcpress.com

Printed and bound in the United States of America by Publishers Graphics, LLC on sustainably sourced paper.

Contents

Series Preface ix
Series Editor xi
Preface xiii
Acknowledgments xv
Editor xvii
Contributors xix

Chapter 1 Developments in Infrared Heating in Food Processing 1

Navin K. Rastogi

Chapter 2 Ohmic Heating and Bioactive Compounds 31

Jorge J. Moreno, Pamela Zúñiga, Erick Jara,
María P. Gianelli, Karla Mella, and Guillermo Petzold

Chapter 3 Encapsulation as a Carrier System to Enrich Foods with
Antioxidants 61

Aline Schneider Teixeira and Lorena Deladino

Chapter 4 Electric Field Treatment in the Retention of Vitamins in Fruit
Pulps and Beverages Processing 79

Giovana D. Mercali and Júlia R. Sarkis

Chapter 5 Effect of Combined High Pressure–Temperature Treatments
on Bioactive Compounds in Fruit Purées 105

Snehasis Chakraborty, Nishant R. Swami Hulle,
Kaunsar Jabeen, and Pavuluri Srinivasa Rao

Chapter 6 Sonication Processing on Bioactive Compound of Fluid Foods 131

Francisco J. Barba

Chapter 7 Multilayer Nanocapsules as a Vehicle for Release of Bioactive
Compounds 149

Ana C. Pinheiro, Ana I. Bourbon, Hélder D. Silva,
Joana T. Martins, and António A. Vicente

Chapter 8 Freeze Concentration as a Technique to Protect Valuable Heat-Labile Components of Foods 183

Guillermo Petzold, Patricio Orellana, Jorge J. Moreno, Julio Junod, and Graciela Bugueño

Chapter 9 Influence of Osmotic Pretreatment and Drying Air Properties on Bioactive Compounds of Fruits 195

Patchimaporn Udomkun, Marcus Nagle, Dimitrios Argyropoulos, and Joachim Müller

Chapter 10 Effect of Hydrostatic High Pressure Treatments on Bioactive Compounds of Vegetable Products 219

Francisco González-Cebrino, Jesús J. García-Parra, and Rosario Ramírez

Chapter 11 Bioactive Compound Encapsulation and Its Behavior during *In Vitro* Digestion 241

Cristian Ramírez, Helena Nuñez, and Ricardo Simpson

Chapter 12 Electrospinning as a Novel Delivery Vehicle for Bioactive Compounds in Food Nanotechnology 259

Behrouz Ghorani, Ali Alehosseini, and Nick Tucker

Index 293

Series Preface

CONTEMPORARY FOOD ENGINEERING

Food engineering is a multidisciplinary field of applied physical sciences combined with the knowledge of product properties. Food engineers provide the technological knowledge transfer essential to the cost-effective production and commercialization of food products and services. In particular, food engineers develop and design processes and equipment to convert raw agricultural materials and ingredients into safe, convenient, and nutritious consumer food products. However, food engineering topics are continuously undergoing changes to meet diverse consumer demands, and the subject is being rapidly developed to reflect market needs.

In the development of food engineering, one of the many challenges is to employ modern tools and knowledge, such as computational materials science and nanotechnology, to develop new products and processes. At the same time, improving food quality, safety, and security continues to be critical issues in food engineering studies. New packaging materials and techniques are being developed to provide more protection to foods, and novel preservation technologies are emerging to enhance food security and defense. In addition, process control and automation are among the top priorities identified in food engineering. Advanced monitoring and control systems are developed to facilitate automation and flexible food manufacturing processes. Furthermore, energy saving and minimization of environmental problems continue to be important food engineering issues, and significant progress is being made in waste management, efficient utilization of energy, and reduction of effluents and emissions in food production.

The *Contemporary Food Engineering Series* addresses some of the recent developments in food engineering. The series covers advances in classical unit operations in engineering applied to food manufacturing as well as to topics such as progress in the transport and storage of liquid and solid foods; heating, chilling, and freezing of foods; mass transfer in foods; chemical and biochemical aspects of food engineering and the use of kinetic analysis; dehydration, thermal processing, nonthermal processing, extrusion, liquid food concentration, membrane processes, and applications of membranes in food processing; shelf life and electronic indicators in inventory management; sustainable technologies in food processing; and packaging, cleaning, and sanitation. The books in this series are aimed at professional food scientists, academics researching food engineering problems, and graduate-level students.

The editors of these books are leading engineers and scientists from different parts of the world. All of them were asked to present their books to address the market's needs and pinpoint cutting-edge technologies in food engineering.

All chapters have been contributed by internationally renowned experts who have both academic and professional credentials. The contributors have attempted to provide critical, comprehensive, and readily accessible information on the art and science of a relevant topic in each chapter, with reference lists for further information. Therefore, each book serves as an essential reference source for students and researchers in universities and research institutions.

Da-Wen Sun
University College Dublin

Series Editor

Da-Wen Sun, born in southern China, is a global authority in food engineering research and education; he is a member of the Royal Irish Academy (RIA), which is the highest academic honor in Ireland; he is also a member of Academia Europaea (The Academy of Europe) and a fellow of the International Academy of Food Science and Technology. He has contributed significantly to the field of food engineering as a researcher, as an academic authority, and educator.

His main research activities include cooling, drying, and refrigeration processes and systems, quality and safety of food products, bioprocess simulation and optimization, and computer vision/image processing and hyperspectral imaging technologies. His many scholarly works have become standard reference materials for researchers, especially in the areas of computer vision, computational fluid dynamics modeling, vacuum cooling, and related subjects. Results of his work have been published in over 800 papers, including more than 390 peer-reviewed journal papers (Web of Science h-index = 62). He has also edited 14 authoritative books. According to Thomson Scientific's *Essential Science Indicators*[SM], based on data derived over a period of 10 years from Web of Science, there are about 4500 scientists who are among the top 1% of the most cited scientists in the category of agriculture sciences; for many years, Professor Sun has consistently been ranked among the top 50 scientists in the world (he was at the 25th position in March 2015, and at the 1st position if ranking is based on "Hot Papers," and at the 2nd position if ranking is based on "Top Papers" or "Highly Cited Papers").

He earned a first class in both bachelor's (honors) and master's degree programs in mechanical engineering, and a PhD in chemical engineering in China before working in various universities in Europe. He became the first Chinese national to be permanently employed in an Irish university when he was appointed college lecturer at the National University of Ireland, Dublin (University College Dublin, UCD), in 1995, and was then progressively promoted in the shortest possible time to senior lecturer, associate professor, and full professor. Dr. Sun is now a professor of food and biosystems engineering and the director of the UCD Food Refrigeration and Computerized Food Technology Research Group.

As a leading educator in food engineering, Dr. Sun has trained many PhD students who have made their own contributions to the industry and academia. He has also frequently delivered lectures on advances in food engineering at academic institutions worldwide, and delivered keynote speeches at international conferences. As a recognized authority in food engineering, he has been conferred adjunct/visiting/consulting professorships from the top 10 universities in China, including Zhejiang University, Shanghai Jiaotong University, Harbin Institute of Technology, China Agricultural University, South China University of Technology, and Jiangnan University. In recognition of his significant contribution to food engineering worldwide and for his outstanding leadership in the field, the International Commission of Agricultural and Biosystems Engineering (CIGR) awarded him the CIGR Merit Award in 2000; in 2006, the Institution of Mechanical Engineers based in the United Kingdom named him Food Engineer of the Year 2004. In 2008, he was awarded the CIGR Recognition Award in honor of his distinguished achievements as one of the top 1% among agricultural engineering scientists in the world. In 2007, he was presented with the only AFST(I) Fellow Award by the Association of Food Scientists and Technologists (India), and in 2010, he was presented with the CIGR Fellow Award; the title of Fellow is the highest honor at the CIGR and is conferred to individuals who have made sustained, outstanding contributions worldwide. In March 2013, he was presented with the "You Bring Charm to the World" award by Hong Kong–based Phoenix Satellite Television with other award recipients, including the 2012 Nobel Laureate in Literature and the Chinese Astronaut Team for Shenzhou IX Spaceship. In July 2013, he received the Frozen Food Foundation Freezing Research Award from the International Association for Food Protection (IAFP) for his significant contributions to improving food-freezing technologies. This was the first time that this prestigious award was presented to a scientist outside the United States.

He is a fellow of the Institution of Agricultural Engineers and a fellow of Engineers Ireland (the Institution of Engineers of Ireland). He is the editor-in-chief of *Food and Bioprocess Technology—An International Journal* (2012 impact factor = 4.115), former editor of *Journal of Food Engineering* (Elsevier), and a member of the editorial boards for a number of international journals, including the *Journal of Food Process Engineering*, *Journal of Food Measurement and Characterization*, and *Polish Journal of Food and Nutritional Sciences*. He is also a chartered engineer.

On May 28, 2010, he was awarded with the membership of the RIA, which is the highest honor that can be attained by scholars and scientists working in Ireland. At the 51st CIGR General Assembly held during the CIGR World Congress in Quebec City, Canada, on June 13–17, 2010, he was elected the CIGR president for 2013–2014. On September 20, 2011, he was elected to Academia Europaea (The Academy of Europe), which functions as the European Academy of Humanities, Letters and Sciences, and is one of the most prestigious academies in the world; election to the Academia Europaea represents the highest academic distinction.

Preface

Development of functional foods has progressed considerably in the past few years. The technology available to produce food products with enhanced active compounds has grown significantly. Natural foods such as fruits and vegetables represent the simplest form of functional foods and can be a good source of functional compounds. *Innovative Processing Technologies for Foods with Bioactive Compounds* include fundamentals in food processing and the innovation made during the past few years. A group of emerging technologies has grown significantly because of the interest in the development of food products with bioactive compounds. Each chapter is designed as a valuable tool for food scientists and food engineering professionals to follow new developments in food processing. The chapters will serve as an essential reference source to undergraduate and postgraduate students and researchers in universities and technological institutions.

This book has 12 chapters, indicating the importance of innovative processing technologies to cater for developments of new foods with bioactive compounds.

Chapter 1 covers the developments in infrared heating in food processing and presents a basic principle governing infrared radiation, opportunities for infrared in food processing (blanching, baking, drying, roasting thawing, microorganism inactivation, and processing of liquid foods), limitations of infrared processing, and equipment for infrared processing.

Chapters 2 and 4 include electric field applications, alone or in combination, for different foods, the retention or enrichment of vegetable matrixes as carriers of bioactive compounds (polyphenols, vitamins, and minerals), enzymatic inactivation, and shelf life.

Chapters 3, 7, 11, and 12 include encapsulation as a carrier system to enrich foods with bioactive compounds; encapsulation of antioxidants (synthetic and natural antioxidants); and techniques, advantages, and disadvantages in relation to the type of antioxidant protected. In addition, multilayer nanocapsules as carriers of bioactive, the impact of multilayers in the release of bioactive compounds and behavior of multilayer nanocapsules in the gastrointestinal tract. In addition, these chapters will be review the encapsulation importance of bioactive compounds during chewing and the starting point of digestion and the inclusion of microcapsules in food matrices. Finally, the encapsulation of bioactive compounds into fibers is often achieved by electrospinning. This technology is based on packaging of bioactive compounds in micro or nanoscaled particles that isolate them and control their release upon applying specific conditions.

Chapters 5 and 10 explain the effect of high pressure processing (HPP) and high pressure thermal (HPT) on bioactive compound stability of fruits and vegetables, such as: vitamins, flavonoids, carotenoids, total phenolics, and antioxidant capacity. Furthermore, these chapters also explain the storage stability of bioactive compounds after HPP treatments.

Other emerging technologies such as sonication provide a potential development in non thermal processing on bioactive compound of fluid foods such as vitamin C, total phenolics, and total antioxidant capacity (Chapter 6). On the other hand, Chapter 8 explicates the advantages of freeze concentration over heat treatments as a technique to concentrate liquid foods and protect valuable heat labile components of foods.

Chapter 9 is devoted to osmotic pretreatment and drying air properties on bioactive compounds of fruits, including the concentration and type of osmotic agents, agitation and material geometry, and temperature and effect of drying air properties such as drying temperature, air velocity, and humidity.

This book has been contributed by world-renowned researchers who have presented thorough research results and have included a list of recent scientific publications. Therefore, this should provide valuable information for further R&D in food processing.

Jorge J. Moreno
Universidad del Bío-Bío

Acknowledgments

I thank Karla Mella for her assistance in the formatting and styling of the manuscript. My final thanks go to all the contributors for enriching the manuscript with their scientific, technical, and engineering experience.

Editor

Jorge J. Moreno was born in Chile. Professor Moreno is an internationally known figure in food engineering research. His main research activities include emerging technologies, ohmic heating, vacuum impregnation, osmotic dehydration, bioactive compounds, and physical and microstructural properties. Especially, his innovative work on osmotic dehydration combined with ohmic heating and vacuum pulse on fruits to accelerate mass transfer process, and his application to retention of polyphenol compounds and aroma of minimally processed fruits, and the obtaining of fruit snack with functional properties using vacuum impregnation technology. Results of his work have been published in several journals and books.

Dr. Moreno earned his bachelor's degrees in food technology and in food engineering, respectively, from the University of La Serena, Chile; he earned a PhD degree in food science and technology from Polytechnic University of Valencia, Spain. Dr. Moreno is now a professor and the director of the Emerging Technologies and Bioactive Compounds in Food Research Group at the Food Engineering Department, Universidad del Bío-Bío, Chile. He has postdoctoral research experience in the United States (1 year) in the enology area of the Department of Food Science and Technology, Oregon State University, Corvallis, Oregon. He uses a combination of emerging technologies to add value to the raw materials and to obtain minimally processed products.

As an educator in food engineering, Dr. Moreno has significantly contributed to the field of food engineering. He has trained many master's and PhD students who have made their own contributions to the industry and academia. He has also delivered speeches in international conferences and has reviewed several journal papers.

Contributors

Ali Alehosseini
Department of Food Nanotechnology
Research Institute of Food Science and Technology
Mashhad, Iran

Dimitrios Argyropoulos
Food Scientist and Technologist
International Institute of Tropical Agriculture
Central Africa Hub Site de Kalambo, UCB
Province du Sud-Kivu, DR Congo

Francisco J. Barba
Nutrition and Food Science Area
Universitat de València
València, Spain

Ana I. Bourbon
Centre of Biological Engineering
University of Minho
Braga, Portugal

Graciela Bugueño
Food Engineering Department
Emerging Technologies of Bioactive Compounds in Food
Universidad del Bío-Bío
Chillán, Chile

Snehasis Chakraborty
Agricultural and Food Engineering Department
Indian Institute of Technology Kharagpur
West Bengal, India

Lorena Deladino
Centro de Investigación y Desarrollo en Criotecnología de los Alimentos
Buenos Aires, Argentina

Jesús J. García-Parra
Centro de Investigaciones Científicas y Tecnológicas de Extremadura
Technological Agri-Food Institute
Badajoz, Spain

Behrouz Ghorani
Department of Food Nanotechnology
Research Institute of Food Science and Technology
Mashhad, Iran

María P. Gianelli
Food Engineering Department
Emerging Technologies of Bioactive Compounds in Food
Universidad del Bío-Bío
Chillán, Chile

Francisco González-Cebrino
Centro de Investigaciones Científicas y Tecnológicas de Extremadura
Technological Agri-Food Institute
Badajoz, Spain

Kaunsar Jabeen
Agricultural and Food Engineering Department
Indian Institute of Technology Kharagpur
West Bengal, India

Erick Jara
Food Engineering Department
Emerging Technologies of Bioactive
 Compounds in Food
Universidad del Bío-Bío
Chillán, Chile

Julio Junod
Food Engineering Department
Emerging Technologies of Bioactive
 Compounds in Food
Universidad del Bío-Bío
Chillán, Chile

Joana T. Martins
Centre of Biological Engineering
University of Minho
Braga, Portugal

Karla Mella
Food Engineering Department
Emerging Technologies of Bioactive
 Compounds in Food
Universidad del Bío-Bío
Chillán, Chile

Giovana D. Mercali
Food Science Department
Institute of Food Science and Technology
Federal University of Rio Grande do Sul
Porto Alegre, Brazil

Jorge J. Moreno
Food Engineering Department
Emerging Technologies of Bioactive
 Compounds in Food
Universidad del Bío-Bío
Chillán, Chile

Joachim Müller
Food Scientist and Technologist
 International Institute of Tropical
 Agriculture
Central Africa Hub Site de Kalambo,
 UCB
Province du Sud-Kivu, DR Congo

Marcus Nagle
Food Scientist and Technologist
 International Institute of Tropical
 Agriculture
Central Africa Hub Site de Kalambo,
 UCB
Province du Sud-Kivu, DR Congo

Helena Nuñez
Department of Chemical and
 Environmental Engineering
Universidad Técnica Federico Santa
 María
Valparaíso, Chile

Patricio Orellana
Food Engineering Department
Emerging Technologies of Bioactive
 Compounds in Food
Universidad del Bío-Bío
Chillán, Chile

Guillermo Petzold
Food Engineering Department
Emerging Technologies of Bioactive
 Compounds in Food
Universidad del Bío-Bío
Chillán, Chile

Ana C. Pinheiro
Centre of Biological Engineering
University of Minho
Braga, Portugal

Cristian Ramírez
Department of Chemical and
 Environmental Engineering
Universidad Técnica Federico Santa
 María
Valparaíso, Chile

Rosario Ramírez
Centro de Investigaciones Científicas
 y Tecnológicas de Extremadura
Technological Agri-Food Institute
Badajoz, Spain

Contributors

Navin K. Rastogi
Department of Food Engineering,
Central Food Technological Research Institute
Mysore, India

Júlia R. Sarkis
Food Science Department
Institute of Food Science and Technology
Federal University of Rio Grande do Sul
Porto Alegre, Brazil

Hélder D. Silva
Centre of Biological Engineering
University of Minho
Braga, Portugal

Ricardo Simpson
Department of Chemional and Enviromental Engineering
Centro Regional para el Estudio de Alimentos Saludables
Conicyt Regional Gore Valparaíso (R06I1004)
Universidad Técnica Federico Santa María
Valparaíso, Chile

Pavuluri Srinivasa Rao
Agricultural and Food Engineering Department
Indian Institute of Technology Kharagpur
West Bengal, India

Nishant R. Swami Hulle
Agricultural and Food Engineering Department
Indian Institute of Technology Kharagpur
West Bengal, India

Aline Schneider Teixeira
Centro de Investigación y Desarrollo en Criotecnología de los Alimentos
Buenos Aires, Argentina

Nick Tucker
School of Engineering
University of Lincoln
Lincoln, United Kingdom

Patchimaporn Udomkun
Food Scientist and Technologist
International Institute of Tropical Agriculture
Central Africa Hub Site de Kalambo, UCB
Province du Sud-Kivu, DR Congo

António A. Vicente
Centre of Biological Engineering
University of Minho
Braga, Portugal

Pamela Zúñiga
Food Engineering Department
Emerging Technologies of Bioactive Compounds in Food
Universidad del Bío-Bío
Chillán, Chile

1 Developments in Infrared Heating in Food Processing

Navin K. Rastogi

CONTENTS

1.1 Introduction 1
1.2 Basic Principles Governing IR Radiation 4
    1.2.1 Planck's Law 4
    1.2.2 Wien's Displacement Law 5
    1.2.3 Stefan–Boltzmann Law 5
1.3 Opportunities for IR in Food Processing 5
    1.3.1 IR Blanching 5
    1.3.2 IR Baking 6
    1.3.3 IR Drying 8
    1.3.4 IR Roasting 8
    1.3.5 IR Thawing 15
    1.3.6 IR Microbial Inactivation 16
1.4 IR Processing of Liquid Foods 16
1.5 Limitations of IR Processing 18
1.6 Equipment for IR Processing 18
1.7 Concluding Remarks 23
Acknowledgments 23
References 24

1.1 INTRODUCTION

Consumers' expectations of convenience, variety, adequate shelf life and caloric content, reasonable cost, and environmental soundness have always triggered the need for advancement and modification of existing food-processing techniques and adoption of further novel processing technologies. Application of infrared (IR) heating in food processing is one such alternative. Food and agricultural sector adopts energy-efficient, less water-intensive, and environment-friendly technologies; the applications of IR technology have gained more attention as alternatives to other processing technologies.

    IR radiation is an electromagnetic radiation, which falls between the region of visible light (0.38–0.78 µm) and microwaves (1–1000 mm). It is transmitted as a wave and gets converted into heat while impinging on the food surface. Based on the wavelength, it can be divided into three regions near- (0.78–1.4 µm), mid- (1.4–3.0 µm), and far-IR

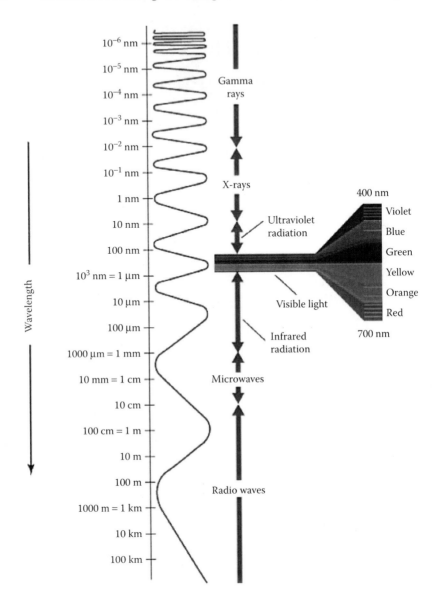

FIGURE 1.1 Electromagnetic wave spectrum.

(3.0–1000 μm) (Figure 1.1). The wavelength of the radiation is determined by the temperature of the emitting body—the higher the temperature, the shorter the wavelength. IR energy can be focused, directed, and reflected the same as light, without introducing more heat into the workplace. In general, the food substances absorb far-IR energy most efficiently through the mechanism of changes in the molecular vibrational state, which can lead to radiative heating. Water and organic compounds (such as proteins and starches) are the main components of food, which absorb far-IR energy at wavelengths greater than 2.5 μm (Decareau, 1985; Sakai and Hanzawa, 1994).

Generally, far-IR radiation is more advantageous for food processing because most food components absorb radiative energy in this region (Sandu, 1986). IR heating provides efficient heat transfer, which reduces processing time and energy cost. At the same time, air in contact with the equipment is not heated; therefore, ambient temperature can be kept at normal levels. During IR heating, surface irregularities have smaller effect on rate of heat transfer, resulting in more uniform heating.

When IR radiation falls on the exposed material, it penetrates and the energy of radiation converts into heat (Ginzburg, 1969). The depth of penetration depends on the composition and structure of the fruits and also on the wavelengths of IR radiation. When a material is exposed to the IR radiation, it is heated intensely, and the temperature gradient in the material reduces within a short period. It results in high rate of heat transfer compared to conventional drying and the product is more uniformly heated rendering better-quality characteristics. In the case of conventional drying, the product is subjected to heating at the surface of the food by convection from circulating hot air, and further transfer of heat to the core takes place by conduction, which results in case hardening of the material and hinders mass transfer. The schematic representation of the difference in heat and mass transfer during conventional and IR heating is shown in Figure 1.2.

The advantages of IR radiation on food processing compared to those of conventional heating include short processing time, increased energy efficiency, uniform product temperature, better-quality finished products, high degree of process control parameters, high heat transfer coefficient, space saving, and environment friendly (Dostie et al., 1989; Navari et al., 1992; Sakai and Hanzawa, 1994; Mongpraneet et al., 2002a). IR heating has been applied in drying, baking, roasting, blanching, pasteurization, and thawing of food products.

There are a number of reviews in the literature discussing the capability and limitations of IR processing (Sakai and Hanzawa 1994; Skjoldebrand, 2001, 2002; Sakai and Mao, 2006; Krishnamurthy et al., 2008a, b, 2009; Rastogi 2012a, b, 2015). Many of the earlier reviews mainly focused on the general aspects of IR processing. This chapter brings to researchers and professionals the latest research results, developments, novel engineering approaches, updated knowledge, and novel applications; it explores the potential of IR heating technology for food and agricultural product processing to meet consumers' needs.

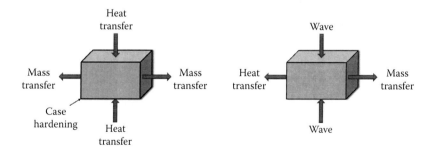

FIGURE 1.2 Schematic representation of conventional and infrared heating. (From Hebbar, U.H. and Rastogi, N.K., *J. Food Eng.*, 47, 1, 2001. With permission.)

1.2 BASIC PRINCIPLES GOVERNING IR RADIATION

IR radiation is a form of energy, which is specifically known as electromagnetic waves arising from the movement of electrons in atoms and molecules. The average or bulk properties of electromagnetic radiation interacting with matter are systematized in a simple set of rules called *radiation laws*. These laws apply when the radiating body is a blackbody radiator. A hypothetical body that completely absorbs all radiant energy falling upon it, reaches some equilibrium temperature, and then reemits that energy as quickly as it absorbs it. The sum of the IR radiation that impinges on any surface has a spectral dependence because the energy coming out of an emitter consists of different wavelengths, and the fraction of the radiation in each band is dependent upon the temperature and emissivity of the emitter. The temperature of the IR heating elements governs the wavelength at which the maximum radiation occurs. The basic laws for blackbody radiation are described in the Sections 1.2.1 and 1.2.3 (Skjoldebrand, 2001; Krishnamurthy et al., 2008a).

1.2.1 PLANCK'S LAW

Planck's law gives the spectral blackbody emissive power distribution, $E(\lambda, T)$, of the radiation emitted by unit surface area into a fixed direction (solid angle) from the blackbody as a function of wavelength for a fixed temperature. The spectral characteristics of the blackbody radiation from objects at different temperatures indicate that the curves give the maximum possible radiation that can be emitted at a selected temperature (Figure 1.3). A blackbody produces maximum intensity according to Planck's law, which can be expressed through the following equation:

$$E(\lambda, T) = \frac{2hc^2}{\lambda^5} \frac{1}{\exp(hc/\lambda kT) - 1} \tag{1.1}$$

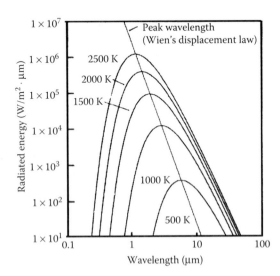

FIGURE 1.3 Spectral characteristics of blackbody radiation at different temperatures. (From Sakai, N. and Hanzawa, T., *Trends Food Sci. Technol.*, 5, 357, 1994. With permission.)

where:
c is the speed of light (3×10^{10} cm/s)
h and k are the Planck (6.625×10^{-27} erg-s) and Boltzmann (1.38×10^{-16} erg/K) constants, respectively
λ (cm) and T (K) are the wavelength and temperature, respectively

1.2.2 Wien's Displacement Law

Wien's displacement law states that the wavelength of the most intense radiation (λ_{max}) emitted by a blackbody depends only on its temperature (T) according to the following formula:

$$\lambda_{max} = \frac{a}{T} \quad (1.2)$$

where:
a is 2897 μm.K
λ_{max} and T are in μm and Kelvin, respectively

Wien's law explains the shift of the peak to shorter wavelengths as the temperature increases.

1.2.3 Stefan–Boltzmann Law

The Stefan–Boltzmann law provides the total energy being emitted at all wavelengths (E_T) by the blackbody (which is the area under the Planck law curve, Figure 1.3) at a specific temperature from an IR source. Radiant heaters are not perfect radiators and foods are not perfect absorbers, although they do emit and absorb a constant fraction of the theoretical maximum. To take account of this, the concept of grey bodies is used, and the Stefan–Boltzmann equation is modified to (Skjoldebrand, 2001)

$$E_T = \varepsilon \sigma A T^4 \quad (1.3)$$

where ε, σ, and A are the emissivity (varies from 0 to 1), Stefan–Boltzmann constant (5.670×10^{-8} W/m²/K⁴), and surface area (m²), respectively. Emissivity varies with the temperature of the body and the wavelength of the radiation emitted.

1.3 OPPORTUNITIES FOR IR IN FOOD PROCESSING

IR technology has long been underestimated in the food field, despite its great potential. It is generally applied for the dehydration of vegetables, fish, pasta, and rice; heating flour; frying meat; roasting cereals; roasting coffee and cocoa; and baking biscuits and bread. This technique has also been used for thawing and surface pasteurization of bread and packaging materials.

1.3.1 IR Blanching

Blanching represents an essential step to inactive enzymes that may cause color change and quality deterioration in further processing and storage. Conventional methods of blanching are energy-intensive unit operations with low energy efficiency and long processing times, and they result in inferior product quality.

IR-blanched peas were demonstrated to contain comparable ascorbic acid retention but better taste and flavor than hot water-blanched samples (Van Zuilichem et al., 1985). Similarly, IR-blanched endive and spinach had firmer texture than steam or hot water-blanched product (Ponne et al., 1994). Far-IR-treated carrot slices had higher tissue strength than the hot water-blanched sample, preserving most of the characteristic texture of the raw tissue due to the less extent of tissue damage caused by IR heating. The cell damage due to IR heating of carrot cells was only up to a depth of 0.5 mm from the surface, while the texture, vitamins, and minerals in undamaged internal tissues were well preserved (Gomez et al., 2005).

A combination of far-IR dry-blanching and dehydration was used to produce high-quality partially dehydrated fruits and vegetables such as pears, carrots, sweet corn kernels, french fries, and apples (Zhongli and McHugh, 2006). The combinations of ascorbic acid and calcium chloride dipping treatment resulted in reduced enzymatic browning of apple cubes and firmer samples after the far-IR dry-blanching process. Surface color changes of product were mainly due to enzymatic browning, which occurred during the process (Yi et al., 2007; Yi and Zhongli, 2009). IR dry-blanching and dehydration method leads to a simpler process and better product quality, and results in better energy and processing efficiency than it is achieved by the conventional two-step process (Zhu and Pan, 2009).

Application of IR radiation for blanching and drying showed a significant increase in total phenols content as compared to fresh ones as well as preserving the green color of fresh olive leaves leading to the enhancement of their luminosity. This method may be used for preserving olive leaves before their use in food or cosmetic applications (Boudhrioua et al. 2009). Bingol et al. (2012) demonstrated the suitability of using IR heating as a dry-blanching pretreatment prior to frying to reduce the oil uptake in french fry production. Besides, IR heating resulted in complete inactivation of polyphenol oxidase enzyme was achieved. The sensory evaluation revealed that IR dry-blanched sample was preferred in terms of taste, texture, color, and appearance.

1.3.2 IR Baking

IR heating has been used in the baking industry to shorten processing times, enhance product quality, and reduce energy consumption (Sumnu and Ozkoc, 2010). Skjoldebrand (2001) stated that IR baking can be divided into three periods: surface heating zone, evaporation zone (transfer of moisture occurs from the central parts), and final stage of baking (central temperature increases to 98°C–99°C). IR baking can be combined with other heating technologies such as microwave heating or hot air heating to improve the heating rate and moisture distribution inside foods (Zhang et al., 2005; Datta et al., 2007). The baking time reduction is due to the more effective heat transfer to the surface than convection or conduction heating. The loss in weight was lower leading to higher water content in the center during baking, causing better and longer shelf life. The energy consumption is also comparable. The rapid surface heating characteristics of IR radiation aid quick removal of moisture from product surface, which promotes color and crust formation. Some of the important findings in this area of bakery products have been summarized in Table 1.1.

TABLE 1.1
Main Findings with Regard to the Effect of IR on Bakery Products

Product	Salient Results	References
Baguettes	A combination of near-IR radiation with jet impingement resulted in increased rate of crust color development and reduced heating time of par-baked baguettes during post-baking compared with heating in a conventional household oven.	Olsson et al. (2005)
Bread	In case of bread baking in near-IR oven, the crust formation and the baking of the crumb took place simultaneously.	Skjoeldebrand and Andersson (1989)
	IR heating at longer wavelength resulted in dry crusts and faster coloring rate of food samples than those emitting shorter wavelength radiations due to the higher rate of temperature increase at the food surface.	Sato et al. (1992)
	Two-stage procedure involving heating in a microwave oven followed by exposure to IR energy resulted in baking and crust formation of bakery products.	Levinson (1992)
	Near-IR-assisted microwave-dried crumbs generally had similar color values compared to conventionally dried samples.	Tireki et al. (2006)
	The addition of xanthan-guar blend resulted in improved bread quality in terms of low hardness values, high specific volume, and porosity studied on the quality of breads baked in IR-microwave combination oven. Retrogradation enthalpy and total mass crystallinity values of bread samples showed delay in staling.	Keskin et al. (2007) and Ozkoc et al. (2009)
Cakes	A combination of microwave and IR oven resulted in surface color development, reduced weight loss and firmness, and increased volume. Besides, it reduces the baking time by 75% compared to conventional baking.	Sumnu et al. (2005)
	Dielectric constant and loss factor of cake samples were dependent on formulation, baking time, and temperature. An increase in baking time, temperature, and fat content resulted in decrease in dielectric constant and loss factor. The cakes containing whey protein baked in microwave and near-IR-microwave combination ovens were found to be the firmest cakes.	Sakiyan et al. (2007a, b)
	Sponge cake baked in a far-IR oven after 7 days storage was softer as compared to the cake baked in an electric oven. But, no significant differences in the volume, water activity, staling rate, or sensory scores were observed.	Yung et al. (2008)
	Gluten-free cakes made from rice flour baked in a microwave-IR combination oven formulated with xanthan gum had better-quality characteristics than cakes containing the xanthan-guar gum blend.	Turabi et al. (2008)
Tortillas	Loss of moisture during baking of the tortillas formed by hot-pressing and baked in IR oven was significantly lower than that of tortillas baked by traditional and commercial methods. IR-baked tortillas showed good characteristics of rollability, puffing, layering, color, and texture.	Martinez et al. (1999)
	Simultaneous application of IR radiation to both sides of the tortilla quickly cooked both the surfaces and formed capping layers that retained a high degree of moisture within the product and prevented dehydration during the cooking cycle.	Gonzalez et al. (1996) and Luz et al. (1996)

1.3.3 IR Drying

Drying is a very important preservation method that is used in the food industry. The basic objective in drying food products is the removal of water from fresh products, thereby achieving a moisture level at which microbial spoilage is avoided. IR drying has gained popularity as an alternative drying method for a variety of agricultural products. Compared to hot air drying, IR radiation drying offers many advantages, such as greater energy efficiency, heat transfer rate, and heat flux, which results in reduced drying time and higher drying rate. The IR heating allows more uniform heating of fruits compared to other drying methods, resulting in better-quality dehydrated foods (Sakai and Hanzawa, 1994; Nowak and Lewicki, 2004). Application of hot air-assisted IR drying for high moisture content materials is reported to be beneficial as it provides a synergistic effect (Ginzburg, 1969). IR radiation with hot air fluidization provides an effective way to increase the drying rate and to prevent the nonuniform heat and mass transfer compared to the conventional drying methods. The intermittent IR drying with energy input of 10 W/m^2 becomes equivalent to convective drying in which the heat transfer coefficient would be as high as 200 $W/m^2\,K$ (Ratti and Mujumdar, 1995). Some of the important finding in this area of fruits and vegetables as well as cereals, grains, and seeds processing have been summarized in Tables 1.2 through 1.4, respectively.

1.3.4 IR Roasting

Roasting is a process that leads to an improvement in flavor, color, texture, and appearance. Compared to conventional drum roasting, IR roasting requires lower investment cost, higher production rates, and more compact design of the roaster configuration. Therefore, IR roasting has been successfully applied to commercial coffee and tea roasting (Sumnu and Ozkoc, 2010). The involvement of far-IR irradiation during green tea leaves manufacturing process in the roasting and drying step resulted in high-quality green tea. It increased the total phenols, flavanols, epigallocatechin, epigallocatechin gallate contents, as well as ascorbic acid, caffeine contents, and nitrite scavenging activities (Kim et al., 2006). Besides, it increased sweetness, umami, and aroma of green tea, while bitterness and astringency were decreased as compared to control (Park et al., 2009). A combination of IR, microwave, conduction, convection, and latent steam heating to roast green coffee beans resulted in energy-efficient process, resulting in more flavorful roasted coffee (Poss, 2007). A coffee bean roaster consisted of an oven with IR ceramic heating plate with a gate attached with a rotary sliding unit, which roasted the coffee beans while rolling them was developed (Chung, 2008).

Yang et al. (2010) developed a pilot-scale new sequential IR and hot air roaster for improved microbial reduction and reduced time for roasting of almond. It consisted of four IR emitters and the metal screen tray. The almond samples are located at distances of 16 and 12.5 cm from the top and bottom emitters, respectively. The roasting process can be easily implemented in the industry by adding an IR preheating device in front of regular hot air roasters.

TABLE 1.2
Main Findings with Regard to the Effect of IR on Dehydration of Fruits

Product	Salient Results	References
Apple	The shrinkage of apple samples during IR drying was found to be directly proportional to the thickness of the material and inversely proportional to the intensity of IR radiation.	Wesolowski and Glowacki (2003)
	The use of near-IR drying offered the ease of adjusting the temperature of the material being dried. It took half of the drying time compared to convective process. The color parameter, rehydration capacity, and mechanical properties were dependent on final material temperature.	Nowak and Lewicki (2004, 2005)
	Various empirical and diffusional models for mid-IR drying of apple in the temperature range of 50°C–80°C were studied.	Togrul (2006)
Banana	A combination of IR drying with hot air pre-drying was found to show 20% reduction in the drying time to remove the same amount of residual moisture as compared to the IR drying alone.	Sun et al. (2007)
	Novel drying technology consisting of low-pressure superheated steam drying with far-IR radiation resulted in more crispness and acceptable color in dried banana slices.	Chatchai et al. (2007)
	During freeze-drying, partially IR-dried banana slices were found to have higher drying rate as compared to the hot air pre-dried sample. IR pretreated samples showed collapse of cellular tissue in the surfaces and center of the slices.	Zhongli et al. (2008b)
	Far-IR radiation resulted in increased porosity of the dried bananas similar to in case of vacuum drying. Further, increase in drying temperature led to an increase in the final porosity of the dried product.	Leonard et al. (2008)
	Combined far-IR radiation and vacuum drying was used to produce fat-free banana-based snacks. A mathematical model was developed to predict the moisture content and temperature of banana.	Thanit et al. (2009)
Blueberries and Strawberries	The use of catalytic-IR drying in case of fresh and sugar-infused blueberries resulted in firmer-texture products in less drying time as compared to hot air drying. NaOH pretreatment increased moisture diffusivity and reduced the number of broken berries.	Junling et al. (2008) and Shi et al. (2008)
	Integration of combined catalytic-IR and freeze-drying process produced high-quality crispy strawberries pieces at less cost having more desirable color and more shrinkage.	Shih et al. (2008)
Cashew	Mass transfer characteristic during far-IR heating of cashew kernel was studied. It was found to be suitable for testa removal for cashew processing, which otherwise would take long time in convective dryers.	Hebbar and Rastogi (2001) and Hebbar et al. (1999)
Fruit Bars	Frozen restructured whole apple and strawberry bars were manufactured by partial dehydration, using IR heating, followed by restructuring and freezing. IR drying reduced the moisture in the fruits quickly and also caused partial degradation of total phenolic and vitamin C.	Hongping et al. (2013)
Figs	IR heating of fresh figs was modeled by the computational fluid dynamics approach. Models were proposed to investigate the progress of surface temperature distribution during IR heating. The thermal deactivation of microorganisms was also investigated during the IR heating process.	Tanaka et al. (2012)

(Continued)

TABLE 1.2 (*Continued*)
Main Findings with Regard to the Effect of IR on Dehydration of Fruits

Product	Salient Results	References
Grapes	Far-IR treatment of grapes was found to increase the levels of antioxidative and phenolic compounds in cv. Campbell Early, but not in cv. Thompson Seedless.	Seok et al. (2009)
	Effects of carbonic maceration, dipping in alkaline emulsion of ethyl oleate solution and dipping in ethyl oleate solution and then freezing at −18°C for 12 h on IR drying kinetics of red grapes, and properties of raisins were investigated. The results indicated that the raisin of carbonic maceration-treated samples had the shorter production time, the highest total phenol content, the best oxidation resistance ability, and the best rehydration ratio. The carbonic maceration technique before drying can greatly increase the drying rate and improve the quality of products; meanwhile, it is free of any chemical reagents, which is friendly to the environment.	Yuxin et al. (2014)
Kiwi Fruit	A combination drying consisting of osmotic dehydration and med-short IR drying of kiwifruit slices mainly occurred in the falling rate drying period. Two different falling rate periods were obtained.	Zeng et al. (2014)
	Drying time decreased with the increase of the drying temperature and IR power. The drying temperature had a significant influence on the drying rate; it was less affected by the power. Weibull distribution function was fitting well with the data.	Mu et al. (2014a, b)
Peach	The effect of different IR power levels (83–209 W) was studied on drying kinetics of peach slices. Drying characteristics of peach slices were greatly influenced by IR power.	Doymaz (2014a)
Pear	Application of ultrasound along with IR heating resulted in reduced drying time. The hardness of samples gradually decreased with an increase in ultrasound intensity.	Dujmic et al. (2013)
Pineapple	Intermittent IR treatment combined with hot air drying and osmotic pre-treatment reduced overall color change, besides maintaining high drying rates.	Tan et al. (2001)
	Developed the mathematical model for pineapple rings during combined far-IR and air convection drying to investigate the evolutions of moisture content and qualities. The kinetics of dried pineapple qualities such as color, shear force ratio, and shrinkage during drying also were studied. The diffusion coefficients were found to increase with increasing intensity of IR and air temperature.	Kamon et al. (2012)
Pistachio Nuts	Combining IR treatment with fluidization technique can be a novel method in drying of pistachio nuts. IR radiation with hot air fluidization provides an effective way to increase the drying rate and to prevent the nonuniform heat and mass transfer compared to the conventional drying methods.	Amiri et al. (2014)
Watermelon Seeds	During IR drying, the power level significantly affected the drying and rehydration characteristics of watermelon seeds. Drying time was reduced from 170 to 40 min as the IR power level increased from 83 to 209 W.	Doymaz (2014b)

TABLE 1.3
Main Findings with Regard to the Effect of IR on Dehydration of Vegetables

Product	Salient Results	References
Carrot	Catalytic IR had higher drying rates, shorter drying times, and greater effective diffusivities than hot air drying. Carrot slices dried with catalytic IR had better rehydration characteristics than the samples dried with hot air drying. Thickness had a significant impact on overall color change of carrot slices.	Bengang et al. (2014)
	IR blanching and IR-assisted hot air drying of carrot slices were compared with conventional blanching and drying techniques in terms of processing time, retention of vitamin C, and rehydration characteristics. Intermittent heating of carrot slices using IR radiation (180°C–240°C, 8–15 min) resulted in desired level of enzyme inactivation. Although the time required for IR blanching was higher, but retention of water soluble vitamin C was also higher as compared to water and steam blanching. IR blanching reduced the moisture content and the samples were dried in hybrid mode that took less time compared to water-blanched hot air-dried samples.	Vishwanathan et al. (2013)
Green Beans	Drying characteristics, rehydration, and color of bean slices were greatly influenced by IR power. The drying data were fitted with five thin-layer drying models available in the literature. Effective moisture diffusivity was calculated in the range of 6.57×10^{-10} to 4.49×10^{-9} m^2/s. Activation energy was estimated by a modified Arrhenius-type equation and found to be 11.38 kW/kg.	Doymaz et al. (2015)
Miang Leaves	Far-IR vacuum was applied for processing tea from Miang leaves. When the temperature increased from 50°C to 65°C and the time from 60 to 120 min, the amount of epicatechin, epicatechin gallate, epigallocatechin gallate, and total catechins increased, while the moisture content and water activity decreased, compared to control sample. The physicochemical properties of dried Miang leaves were also influenced by time and temperature. Drying conditions of 65°C for 120 min are recommended for optimization of drying.	Sathira et al. (2015)
Mint Leaves	The optimum conditions for drying mint leaves were found to be 18 and 38 min at drying temperatures of 60°C–80°C. The lightness and yellowness of the dried mint leaves were significantly increased when compared with fresh samples. The high drying temperature yielded in darker, less green, and more yellow mint leaves.	Ertekin and Heybeli (2014)
Mushroom	Microwave-vacuum drying and microwave-vacuum combined with IR drying were used for drying. Drying rate increased with lower absolute pressure, higher microwave power, and higher IR power. Microwave-vacuum combined with IR drying provided better quality of product in terms of color of dried mushroom, rehydration ratio, and texture of rehydrated ones.	Hataichanok et al. (2014)

(Continued)

TABLE 1.3 (*Continued*)
Main Findings with Regard to the Effect of IR on Dehydration of Vegetables

Product	Salient Results	References
	A combination of freeze drying (for 4 h) followed by mid-IR saved 48% time compared to freeze drying while keeping the product quality at an acceptable level. The mid-IR and freeze drying combination was found to be inferior compared to the freeze drying and mid-IR combination as the former tended to produce products with a collapsed surface layer and poor rehydration capability. The combination of mid-IR with freeze drying had a significant effect on aroma retention and caused an increase of sulfur compounds such as dimethyl, trisulfide, and 1,2,4-trithiolane.	Hong et al. (2015)
Onion	Drying curves of Welsh onion were divided into three regions: a rising rate, a constant rate, and a falling rate period. Increase in radiation intensities increased material temperature and significantly affected the chlorophyll content.	Mongpraneet et al. (2002a, b)
	Exponential model and an approximation of the diffusion model fitted well for far-IR radiation drying of onion slices.	Wang (2002)
	A hybrid dryer combining IR and convective drying for drying of onion slices was developed. Onion slices had better rehydration ratio as compared to conventionally dried sample.	Sharma et al. (2005a, b)
	Combined IR and hot air drying of onion resulted in a shorter drying time and better quality of onion slice (in terms of color and pyruvic acid content) as compared to individual treatments.	Praveen Kumar et al. (2005, 2006)
	In case of IR-convective drying, moisture diffusion coefficient values increased. Further, the values were found to increase with the level of IR radiation and air velocity.	Pathare and Sharma (2006)
	Catalytic IR heating with or without air recirculation had higher drying rates than forced air convection heating. Besides, catalytic IR processed sample had lower yeast and mould counts, but no difference was seen for the aerobic plate counts and coliform counts.	Gabel et al. (2006)
Potato and Yam	During far-IR drying of potato, diffusion coefficient was found to increase with an increase in radiation intensity and slice thickness. Drying rates were dependent on radiation intensity and air velocity, but independent of relative humidity.	Afzal and Abe (1998) and Afzal et al. (1999)
	Combined IR heating and freeze-drying in case of sweet potatoes reduced the processing time by less than half.	Yeu et al. (2005, 2007)
Red Bell Pepper	Dry-blanching of red bell pepper slices using IR and microwave radiations was compared with conventional water and steam blanching methods. Processing conditions (time and temperature) were standardized on the basis of degree of enzyme inactivation. Retention of water-soluble nutrients, prevention of solid loss, and elimination of waste water generation are some of the advantages of dry-blanching.	Jeevitha et al. (2013)

(*Continued*)

TABLE 1.3 (*Continued*)
Main Findings with Regard to the Effect of IR on Dehydration of Vegetables

Product	Salient Results	References
Sea Cucumber	Far-IR drying dramatically shortens the drying time of sea cucumber compared to air drying. The quality of IR-dried product was better than that of the air drying, because the surface hardening of the sea cucumber was minimized by the IR heating. Far-IR might be a good technology to replace the solar drying as well as the air drying to produce the dried sea cucumber.	Ji et al. (2014)
Korean Vegetable (*Pimpinella bracycarpa*)	Lower temperature far-IR processing resulted in higher sensory and color score. Rehydration rates increased with drying temperature.	Myung et al. (2000)
Tomato	IR dry-peeling of tomatoes has emerged as a nonchemical alternative to conventional peeling methods using hot lye or steam. Successful peel separation was dependent on the delivery of a sufficient amount of thermal energy onto the tomato surface in a very short duration. The transient heat transfer during IR dry-peeling was explained by developing a computer simulation model. A predictive mathematical model simulated heat transfer in a tomato undergoing double-sided IR heating in a dry-peeling process.	Xuan and Zhongli (2014a, b)
	Compared to conventional lye peeling, IR dry-peeling took less heating time, lower peeling loss, thinner thickness of peeled-off skin, and slightly firmer texture of peeled products while similar ease of peeling was achieved. Besides, IR heating increased the Young's modulus of tomato peels and reduced the peel adhesiveness, indicating the tomato peels to loosen, become brittle, and crack more easily. In addition, IR heating resulted in melting of cuticular membrane, collapse of several cellular layers, and severe degradation of cell wall structures, which in turn caused peel separation.	Xuan et al. (2014)
	IR power level was found to affect the drying time and rehydration capacity. Drying time was reduced from 450 to 240 min as the IR power level increased from 125 to 188 W. IR drying has appeared as one of the potential alternatives to the traditional drying methods in obtaining high-quality dried products.	Doymaz (2014c)
	Drying time was prolonged with increasing air velocity while it was shortened with increasing IR intensity. The lowest energy consumption occurred at the air velocity of 1.0 m/s and at IR intensity of 2640 W/m^2. The contents of vitamin C and beta-carotene in IR-dried tomatoes were decreased, while there was an increase in lycopene content. It was observed that IR drying of tomato provided good nutrient retention and low cost of energy.	Kocabiyik et al. (2015)

TABLE 1.4
Main Findings with Regard to the Effect of IR on Dehydration of Cereals, Grains, and Seeds

Product	Salient Results	References
Barley	Gas-fired IR heating of grains (barley, kidney beans, green peas, black beans, lentils, and pinto beans) alters the physical, mechanical, chemical, and functional properties due to their possible cracking. The trypsin inhibitor activity was also reduced.	Fasina et al. (1998, 1999, 2001)
	IR irradiation decreased moisture contents and destroyed antinutritional factors in beans and cereals. IR treatment resulted in destruction of tannin in sorghum, aflatoxins in corn and sorghum, and trypsin inhibitor activity in common beans.	Keya and Sherman (1997)
	Combination of convective and far-IR heating process of barley reduced the total energy required as compared to convection drying.	Afzal et al. (1999) and Afzal and Abe (2000)
	Far-IR drying of barley kernel resulted in size expansion along with an increase in the yellowness on the surface due to decomposition of carotenoids.	Konopka et al. (2008)
Canola Seeds	IR-processed canola seed resulted in lower crude fiber content and yielded in dehulling without affecting the crude oil quality.	McCurdy (1992)
Hazelnut	Combination of microwave-IR oven for roasting of hazelnut resulted in the product of comparable quality with conventionally roasted ones with respect to color, texture, moisture content, and fatty acid composition.	Uysal et al. (2009) and Ozdemir and Devres (2000)
Sesame Seeds	Near-IR roasting of sesame seeds increased the oxidative stability of sesame oil synergistically with tocopherols due to degradation of lignan sesamolin to sesamol. No difference in the functional properties of defatted flours obtained from either IR roasted or conventionally roasted sesame seed was observed.	Kumar et al. (2009)
Paddy (Rice)	A vibration-aided IR dryer of laboratory scale for drying paddy was developed and frequency as well as amplitude of the vibrations was optimized. Drying time was lower when higher radiation intensity was used.	Das et al. (2004a, b, 2009) and Naret et al. (2007)
	IR treatment followed by tempering and slow cooling resulted in simultaneous drying and disinfestation along with high rice milling quality.	Zhongli et al. (2008a)
	An increase in IR drying time was found to increase the yield of head rice and decreased the whiteness.	Juckamas and Seree (2009)
	Optimize the operating parameters of IR dryer to achieve high heating rate, fast drying, and good quality of end products. IR heating of rough rice to about 60°C followed by tempering and natural cooling could be an effective approach for designing IR rough rice dryers. It can be used as an energy saving drying method with improved drying efficiency, space saving, clean working environment, and superior product quality compared with the conventional heated air drying method.	Khir et al. (2014)

(Continued)

TABLE 1.4 (*Continued*)
Main Findings with Regard to the Effect of IR on Dehydration of Cereals, Grains, and Seeds

Product	Salient Results	References
	The mechanical quality aspects of paddy kernels dried using combined hot-air/IR thin-layer drying conditions were evaluated in terms of percentage of cracked kernels and also required failure force obtained from bending tests. In addition to the product quality aspects, the specific energy consumption was estimated. Application of a low-intensity IR radiation (2000 W/m^2), together with lower values of inlet air temperature (30°C) and moderate values of inlet air velocity (0.15 m/s), can effectively improve the final quality of paddy (as a heat-sensitive product) with a reasonable specific energy consumption.	Zare et al. (2015)
Parboiled Rice	A comparison of hot air, IR and a combination of hot air and IR drying revealed that the effective diffusion coefficient values of hot air and IR drying were relatively depended on temperature. Head rice yield using hot air and IR drying was highest. A combined hot air and IR drying offers a great potential for preserving parboiled rice and proved to be more efficient method compared to other drying methods.	Bualuang et al. (2013)
Dry Peas	IR heating of dry peas induced no change in dehulling and air classification characteristics. Bitterness and protein solubility of raw pea protein was reduced; water hydration rate was increased and cooking time was reduced. It can be used as an effective technique for instantizing split peas.	McCurdy (1992) and Cenkowski and Sosulski (1998)
Peanut	Far-IR radiation resulted in higher antioxidative activity of extracts from peanut shells as compared to heat treated sample. The antioxidative activity was found to increase with an increase in exposure time.	Rim et al. (2005)
Pinto Beans	IR heating on pinto beans (*Phaseolus vulgaris*) improved rehydration rate, increased degree of swelling, and increased cooking time.	Abdul et al. (1990)

1.3.5 IR Thawing

In the IR region, the absorption coefficient of ice and water is approximately the same (Sakai and Mao, 2006), and this prevents runaway heating rendering IR heating as a possible thawing method. Generally, the quality of frozen food is affected by thawing rate and final product temperature. Frozen foods heated by far-IR radiation may result in less damage to foods during thawing. Combination of far-IR with air blast thawing has a potential for improving thawed meat quality aspects. An increase in IR dosage added to the thawing loss, but reduced the cooking loss, the water-holding capacity, and shear force (Geun et al., 2009). Frozen tuna was thawed using far-IR heating without drip losses and discoloration. Evolution of surface temperature due to absorption of IR energy is a significant parameter to control in thawing using IR energy. Liu et al. (1999) investigated the far-IR conditions on the temperature distribution of frozen tuna during thawing. A method for the preparation of packaged chestnuts involving aging, roasting, peeling, washing, freezing, far-IR thawing, and pasteurizing was developed. Far-IR thawing (170°C–230°C for 30–60 min) was used to inhibit microorganisms and prevent the formation of particular flavors, aromas, and deformities during processing (Hee, 2006).

1.3.6 IR Microbial Inactivation

IR heating can be used to inactivate bacteria, spores, yeast, and mold in both liquid and solid foods. IR radiation (70°C for 5 min) was found to be effective in reducing the growth of yeasts and fungi on cheese surface without affecting the product quality, resulting in a shelf life of 3–4 weeks at 4°C (Rosenthal et al., 1996). Two-stage IR treatment of wheat obtained a 1.56 $\log_{10}$ CFU/g reduction. The first irradiation treatment helped in activation of spores into vegetative cells and the second irradiation effectively inactivated spore formers (Hamanaka et al., 2000). The bacterial counts on the surface of eggs were reduced by IR radiation without significantly raising interior temperature (James et al., 2002).

The inactivation of *Escherichia coli* followed first-order reaction kinetics during IR heating. The higher death rate constant for far-IR heating compared to thermal conductive heating indicated that far-IR is potentially more efficient than conductive heating for pasteurization (Sawai et al., 2003, 2006).

Selective far-IR heating (5.88–6.66 μm) for inactivation of *Aspergillus niger* and *Fusarium proliferatum* in corn meal resulted in 40% increase in inactivation as compared to normal IR heating (Jun and Irudayaraj, 2003). The higher inactivation of *Bacillus subtilis* for wavelength of 950 nm compared to 1100 and 1150 nm demonstrated that inactivation efficiency is dependent on the radiation spectrum (Hamanaka et al., 2006).

Far-IR radiation was used to prevent fungal spoilage of strawberry during storage because it can achieve rapid and noncontact heating (Tanaka et al., 2007). IR exposure to raw almonds followed by holding at warm temperature for 60 min resulted in higher than 7.5-log reduction in *Salmonella enteric* without any significant change in kernel quality (Brandl et al., 2008). Dried powders (e.g., spice powders) may contain high microbial counts, particularly of bacterial spores. These spores do not germinate in dry condition. But the reconstitution of powders in high-moisture foods established a suitable environment for microbial growth. The reduction in the concentration of *Bacillus cereus* spores during near-IR heating of paprika powder was reduced by 4.5 $\log_{10}$CFU/g within 6 min and the final spore concentration remained approximately 2 $\log_{10}$CFU/g (Staack et al., 2008).

1.4 IR PROCESSING OF LIQUID FOODS

In case of fluid foods, IR heating mainly heats only a thin layer from the surface, which can be rapidly cooled after the treatment, and thus provides less change in the quality of food material because of negligible heat conduction (Hamanaka et al., 2000). IR can heat up to only a few millimeters below the surface of the sample. As the sample volumes increases, the total energy absorbed becomes limited. Some of the important findings for the application of liquid food processing have been summarized in Table 1.5.

TABLE 1.5
Main Findings with Regard to the Effect of IR on Liquid Food Products

Product	Salient Results	References
Microbes in Suspension	Far-IR radiation is easily absorbed by water and organic materials, which are the main components of food, and thus offers considerable potential for efficient pasteurization. Far-IR irradiation resulted in activation of *Bacillus subtilis* spores over a temperature range in which thermal conductive heating had no effect on spore viability. Temperature distribution within the far-IR-treated microbial suspension suggested that the temperature of the surface region was significantly higher than the bulk temperature.	Sawai et al. (1997b) and Hashimoto et al. (1992)
	E. coli cells in phosphate-buffered saline irradiated with far-IR energy were injured and killed even under the condition where the bulk temperature of the suspension was maintained below the lethal temperature. The survival ratio of *E. coli* was found to decrease with an increase in irradiation.	Sawai et al. (2000)
	Far-IR heating can decrease bacterial contamination levels more quickly and maintain greater enzyme activity levels than conductive heating. It has a potential for maintaining the required level of pasteurization at lower temperatures than conductive heating while maintaining the α-amylase activity and less change in lipase activity.	Sawai et al. (2003)
Orange Juice	A comparison of kinetics of degradation of vitamin C in orange juice during IR heating and conventional heating resulted in higher k-value or lower D-values in case of IR heating, which indicated higher degradation of vitamin C than conventional heating.	Vikram et al. (2005)
Milk	IR heating was demonstrated to be a potential for effective inactivation of *Staphylococcus aureus* in milk. To ensure the sterility efficacy, the heating patterns of milk samples under IR radiation were simulated using computational fluid dynamics.	Krishnamurthy et al. (2008b)
	The unprocessed honey tends to ferment within a few days of storage at ambient temperature because of its high moisture content and yeast count. To prevent fermentation, honey is heat processed before storage. IR heating achieved the desired results in a relatively shorter period offering advantages over the conventional method. Conventional heating for 5 min resulted in a product temperature of 85°C, which resulted in 220% increase of hydroxymethylfurfural content and 37% drop in enzyme activity. IR heating was reported to be adequate to obtain a commercially acceptable product, which met all the quality requirements in terms of hydroxymethylfurfural ($\leq$40 mg/kg), diastase activity (DN $\geq$ 8), moisture content (19.8%), and yeast count (200–300 CFU/mL).	Hebbar et al. (2003)
Beer	In order to extend the shelf life of beer, it can be either thermally pasteurized or subjected to a sterile micro-porous filtration. The short-term exposure of beer to near-IR treatment strongly suppressed the propagation of yeast and inactivated bacteria. The other advantages were in-pack processing, low energy consumption, low prices, and cheaper than the commercial methods currently used by breweries worldwide, besides offering the good-quality product.	Vasilenko (2001)

1.5 LIMITATIONS OF IR PROCESSING

In case of IR heating, the layer, close to the IR source, dried more rapidly compared to the one that was deep inside. Since IR energy is absorbed on the surface, it allows the outer layer to be dried more quickly. When the thickness of the bed is high, external agitation-like vibration would be helpful to turn the bed, so that each part can receive uniform radiation. Use of ultrasound in combination with IR heating may have overcome this limitation partially.

The absorption of total energy in case of IR heating is limited because IR radiation cannot penetrate deep inside the food and heats up only a few millimeters below the surface of the sample. Furthermore, the absorbed energy is then transferred by conduction to other areas within the food material. The increase in the sample volume limits the conduction and hence a limited total energy can only be absorbed.

The penetrative radiation energy does not make significant contribution to internal heating. A combination of IR heating with microwave or other common conductive and convective modes of heating may be used to achieve optimum use of energy. The utilization of IR heating for prolonged exposure of biological materials may cause fracturing of foods and may impair quality.

1.6 EQUIPMENT FOR IR PROCESSING

Owing to simplicity of construction and operation of IR equipment, IR heat processing is becoming an important heat treatment in the food industry. The other features are its fast transient response, significant associated energy savings over other thermal processes, and easy accommodation with convective, conductive, and microwave heating (Sandu, 1986). IR ovens have a modular design that can fit easily into most production lines, take up less floor space than convection ovens, and need minimal maintenance. The information regarding the equipment and techniques used in IR processing is either proprietary or available in the form of patent. A number of designs of IR heaters are also available in the literature for different products such as dehydration of vegetables or cereals. A few have been discussed in the following section.

An integrated mid-IR and hot air drying system for the dehydration of vegetables consisting of three insulated chambers, fitted with quartz IR heaters on either side of the wire mesh conveyor, was developed. A through flow hot air heating system was provided for convective heating. The combined IR drying reduced the processing time significantly (48%) and consumed less energy (63%) for water evaporation as compared to hot air drying (Hebbar et al., 2004, Figure 1.4).

The dryer consisted of mainly two basic units: a heating unit and a vibrating unit, which were developed for the IR drying of the paddy. This unit had two IR lamps (250 W) connected in series with provision for varying the electrical input to the lamp and thereby changing the radiation intensity. The whole setup was mounted over the vibrating unit. The vibrating unit had the provision for vibrating the drying tray on which the paddy sample was placed (Das et al., 2004a, Figure 1.5).

Mongpraneet et al. (2002a) developed an experimental far-IR dryer with vacuum drying chamber. The material was dried by placing it in the vacuum drying chamber, and then simultaneously reducing the pressure by means of the aspirator while

Developments in Infrared Heating in Food Processing

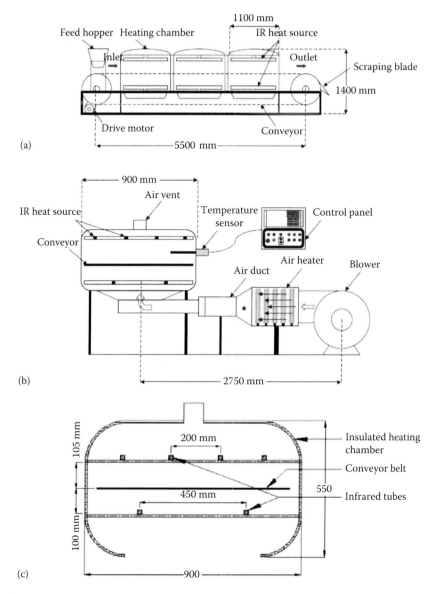

FIGURE 1.4 Combined infrared and hot air dryer. (a) Front view; (b) side view; and (c) cross-sectional view. (From Hebbar, U.H. et al., *J. Food Eng.*, 65, 557, 2004. With permission.)

starting the heater. Vaporization was promoted even at low temperature by the application of vacuum. A stainless steel heating coil comprised the heating element and was effectively made into a fluororesin plate heater by covering the surface in fluororesin, which had excellent radiation efficiency for far-IR. The coated fluorine resin plastic board IR heater was operated at 100 V, with a maximum power of 150 W (Figure 1.6).

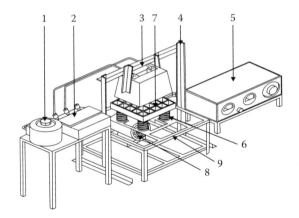

FIGURE 1.5 Vibration-aided IR dryer (1—variac; 2—wattmeter; 3—IR lamps; 4—main frame; 5—motor speed regulating unit; 6—helical springs; 7—drying tray; 8—motor; 9—base plate). (From Das, I. et al., *J. Food Eng.*, 64, 129, 2004a. With permission.)

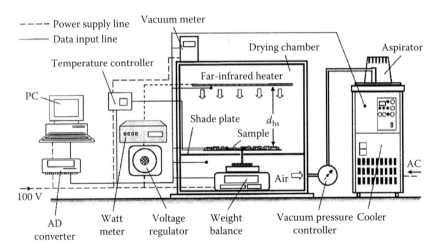

FIGURE 1.6 A schematic view of far-infrared dryer with vacuum extractor. (From Mongpraneet, S. et al., *J. Food Eng.*, 55, 147, 2002a. With permission.)

Fasina et al. (1999) developed a laboratory-scale IR heating system for the processing of grains consisted of vibratory conveyor and vibratory feeder to turn the seeds frequently as they passed beneath the IR burner (Figure 1.7). During operation, the grain seeds were fed onto the vibratory conveyor via the vibratory feeder. The vibration of the conveyor turned the seeds frequently as they passed beneath the IR burner to ensure that the seeds received uniform IR energy on all surfaces. The propane-fired burners were lit and allowed to reach a steady-state operating condition. The rate of flow of sample when passing under the IR heater was adjusted such that the seeds attained the desired surface temperature while exiting the conveyor. Depending on their initial moisture contents (12.2%–26.5%, w.b.), samples were heated to surface temperatures in the range of 105°C–150°C.

Developments in Infrared Heating in Food Processing

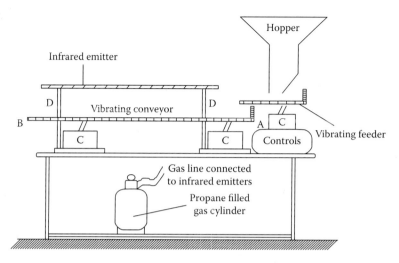

FIGURE 1.7 Schematic Diagram of the laboratory-scale infrared system (AB—vibrating conveyor, C—vibrating feeder, D—stand supporting infrared emitters and side plates). (From Fasina, O.O. et al., *J. Food Proc. Preserv.*, 23, 135, 1999. With permission.)

Vikram et al. (2005) used an IR heater for the processing of orange juice, which consisted of a heating chamber with IR modules (250 W), which was equipped with reflectors to direct IR waves on to a platform. A temperature controller maintained the set temperature during processing. The provision was provided to adjust the distance between the platform and the IR source in order to vary the intensity (Figure 1.8). Hebbar et al. (2003) employed a near-IR batch oven for honey processing, which was fitted with IR lamps (1.0 kW, peak wavelength 1.1–1.2 μm). The distance between the sample and the source was fixed in such a way, which ensured uniform power intensity of 0.2 W/cm^2.

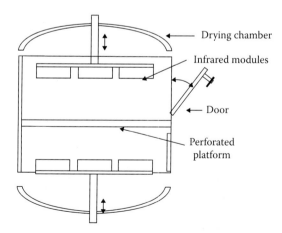

FIGURE 1.8 Batch-type infrared heater. (From Vikram, V.B. et al., *J. Food Eng.*, 69, 31, 2005. With permission.)

Krishnamurthy et al. (2008b) employed a lab-scale, custom-made IR heating system with a cone-shaped waveguide for IR processing of milk. The IR heating system had six ceramic IR lamps (500 W) with a cast-in K-type thermocouple. All the lamps were fixed inside the closing on top of the waveguide and arranged symmetric to the central axis of the waveguide (Figure 1.9).

Hashimoto et al. (1991) and Sawai et al. (2003) used a far-IR heater consisting of a cylinder and reflector. The reflector was placed at the top of irradiation chamber made up of an aluminum plate. During irradiation, the enzyme solution was agitated with the help of a rotary shaker (Figure 1.10).

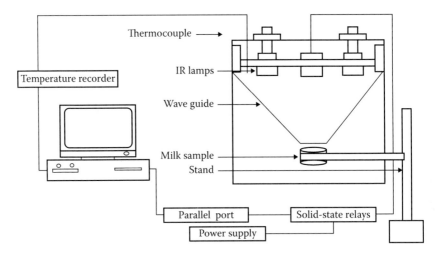

FIGURE 1.9 Schematic view of the lab-scale infrared heating system. (From Krishnamurthy, K. et al., *J. Food Proc. Eng.*, 31, 798, 2008b. With permission.)

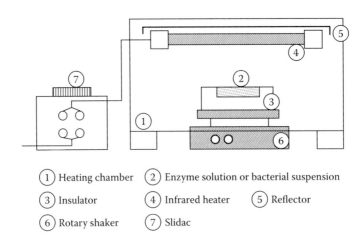

FIGURE 1.10 Schematic apparatus for FIR heating. (From Sawai, J. et al., *Int. J. Food Sci. Technol.*, 38, 661, 2003. With permission.)

1.7 CONCLUDING REMARKS

The significant attractive merits of IR heating over conventional heating methods for various food-processing applications include high heat delivery rate, absence of heating medium, reduced processing time, improved energy efficiency, and enhanced product quality and safety without significant negative environmental footprint. The development and commercialization of IR-based food-processing technologies could open new avenues to deliver desirable foods to consumers while reducing the consumption of natural resources.

IR heating is fast and produces heating inside the material being dried, but its penetrating powers are limited, and hence it is difficult to achieve optimum energy usage and efficient practical applicability of IR heating in the food-processing industry. Sometimes, prolonged exposure of a biological material to IR heat results in swelling and ultimately fracturing of the material. IR heating has attracted a lot of attention for surface heating applications such as prevention of growth of yeasts and fungi on cheese surfaces, peeling of tomatoes and cashew, pasteurization of the surface of eggs, arresting fungal spoilage of strawberry during storage, and dry pasteurization of raw almonds.

The use of IR spectrum, after passing through suitable filters to allow radiation within a specific spectral range for specific application, is referred to as *selective heating*. It was found to be more effective for microbial inactivation. Such a controlled radiation can stimulate the maximum optical response of the target object when the emission band of IR and the peak absorbance band of the target object are coinciding. Specific applications of IR radiation for selective heating of foods could be very useful and open up avenues for future research in this area. Sequential or combination of IR with other treatments also offers diverse food-processing merits, including improved process efficiency, product characteristics, and safety.

The effect of irradiation on nutritional or sensorial characteristic and physicochemical properties as well as interaction of food components under IR radiation may further justify the use of IR radiation as a future food-processing option. More knowledge about the interaction between processes and products needs to be gained.

Good background knowledge may combine the techniques in the most optimal way. More knowledge is important for the success of the heating technique. Its application is certain to grow as food equipment manufacturers begin to realize its full potential. The IR technology should probably be used at the start of the heating procedure or a stepwise combination of different power levels should be designed.

ACKNOWLEDGMENTS

The author is very grateful to Prof. Ram Rajasekharan, Director, Central Food Technological Research Institute, Mysore, India, for constant encouragement. Thanks are also due to Mr. K. Shesha Narayana for his help in providing useful information.

REFERENCES

Abdul, K.R., Bargman, T.J., and Rupnow, J.H. Effect of infrared heat processing on rehydration rate and cooking of *Phaseolus vulgaris* (var. pinto). *J. Food Sci.* 55, (1990): 1472–1473.

Afzal, T.M. and Abe, T. Diffusion in potato during far-infrared radiation drying. *J. Food Eng.* 37(4), 1998: 353–365.

Afzal, T.M. and Abe, T. Simulation of moisture changes in barley during far infrared radiation drying. *Comp. Electronics Agric.* 26, (2000): 137–145.

Afzal, T.M., Abe, T., and Hikida, Y. Energy and quality aspects of combined FIR-convection drying of barley. *J. Food Eng.* 42, (1999): 177–182.

Amiri, C.R., Majid, T.S., and Rahimi, S.F. Modeling infrared-convective drying of pistachio nuts under fixed and fluidized bed conditions. *J. Food Proc. Preserv.* 38(3), (2014): 1224–1233.

Bengang, W., Haile, M., Wenjuan, Q., Bei, W.X.Z., Peilan, W., Juan, W., Atungulu, G.G., and Zhongli, P. Catalytic infrared and hot air dehydration of carrot slices. *J. Food Proc. Eng.* 37(2), (2014): 111–121.

Bingol, G., Zhang, A., Zhongli, P., and McHugh, T.H. Producing lower-calorie deep fat fried French fries using infrared dry-blanching as pretreatment. *Food Chem.* 132, (2012): 686–692.

Boudhrioua, N., Bahloul, N., Slimen, I.B., and Kechaou, N. Comparison on the total phenol contents and the color of fresh and infrared dried olive leaves. *Indust. Crops Prod.* 29, (2009): 412–419.

Brandl, M.T., Zhongli, P., Huynh, S., Yi, Z., and McHugh, T.H. Reduction of *Salmonella enteritidis* population sizes on almond kernels with infrared heat. *J. Food Prot.* 71, (2008): 897–902.

Bualuang, O., Tirawanichakul, Y., and Tirawanichakul, S. Comparative study between hot air and infrared drying of parboiled rice: Kinetics and qualities aspects. *J. Food Proc. Preserv.* 37(6), (2013): 1119–1132.

Cenkowski, S. and Sosulski, F.W. Cooking characteristics of split peas treated with infrared heat. *Trans ASAE.* 41, (1998): 715–720.

Chatchai N., Sakamon, D., Thanit, S., and Somchart, S. Drying of banana slices using combined low pressure superheated steam and far infrared radiation. *J. Food Eng.* 81, (2007): 624–633.

Chung, D. (2008). Coffee bean roaster and method for roasting coffee beans using the same. US Patent 2008/0268119 A1.

Das, I., Das, S.K., and Bal, S. Drying performance of a batch type vibration aided infrared dryer. *J. Food Eng.* 64, (2004a): 129–133.

Das, I., Das, S.K., and Bal, S. Specific energy and quality aspects of infrared (IR) dried parboiled rice. *J. Food Eng.* 62, (2004b): 9–14.

Das, I., Das, S.K., and Bal, S. Drying kinetics of high moisture paddy undergoing vibration assisted infrared (IR) drying. *J. Food Eng.* 95, (2009): 166–171.

Datta, A.K., Sahin, S., Sumnu, G., and Ozge, K.S. Porous media characterization of breads baked using novel heating modes. *J. Food Eng.* 79, (2007): 106–116.

Decareau, R.V. *Microwaves in the Food Processing Industry.* Orlando, FL: Academic Press, (1985).

Dostie, M., Seguin, J.N., Maure, D., Tonthat, Q.A., and Chatingy, R. Preliminary measurements on the drying of thick porous materials by combinations of intermittent infrared and continuous convection heating. In *Drying'89*, A.S. Mujumdar and M.A. Roques (ed.), New York: Hemisphere Press, (1989).

Doymaz, I. Suitability of thin-layer drying models for infrared drying of peach slices. *J. Food Proc. Preserv.* 38(6), (2014a): 2232–2239.

Doymaz, I. Experimental study and mathematical modeling of thin-layer infrared drying of watermelon seeds. *J. Food Proc. Preserv.* 38(3), (2014b): 1377–1384.

Doymaz, I. Mathematical modeling of drying of tomato slices using infrared radiation. *J. Food Proc. Preserv.* 38(1): (2014c): 389–396.

Doymaz, I., Kipcak, A. S., and Piskin, S. Characteristics of thin-layer infrared drying of green bean. *Czech J. Food Sci*, 33(1), (2015): 83–90.

Dujmic, F., Brncic, M., Karlovic, S., Bosiljkov, T., Jezek, D., Tripalo, B., and Mofardin, I. Ultrasound-assisted infrared drying of pear slices: Textural issues. *J. Food Proc. Eng.* 36(3), (2013): 397–406.

Ertekin, C. and Heybeli, N. Thin-layer infrared drying of mint leaves. *J. Food Proc. Preserv.* 38(4), (2014): 1480–1490.

Fasina, O., Tyler, B., Pickard, M., Guo, H.Z., and Ning, W. Effect of infrared heating on the properties of legume seeds. *Int. J. Food Sci. Technol.* 36, (2001): 79–90.

Fasina, O.O., Tyler, R.T., and Pickard, M.D. Modelling the infrared radiative heating of agricultural crops. *Drying Technol.* 16, (1998): 2065–2082.

Fasina, O.O., Tyler, R.T., Pickard, M.D., and Zheng, G.H. Infrared heating of hulless and pearled barley. *J. Food Proc. Preserv.* 23, (1999): 135–151.

Gabel, M.M., Zhongli, P., Amaratunga, K.S.P., Harris, L.J., and Thompson, J.F. Catalytic infrared dehydration of onions. *J. Food Sci.* 71, (2006): E351–E357.

Geun, P.H., Kook, B.S., Mi, J.C., and Sang, G.M. Effects of air blast thawing combined with infrared radiation on physical properties of pork. *Korean J. Food Sci. An. Resource.* 29, (2009): 302–309.

Ginzburg, A.S. *Application of Infrared Radiation in Food Processing.* London, UK: Leonard Hill Books, (1969).

Gomez, G.F., Toledo, R.T., and Sjoholm, I. Tissue damage in heated carrot slices. Comparing mild hot water blanching and infrared heating. *J. Food Eng.* 67, (2005): 381–385.

Gonzalez, H.J., Luz, M.J., Sanchez, S.F., Martinez, B.F., Dios, F.C., and Ruiz, T.M. (1996). Method of cooking corn dough tortillas using infrared radiation. US Patent 5 567 459.

Hamanaka, D., Dokan, S., Yasunaga, E., Kuroki, S., Uchino, T., and Akimoto, K. (2000). The sterilization effects on infrared ray of the agricultural products spoilage microorganisms (part 1). An ASAE Meeting Presentation, Milwaukee, WI, July 9–12, No. 00 6090.

Hamanaka, D., Uchino, T., Furuse, N., Han, W., and Tanaka, S.I. Effect of the wavelength of infrared heaters on the inactivation of bacterial spores at various water activities. *Int. J. Food Microbiol.* 108, (2006): 281–285.

Hashimoto, A., Shimizu, M., and Igarashi, H. Effect of far infrared radiation on pasteurization of bacteria suspended in phosphate-buffered saline. *Kagaku Kogaku Ronbunshu (in Japanese).* 17, (1991): 627–633.

Hashimoto, A., Igarashi, H., and Shimizu, M. Far-infrared irradiation on pasteurization of bacteria on or within wet-solid medium. *J. Chem. Eng.* Japan 25, (1992): 666–671.

Hataichanok, K., Ampawan, T., and Mittal, G.S. Drying characteristics and quality of shiitake mushroom undergoing microwave-vacuum drying and microwave-vacuum combined with infrared drying. *J. Food Sci. Tech.* 51(12) (2014), 3594–3608.

Hebbar, U., Ramesh, A., and Rastogi, N.K. (1999). Detachment of cashew kernel from testa layer using infrared radiation. Indian Patent 1562/DEL/99.

Hebbar, U.H., Nandini, K.E., Lakshmi, M.C., and Subramanian, R. Microwave and infrared heat processing of honey and its quality. *Food Sci. Technol. Res.* 9, (2003): 49–53.

Hebbar, U.H. and Rastogi, N.K. Mass transfer during infrared drying of cashew kernel. *J. Food Eng.* 47, (2001): 1–5.

Hebbar, U.H., Vishwanathan, K.H., and Ramesh, M.N. Development of combined infrared and hot air dryer for vegetables. *J. Food Eng.* 65, (2004): 557–563.

Hee, S.C. (2006). Preparation methods of retort roast chestnuts using far infrared ray thawing. WO 2006/065018A1.

Hong, C.W., Min, Z., and Adhikari, B. Drying of shiitake mushroom by combining freeze-drying and mid-infrared radiation. *Food Bioprod Process*, 94, (2015): 507–517.

Hongping, T., Zhongli, P., Yi, Z., McHugh, T.H., and Yibin, Y. Quality of frozen fruit bars manufactured through infrared partial dehydration. *J. Food Proc. Preserv.* 37(5), (2013): 784–791.

James, C., Lechevalier, V., and Ketteringham, L. Surface pasteurization of shell eggs. *J. Food Eng.* 53, (2002): 193–197.

Jeevitha, G.C., Hebbar, U.H., and Raghavarao, K.S.M.S. Electromagnetic radiation-based dry blanching of red bell peppers: A comparative study. *J. Food Proc. Eng.* 36(5), (2013): 663–674.

Ji, H.M., Moo, J.K., Dong, H.C., Cheol, H.P., and Won, B.Y. Drying characteristics of sea cucumber (*Stichopus japonicas* Selenka) using far infrared radiation drying and hot air drying. *J. Food Proc. Preserv.* 38(4), (2014): 1534–1546.

Juckamas, L. and Seree, W. Drying characteristics and milling quality aspects of paddy dried with gas fired infrared. *J. Food Process Eng.* 32, (2009): 442–461.

Jun, S. and Irudayaraj, J. A dynamic fungal inactivation approach using selective infrared heating. *Trans ASAE*. 46, (2003): 1407–1412.

Junling, S., Zhongli, P., McHugh, T.H., Wood, D., Hirschberg, E., and Olson, D. Drying and quality characteristics of fresh and sugar infused blueberries dried with infrared radiation heating. *LWT-Food Sci. Technol.* 41, (2008): 1962–1972.

Kamon, P., Naret, M., Somchart, S., and Sirithon, S. Modeling of combined far-infrared radiation and air drying of a ring shaped-pineapple with/without shrinkage, *Food and Bioprod. Proc.* 90(2), (2012): 155–164.

Keskin, S.O., Sumnu, G., and Sahin, S. A study on the effects of different gums on dielectric properties and quality of breads baked in infrared microwave combination oven. *Eur. Food Res. Technol.* 224, (2007): 329–334.

Keya, E.L. and Sherman, U. Effects of a brief, intense infrared radiation treatment on the nutritional quality of maize, rice, sorghum and beans. *Food Nutr. Bull.* 18, (1997): 382–387.

Khir, R., Zhongli, P., Thompson, J.F., El Sayed, A.S., Hartsough, B.R., and El-Amir, M.S. Moisture removal characteristics of thin layer rough rice under sequenced infrared radiation heating and cooling. *J. Food Proc. Preserv.* 38(1), (2014): 430–440.

Kim, S.Y., Jeong, S.M., Jo, S.C., and Lee, S.C. Application of far infrared irradiation in the manufacturing process of green tea. *J. Agric. Food Chem.* 54, (2006): 9943–9947.

Kocabiyik, H., Yilmaz, N., Tuncel, N.B., Sumer, S.K., and Buyukcan, M.B. Drying, energy, and some physical and nutritional quality properties of tomatoes dried with short-infrared radiation. *Food Bioprocess Tech.* 8(3), (2015): 516–525.

Konopka, I., Markowski, M., Tanska, M., Zmojda, M., Malkowski, M., and Bialobrzewski, I. Image analysis and quality attributes of malting barley grain dried with infrared radiation and in a spouted bed. *Int. J. Food Sci. Technol.* 43, (2008): 2047–2055.

Krishnamurthy, K., Khurana, H.K., Jun, S., Irudayaraj, J.M., and Demirci, A. Infrared radiation for food processing. In *Food Processing Operations Modeling: Design and Analysis*, S. Jun and J.M. Irudayaraj (eds.), Boca Raton, FL: CRC Press, pp. 115–142, (2009).

Krishnamurthy, K., Khurana, K., Soojin, J., Irudayaraj, J., and Demirci, A. Infrared heating in food processing: An overview. *Comp. Rev. Food Sci. Food Safety*. 7, (2008a): 2–13.

Krishnamurthy, K., Soojin, J., Irudayaraj, J., and Demirci, A. Efficacy of infrared heat treatment for inactivation of *Staphylococcus aureus* in milk. *J. Food Proc. Eng.* 31, (2008b): 798–816.

Kumar, C.M., Rao, A.G.A., and Singh, A.S. Effect of infrared heating on the formation of sesamol and quality of defatted flours from *Sesamum indicum* L. *J. Food Sci.* 74, (2009): H105–H111.

Leonard, A., Blacher, S., Nimmol, C., and Devahastin, S. Effect of far infrared radiation assisted drying on microstructure of banana slices: An illustrative use of X ray microtomography in microstructural evaluation of a food product. *J. Food Eng.* 85, (2008): 154–162.

Levinson, M.L. (1992). Two stage process for cooking/browning/crusting food by microwave energy and infrared energy. WO 92/14369 A1.
Liu, C.M., Sakai, N., and Hanzawa T. Three dimensional analysis of heat transfer during food thawing by far-infrared radiation. *Food Sci. Technol. Res.* 5(3), (1999): 294–299.
Luz, M.J., Gonzalez, H.J., Sanchez, S.F., Ruiz, T.M., Dios, F.C.J., and Martinez, B.F. (1996). Method for cooking wheat flour products by using infrared radiation. US 5 589 210.
Martinez, B.F., Morales, S.E., Chang, Y.K., Herrera, G.A, Martinez, M.J.L., Banos, L, Rodriguez, M.E., and Flores, M.H.E. Effect of infrared baking on wheat flour tortilla characteristics. *Cereal Chem.* 76, (1999): 491–495.
McCurdy, S.M. Infrared processing of dry peas, canola, and canola screenings. *J. Food Sci.* 57, (1992): 941–944.
Mongpraneet, S., Abe, T., and Tsurusaki, T. Accelerated drying of Welsh onion by far infrared radiation under vacuum conditions. *J. Food Eng.* 55, (2002a): 147–156.
Mongpraneet, S., Abe, T., and Tsurusaki, T. Far infrared vacuum and convection drying of Welsh onion. *Trans ASAE.* 45, (2002b): 1529–1535.
Mu, C.Z., Jin, F.B., Qin, Q.C., Xuan, L., Xin, Y, W., and Yi, Jiao. Drying characteristics and kinetics of kiwifruit slice under medium and shortwave infrared radiation. *Mod Food Sci and Technol.* 1(199), (2014a): 153–159.
Mu, C.Z., Jin, F.B., Qin, Q.C., Xuan, L., Xin, Y, W., and Yi, Jiao. Weibull distribution for modeling med-and short-wave infrared radiation drying of kiwifruit slices. *Mod Food Sci and Technol.* 6(201) (2014b): 146–151.
Myung, K.L., Sang, H.K., Seung, S.H., Sang, Y.L., Cha, K.C., Il, J.K., and Deog, H.O. The effect of far infrared ray vacuum drying on the quality changes of *Pimpinella brachycarpa*. *J. Korean Soci. Food Sci. Nutr.* 29, (2000): 561–567.
Naret, M., Adisak, N., Thanid, M., and Somchart, S. Modelling of far-infrared irradiation in paddy drying process. *J. Food Eng.* 78(4) (2007): 1248–1258.
Navari, P., Andrieu, J., and Gevaudan, A. Studies on infrared and convective drying of non-hygroscopic solids. In *Drying 92*, A.S. Mujumdar (ed.), Amsterdam, the Netherlands: Elsevier Science, pp 685–694, (1992).
Nowak, D. and Lewicki, P.P. Infrared drying of apple slices. *Innov. Food Sci. Emerg. Technol.* 5, (2004): 353–360.
Nowak, D. and Lewicki, P.P. Quality of infrared dried apple slices. *Drying Technol.* 23, (2005): 831–846.
Olsson, E.E.M., Tragardh, A.C., and Ahrne, L.M. Effect of near infrared radiation and jet impingement heat transfer on crust formation of bread. *J. Food Sci.* 70, (2005): E484–E491.
Ozdemir, M. and Devres, O. Analysis of color development during roasting of hazelnuts using response surface methodology. *J. Food Eng.* 45, (2000): 17–24.
Ozkoc, S.O., Sumnu, G., Sahin, S., and Turabi, E. Investigation of physicochemical properties of breads baked in microwave and infrared microwave combination ovens during storage. *Eur. Food Res. Technol.* 228, (2009): 883–893.
Park, J.H., Lee, J.M., Cho, Y.J., Kim, C.T., Kim, C.J., Nam, K.C., and Lee, S.C. Effect of far infrared heater on the physicochemical characteristics of green tea during processing. *J. Food Biochem.* 33, (2009): 149–162.
Pathare, P.B., and Sharma, G.P. Effective moisture diffusivity of onion slices undergoing infrared convective drying. *Biosys. Eng.* 93, (2006): 285–291.
Ponne, C.T., Baysal, T., and Yuksel, D. Blanching leafy vegetables with electromagnetic energy. *J. Food Sci.* 59, (1994): 1037–1041.
Poss, G.T. (2007). Roasting coffee beans. US Patent 7 235 764 B2.
Praveen Kumar, D.G., Hebbar, U.H., and Ramesh, M.N. Suitability of thin layer models for infrared hot air drying of onion slices. *LWT-Food Sci. Technol.* 39, (2006): 700–705.
Praveen Kumar, D.G., Hebbar, U.H., Sukumar, D., and Ramesh, M.N. Infrared and hot air drying of onions. *J. Food Proc. Preserv.* 29, (2005): 132–150.

Rastogi, N.K. Applications of infrared heating in food processing. In *Novel Thermal and Nonthermal Technologies for Fluid Foods*, P.J. Cullen, B.K. Tiwari, and V.P. Valdramidis (eds.), Philadelphia, PA: Academic Press, Elsevier, pp. 411–432, (2012a).

Rastogi, N.K. Recent trends and developments in infrared heating in food processing. *Crit. Rev. Food Sci. Nutr.* 52(9), (2012b): 737–760.

Rastogi, N.K. Opportunities and challenges in application of infrared in food processing. In *Electron Beam Pasteurization and Complementary Food Processing Technologies*, S. Pillai, S. Shayanfar, and N. Holden (eds.), Cambridge, UK: Woodhead Publishing, pp. 61–82, (2015).

Ratti, C. and Mujumdar, A.S. Infrared drying. In *Handbook of Industrial Drying*, A.S. Mujumdar (ed.), New York: Marcel Dekker, pp. 567–588, (1995).

Rim, A.R., Jung, E.S., Jo, S.C., and Lee, S.C. Effect of far infrared irradiation and heat treatment on the antioxidant activity of extracts from peanut (*Arachis hypogaea*) shell. *J. Korean Soci. Food Sci. Nutr.* 34, (2005): 1114–1117.

Rosenthal, I., Rosen, B., and Bernstein, S. Surface pasteurization of cottage cheese. *Milchwissenschaft.* 51, (1996): 198–201.

Sakai, N. and Hanzawa, T. Applications and advances in far-infrared heating in Japan. *Trends Food Sci. Technol.* 5, (1994): 357–362.

Sakai, N. and Mao, W. Infrared heating. In *Thermal Food Processing: New Technologies and Quality Issues*, D.W. Sun (ed.), Boca Raton, FL: CRC Press, pp. 493–544, (2006).

Sakiyan, O., Sumnu, G., Sahin, S., and Meda, V. Investigation of dielectric properties of different cake formulations during microwave and infrared microwave combination baking. *J. Food Sci.* 72, (2007a): E205–E213.

Sakiyan, O., Sumnu, G., Sahin, S., and Meda, V. The effect of different formulations on physical properties of cakes baked with microwave and near infrared microwave combinations. *J. Microwave Pow. Electromag. Energy.* 41, (2007b): 20–26.

Sandu, C. Infrared radiative drying in food engineering: A process analysis. *Biotechnol. Prog.* 2, (1986): 109–119.

Sathira, H., Jeong, H.C., Jutatip, A., Chonnipa, P., Chanutchamon, S., Quan, V.V. Suwimol, C., Young, R.H., and Scarlett, C.J. Optimization of far-infrared vacuum drying conditions for Miang leaves (*Camellia sinensis var. assamica*) using response surface methodology. *Food Sci. Biotechnol.* 24(2), (2015): 461–469.

Sato, H., Hatae, K., and Shimada, A. Studies on radiative heating condition of foods. I. Effect of radiant characteristics of heaters on crust formation and coloring processes of food surfaces. *J. Japanese Soci. Food Sci. Technol.* 39, (1992): 784–789.

Sawai, J., Kojima, H., Igarashi, H., Hashimoto, A., Fujisawa, M., Kokugan, T. and Shimizu, M. Pasteurization of bacterial spores in liquid medium by far-infrared irradiation. *J. Chem. Eng. Jpn.* 30, (1997): 170–172.

Sawai, J., Isomura, Y., Honma, T., and Kenmochi, H. Characteristics of the inactivation of *Escherichia coli* by infrared irradiative heating. *Biocont. Sci.* 11, (2006): 85–90.

Sawai, J., Sagara, K., Hashimoto, A., Igarashi, H., and Shimizu, M. Inactivation characteristics shown by enzymes and bacteria treated with far infrared radiative heating. *Int. J. Food Sci. Technol.* 38, (2003): 661–667.

Sawai, J., Sagara, K., Kasai, S., Igarashi, H., Hashimoto, A., Kokugan, T., Shimizu, M., and Kojima, H. Far infrared irradiation induced injuries to *Escherichia coli* at below the lethal temperature. *J. Indust. Microbiol. Biotechnol.* 24, (2000): 19–24.

Seok, H.E., Hyung, J.P., Dong, W.S., Won, W.K., and Dong, H.C. Stimulating effects of far infrared ray radiation on the release of antioxidative phenolics in grape berries. *Food Sci. Biotechnol.* 18, (2009): 362–366.

Sharma, G.P., Verma, R.C., and Pankaj, P. Mathematical modeling of infrared radiation thin layer drying of onion slices. *J. Food Eng.* 71, (2005a): 282–286.

Sharma, G.P., Verma, R.C., and Pankaj, P.B. Thin layer infrared radiation drying of onion slices. *J. Food Eng.* 67, (2005b): 361–366.
Shi, J., Pan, Z., McHugh, T.H., Wood, D., Zhu, Y., Avena Bustillos, R.J., and Hirschberg, E. Effect of berry size and sodium hydroxide pretreatment on the drying characteristics of blueberries under infrared radiation heating. *J. Food Sci.* 73, (2008): E259–E265.
Shih, C., Pan, Z., McHugh, T., Wood, D., and Hirschberg, E. Sequential infrared radiation and freeze drying method for producing crispy strawberries. *Trans ASABE.* 51, (2008): 205–216.
Skjoeldebrand, C. and Andersson, C. A comparison of infrared bread baking and conventional baking. *J. Microwave Pow. Electromag. Energy.* 24, (1989): 91–101.
Skjoldebrand, C. Infrared heating. In *Thermal Technologies in Food Processing*, P. Richardson (ed.), Abington, England: Woodhead Publishing, pp. 208–228, (2001).
Skjoldebrand, C. Infrared processing. In *The Nutrition Handbook for Food Processors*, C.J.K. Henry and C. Chapman (eds.), Abington, England: Woodhead Publishing, pp. 423–433, (2002).
Staack, N., Ahrne, L., Borch, E., and Knorr, D. Effects of temperature, pH, and controlled water activity on inactivation of spores of *Bacillus cereus* in paprika powder by near IR radiation. *J. Food Eng.* 89, (2008): 319–324.
Sumnu, G. and Ozkoc, S.O. Infrared baking and roasting. In *Infrared Heating for Food and Agricultural Processing*, Z. Pand and G.G. Atungulu (eds.), Boca Raton, FL: CRC Press, pp. 203–236, (2010).
Sumnu, G., Sahin, S., and Sevimli, M. Microwave, infrared and infrared microwave combination baking of cakes. *J. Food Eng.* 71, (2005): 150–155.
Sun, J., Hu, X., Zhao, G., Wu, J., Wang, Z., Chen, F., and Liao, X. Characteristics of thin layer infrared drying of apple pomace with and without hot air pre drying. *Food Sci. Technol. Int.* 13, (2007): 91–97.
Tan, M., Chua, K.J., Mujumdar, A.S., and Chou, S.K. Effect of osmotic pre-treatment and infrared radiation on drying rate and color changes during drying of potato and pineapple. *Drying Technol.* 19, (2001): 2193–2207.
Tanaka, F., Chatani, M., Kawashima, H., Hamanaka, D., and Uchino, T. CFD modeling of infrared thermal treatment of figs (*Ficus carica* L.). *J. Food Proc. Preserv.* 35(6), (2012): 821–828.
Tanaka, F., Verboven, P., Scheerlinck, N., Morita, K., Iwasaki, K., and Nicolai, B. Investigation of far infrared radiation heating as an alternative technique for surface decontamination of strawberry. *J. Food Eng.* 79, (2007): 445–452.
Thanit, S., Sakamon, D., Poomjai, S.A., and Somchart S. Mathematical modeling of combined far infrared and vacuum drying banana slice. *J. Food Eng.* 92, (2009): 100–106.
Tireki, S., Sumnu, G., and Esin, A. Production of bread crumbs by infrared assisted microwave drying. *Eur. Food Res. Technol.* 222, (2006): 8–14.
Togrul, H. Suitable drying model for infrared drying of carrot. *J. Food Eng.* 77, (2006): 610–619.
Turabi, E., Sumnu, G., and Sahin, S. Optimization of baking of rice cakes in infrared microwave combination oven by response surface methodology. *Food Bioproc. Technol.* 1, (2008): 64–73.
Uysal, N., Sumnu, G., and Sahin, S. Optimization of microwave infrared roasting of hazelnut. *J. Food Eng.* 90, (2009): 255–261.
Van Zuilichem, D.J., Riet, V.T.K., and Stolp, W. Food engineering and process applications. In *Transport Phenomena*, Elsevier applied science, vol. 1, L.M. Maguer and P. Jelen (eds.), New York: Elsevier, pp. 595–610, (1985).
Vasilenko, V. The pasteurization effect of laser infrared irradiation on beer, Master Brewers Association of the Americas, *Technical Quarterly*, 38, (2001): 211–214.
Vikram, V.B., Ramesh, M.N., and Prapulla, S.G. Thermal degradation kinetics of nutrients in orange juice heated by electromagnetic and conventional methods. *J. Food Eng.* 69, (2005): 31–40.

Vishwanathan, H.K., Girish, K.G., and Hebbar, H.U. Infrared assisted dry-blanching and hybrid drying of carrot. *Food Bioprod. Proc.* 91(2), (2013): 89–94.

Wang, J. A single layer model for far infrared radiation drying of onion slices. *Drying Technol.* 20, (2002): 1941–1953.

Wesolowski, A. and Glowacki, S. Shrinkage of apples during infrared drying. *Pol. J. Food Nutr. Sci.* 12/53, (2003): 9–12.

Xuan, L. and Zhongli, P. Dry-peeling of tomato by infrared radiative heating: Part I. Model development. *Food Bioprocess Technol.* 7(7), (2014a): 1996–2004.

Xuan, L. and Zhongli, P. Dry peeling of tomato by infrared radiative heating: Part II. Model validation and sensitivity analysis. *Food Bioprocess Technol.* 7(7), (2014b): 2005–2013.

Xuan, L., Zhongli, P., Atungulu, G.G., Xia, Z., Wood, D., Delwiche, M., and McHugh, T.H. Peeling of tomatoes using novel infrared radiation heating technology. *Innov. Food Sci. Emerg. Technol.* 21, (2014): 123–130.

Yang, J., Bingol, G., Pan, Z., Brandl, M.T., McHugh, T.H., and Wang, H. Infrared heating for dry-roasting and pasteurization of almonds. *J. Food Eng.* 101(3), (2010): 273–280.

Yeu, P.L., Jen, H.T., and An Erl King, V. Effects of far infrared radiation on the freeze drying of sweet potato. *J. Food Eng.* 68, (2005): 249–255.

Yeu, P.L., Tsai, Y.L., Jen, H.T., and An Erl King, V. Dehydration of yam slices using far-IR assisted freeze drying. *J. Food Eng.* 79, (2007): 1295–1301.

Yi, Z., Zhongli, P., and McHugh, T.H. Effect of dipping treatments on color stabilization and texture of apple cubes for infrared dry blanching process. *J. Food Proc. Preserv.* 31, (2007): 632–648.

Yi, Z. and Zhongli, P. Processing and quality characteristics of apple slices under simultaneous infrared dry blanching and dehydration with continuous heating. *J. Food Eng.* 90, (2009): 441–452.

Yung, S.S., Wen, C.S., Ming, H.C., and Jean, Y.H. Effect of far infrared oven on the qualities of bakery products. *J. Culinary Sci. Technol.* 6, (2008): 105–118.

Yuxin, W., Hongyan, T., Junsi, Y., Kejing, A., Shenghua, D., Dandan, Z., and Zhengfu, W. Effect of carbonic maceration on infrared drying kinetics and raisin qualities of Red Globe (*Vitis vinifera* L.): A new pre-treatment technology before drying. *Innov. Food Sci. & Emerg. Technol.* 26, (2014): 462–468.

Zare, D., Naderi, H., and Ranjbaran, M. Energy and quality attributes of combined hot-air/ infrared drying of paddy, *Drying Technol*, 33(5), (2015): 570–582.

Zeng, M., Bi, J., Chen, Q., Liu, X., Wu, X., and Jiao, Yi. Effect of the combination of osmotic dehydration and medium-short wave infrared radiation on the drying characteristics and kinetics model of kiwifruit slices. *J. Chin. Inst. Food Sci. and Technol.* 14(10), (2014): 83–91.

Zhang, J., Datta, A.K., and Mukherjee, S., Transport processes and large deformation during baking of bread. *AIChE J.* 51(9), (2005): 2569–2580.

Zhongli, P. and McHugh, T.H. (2006). Novel infrared dry blanching (IDB), infrared blanching, and infrared drying technologies for food processing. US Patent 2006/0034981 A1. Filed August 13, 2004, published February 16, 2006.

Zhongli, P., Ragab, K., Godfrey, L.D., Lewis, R., Thompson, J.F., and Salim, A. Feasibility of simultaneous rough rice drying and disinfestations by infrared radiation heating and rice milling quality. *J. Food Eng.* 84, (2008a): 469–479.

Zhongli, P., Shih, C., McHugh, T.H., and Hirschberg, E. Study of banana dehydration using sequential infrared radiation heating and freeze drying. *LWT-Food Sci. Technol.* 41, (2008b): 1944–1951.

Zhu, Y. and Pan, Z. Processing and quality characteristics of apple slices under simultaneous infrared dry-blanching and dehydration with continuous heating. *J Food Eng.* 90(4), (2009): 441–452.

2 Ohmic Heating and Bioactive Compounds

Jorge J. Moreno, Pamela Zúñiga, Erick Jara, María P. Gianelli, Karla Mella, and Guillermo Petzold

CONTENTS

2.1　Introduction 32
2.2　Bioactive Compounds 32
　　　2.2.1　Polyphenols 32
　　　　　　2.2.1.1　Nonflavonoids 33
　　　　　　2.2.1.2　Flavonoids 34
　　　2.2.2　Vitamins 38
　　　　　　2.2.2.1　Liposoluble Vitamins (A, D, E, and K) 39
　　　　　　2.2.2.2　Vitamin C 40
　　　　　　2.2.2.3　B-Complex 40
　　　　　　2.2.2.4　Choline 41
　　　　　　2.2.2.5　Biotin 41
　　　2.2.3　Minerals 41
　　　　　　2.2.3.1　Calcium 42
　　　　　　2.2.3.2　Magnesium 42
　　　　　　2.2.3.3　Iron 43
　　　　　　2.2.3.4　Zinc 43
2.3　Ohmic Heating 43
　　　2.3.1　Minimal Processing and OH 43
　　　2.3.2　OD/VI and OH 46
2.4　Advantages and Limitations of Ohmic Heating 50
2.5　Vegetable Matrix as Carriers of Bioactive Compounds 51
　　　2.5.1　Enrichment with Bioactive Compounds 51
　　　2.5.2　Enzymatic Inactivation 53
2.6　Impact of OH in Protection of Bioactive Compounds 54
2.7　Conclusions and Future Perspectives 55
Acknowledgments 55
References 55

2.1 INTRODUCTION

The evolution of the functional food has had considerable advancement over the past decade. Emergent technologies to produce food with enriched bioactive compounds have grown meaningfully. Natural foods such as fruits and vegetables represent the simplest form of functional foods and can be a good source of functional compounds. The population currently has a high proportion of individuals who are overweight or obese and suffer from heart risks and hypertension due to a reduction in the consumption of healthy foods such as fruits, vegetables, fish, and whole grains. Interest in healthy food has increased the consumption of nutraceutical/functional food. An alternative of healthy, natural and convenient food are fruit products. Innovative technologies for foods with enhanced active compounds include fundamentals in food processing as well as the innovation made during the past few years. A group of emerging technologies has grown significantly because of the interest in the development of foods with active compounds. Technologies such as osmotic dehydration (OD) and vacuum impregnation (VI) at mild temperatures are considered minimal processing techniques because they preserve the fresh characteristics of fruits. On the other hand, ohmic heating (OH) is a thermal process in which heat is internally generated by the passage of an electrical alternating current (AC) through a body, such as a food system, which serves as a source of electrical resistance. In OH treatments, the moderate electric field (MEF) application offers an interesting addition for enhanced diffusion via electric field treatment, and the potential of low field strength and low frequencies for plant membrane permeabilization processes increases with increasing field strength and decreasing frequency. Studies have indicated that the combination of OH and VI has beneficial effects on the acceleration of mass transfer in fruit samples, and the shelf life at 5°C was extended compared to that of control samples.

2.2 BIOACTIVE COMPOUNDS

Foods are complex matrices composed of multiple components that contain macro- and micronutrients needed to meet the normal biological functions of the body. Those components are water, carbohydrates, lipids, and proteins, which are in larger quantities, while other organic and inorganic substances are present in trace amounts. Some compounds in foods, called *bioactive compounds*, provide health benefits in addition to their nutritional value. Polyphenols, vitamins, and minerals are also included in this group.

2.2.1 POLYPHENOLS

Polyphenols are chemical compounds synthesized in plants, which have antioxidant effects, and possess a molecular structure with one or more hydroxyl groups directly attached to an aromatic benzenoid ring (Figure 2.1).

Polyphenols are generally found as esters or glycosides and not as free compounds. As polyphenols are a large number of molecules, they can be classified according to their different chemical structures (Figure 2.2).

FIGURE 2.1 Basic structures of polyphenols. (From Vermerris W. and Nicholson R., *Phenolic Compound Biochemistry*, Springer, Dordrecht, the Netherlands, 2006, pp. 1–34. With permission.)

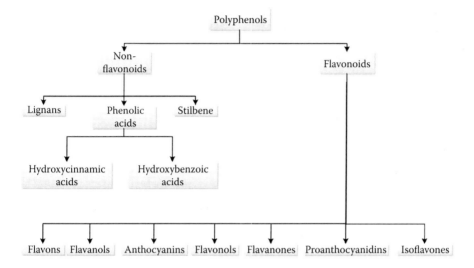

FIGURE 2.2 Classification of polyphenols according to the chemical structure.

2.2.1.1 Nonflavonoids

Non flavonoids are a group of phenolic compounds consisting of a basic structure (phenol) with different branches. This group differs from flavonoids in that they have a single ring in their monomers. Within this group are present lignans, phenolic acids, and the stilbene.

1. *Lignans*: Lignans are molecules present in many foods, especially in seeds such as flaxseed and in smaller quantities in fruits and vegetables. In addition, lignans are present in beverages such as tea, coffee, and wine, and are composed by pinoresitol, podophyllotoxin, and steganacin, which are stable under normal conditions and soluble in water (Landete, 2012). It has been found that lignans have antioxidant activity where studies indicate that consumption of dietary lignans reduces the risk of colon, prostate, and breast cancer and can even prevent cardiovascular diseases (Adlercreutz, 2007).
2. *Phenolic acid*: Phenolic acids have in their basic structure a phenolic ring attached to a carboxylic acid; this group is subdivided into hydroxycinnamic acid and hydroxybenzoic acid. The hydroxycinnamic acids are sensitive to oxidation and pH and are basically composed of an aromatic ring, an aliphatic

group, and a carboxylic acid, which are classified into *p*-coumaric acid, caffeic acid, ferulic acid, 5-hydroxyferulic, and synaptic acid (Figure 2.3). The biological effects of hydroxycinnamic acid in humans are related to their antioxidant functions. It has been reported that hydroxycinnamic acid inhibits cardiovascular disorders, cancer, and neurological disorders such as Alzheimer's and Parkinson's diseases, and have antigenotoxic activity (Ruiza et al., 2015).

Hydroxybenzoic acid is a compound present in tea, wheat, red fruits (raspberry, currant, and strawberry), and some alcoholic beverages such as wine. This group is composed mainly of gallic acid, *p*-hydroxybenzoic acid, and vanillic acid (Figure 2.4). These acids are soluble in water and sensitive to temperature, oxidation, and pH. Hydroxybenzoic acids have therapeutic interest due to the fungitoxicity, antimicrobial activity, and anti-inflammatory properties of salicylates (Parisi et al., 2014).

3. *Stilbene*: Stilbene compounds are produced by many plants, where resveratrol can be highlighted in this group (Figure 2.5), because it has potential health benefits related to the prevention of cardiovascular diseases. It also has effect on obesity, acts as antidiabetic, and has neuroprotective properties (Reinisalo et al., 2015). Other compounds such as pterostilbene have a higher bioavailability and better neuroprotection activity than resveratrol (Chang et al., 2012).

2.2.1.2 Flavonoids

Flavonoids are metabolites generated by plants that possess in their structure two aromatic rings (Figure 2.6) and are present in fruits, vegetables, wine, tea, and cocoa. These compounds are very sensitive to heat, oxidation, light, and pH (Parisi et al., 2014).

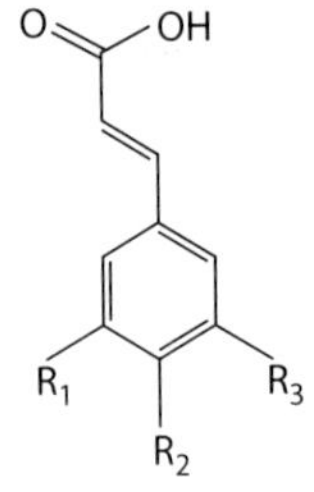

Hydroxycinnamic acid	R_1	R_2	R_3
p-Coumaric acid	H	OH	H
Caffeic acid	OH	OH	H
Ferulic acid	OCH_3	OH	H
Synaptic acid	OCH_3	OH	OCH_3

FIGURE 2.3 Structure of hydroxycinnamic acid modified. (From Rentzsch M. et al., Non-flavonoid phenolic compounds, in *Wine Chemistry and Biochemistry*, C. Polo, and M. V. Moreno-Arribas (eds.), Springer, New York, 2009, pp. 509–521. With permission.)

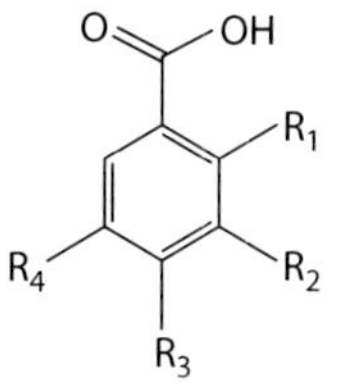

Hydroxybenzoic acid	R₁	R₂	R₃	R₄
Gallic acid	H	OH	OH	OH
Gentisic acid	OH	H	H	OH
p-Hydroxybenzoic acid	H	H	OH	H
Protocatechuic acid	H	OH	OH	H
Salicylic acid	OH	H	H	H
Syringic acid	H	OCH₃	OH	OCH₃
Vanillic acid	H	OCH₃	OH	H

FIGURE 2.4 Structure of hydroxybenzoic acid modified. (From Rentzsch M. et al., Non-flavonoid phenolic compounds, in *Wine Chemistry and Biochemistry*, C. Polo, and M. V. Moreno-Arribas (eds.), Springer, New York, 2009, pp. 509–521. With permission.)

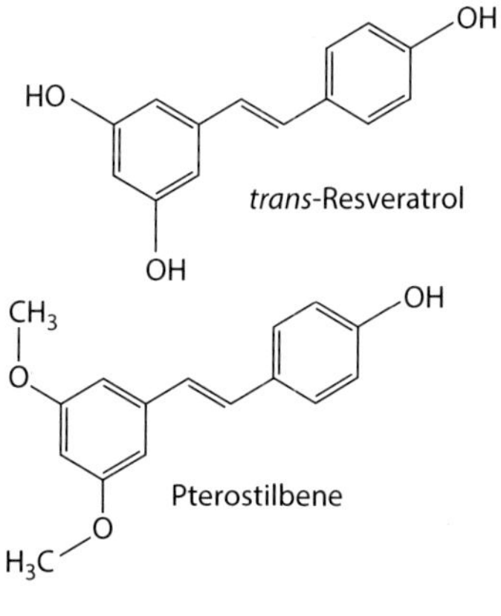

FIGURE 2.5 Structure of *trans*-resveratrol and pterostilbene. (From Reinisalo M. et al., *Oxid. Med. Cell. Long.*, Article ID 340520, 1–24, 2015. With permission.)

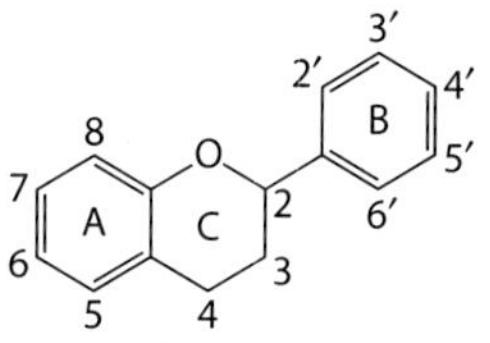

FIGURE 2.6 Generic structures of flavonoids. (From Wolfe, K. L. and Liu, R. H., *J. Agric. Food Chem.*, 56, 8404–8411, 2008. With permission.)

These are recognized mainly for their antioxidant activity, are cardioprotective, and neuroprotective, among other beneficial health effects (Heima et al., 2002).

1. *Anthocyanidins and anthocyanins*: Anthocyanins are a class of flavonoids characterized by a hydronium ion in its structure. These compounds are glycosides of anthocyanidins because they are constituted by an anthocyanidin molecule attached to a sugar. The most common anthocyanins are cyanidin, delphinidin, malvidin, pelargonidin, peonidin, and petunidin (Figure 2.7), which are distributed in the plant kingdom, especially in vegetables, fruits, and flowers, and are responsible for their orange, red, blue, and purple colors (Ali et al., 2015). Some examples of foods that contain anthocyanins are berries, red grapes, and red wine. These compounds have effects on health, including the prevention of cardiovascular disease, inflammatory, antiallergic, antiviral, antibacterial, antioxidant, and anticancer properties and problems associated with aging (Parisi et al., 2014).
2. *Flavanols*: Flavanols or flavan-3-ols are a type of flavonoid that has a saturated carbon ring and a hydroxyl group at the C3 carbon. They are present as monomers catechin, epicatechin, gallocatechin, and epigallocatechin (Figure 2.8) that can be esterified with gallic acid. These also exist as dimers (theaflavin, thearubigins, and procyanidins), which are colorless, or polymers (proanthocyanidins), which go from colorless to colored when are degraded. Proanthocyanidins are also called *condensed tannins* since the chains to be constituted with less than four monomers provide a bitter taste,

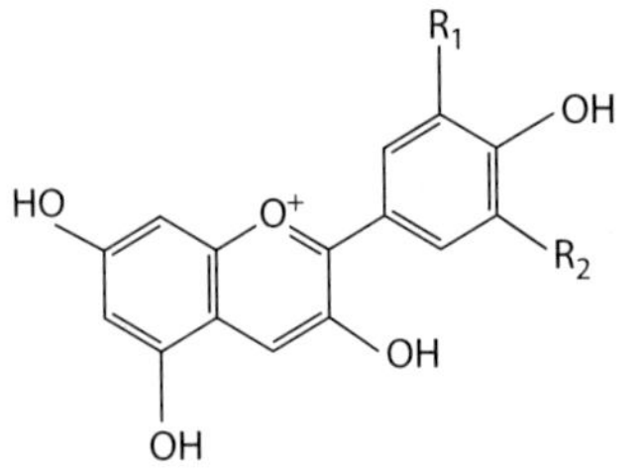

Anthocyanidins	R_1	R_2
Cyanidin	OH	H
Delphinidin	OH	OH
Malvidin	OCH_3	OCH_3
Pelargonidin	H	H
Peonidin	OCH_3	H
Petunidin	OCH_3	OH

FIGURE 2.7 Structures of anthocyanidins. (From Durst, R. and Werolstad, R., Separation and characterization of anthocyanins by HPLC, in: *Handbook of Food Analytical Chemistry*, R. E. Wrolstad et al. (eds.), John Wiley & Sons, New Jersey, 2001, pp. 33–45. With permission.)

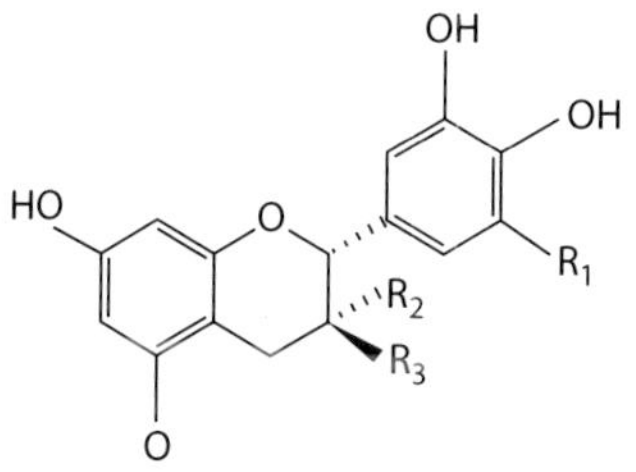

Ohmic Heating and Bioactive Compounds

Flavonols	R₁	R₂	R₃
Catechin	H	H	OH
Epicatechin	H	OH	H
Gallocatechin	OH	H	OH
Epigallocatechin	OH	OH	H

FIGURE 2.8 Structures of flavanols modified. (From Aron, P. and Kennedy, J., *Mole. Nutr. Food Res.*, 52, 79–104, 2008. With permission.)

while being more than four, it delivers an astringent sensation. Flavanols are found in tea, cacao, apples, pears, grapes (skin and seed), and wine (Donovan and Waterhouse, 2003). Some benefits of consumption of these elements are the reduction of the risk of cardiovascular disease, including coronary heart disease, myocardial infarction, and stroke. In addition, these elements act as mitigating the risk factors on metabolic syndrome such as hypertension, vascular endothelial dysfunction, dyslipidemia, and glucose intolerance (Osakabe et al., 2014).

3. *Flavanones:* Flavanones are present in almost all citrus fruits and in smaller quantities in tomatoes and some herbs. The molecules that constitute this group are naringenin, hesperetin, and eriodictyol, which are present in greater amount in the membranes that separate the segments of grapefruit, orange, mandarin, and lemon, compared to the pulp (Chanet et al., 2012). These have anti-inflammatory, antiallergic, antibacterial, antioxidant, anticancer, and hepatoprotective effects (Parisi et al., 2014).

4. *Flavonols:* Flavonoids and flavones 3-hydroxy are a type of flavonoid in their structure having a carbonyl group in position 4 and a hydroxyl group at position 3 of the ring atoms. This compound is photosensitive because it oxidizes with light easily. This group is divided into quercetin, kaempferol, myricetin, and isorhamnetin (Figure 2.9), which are present in onions, kale, broccoli, apples, berries, and tea, among others. The flavonols have antibacterial, antiviral, antifungal, antimutagenic, antiproliferative, anticancer, anti-aging, anti-inflammatory and anti-angiogenic effects (Ensafia et al., 2015).

5. *Flavones*: Flavones are a group of flavonoid characterized by having a double bond in the heterocycle in position 2–3; carbon 4 in carbonilon groups is soluble in water and ethanol. The most important flavones are apigenin

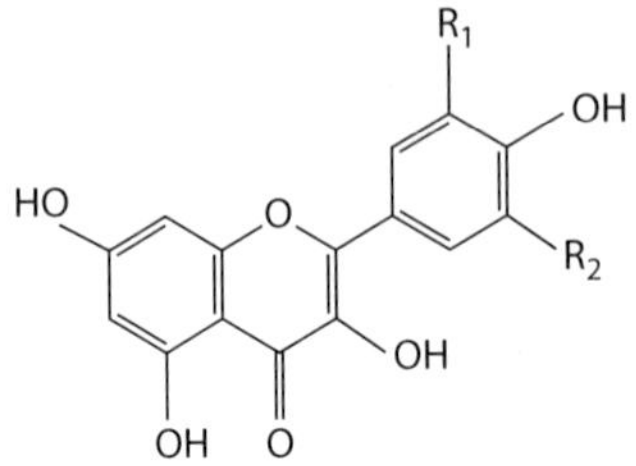

Flavonols	R₁	R₂
Quercetin	OH	H
Kaempferol	H	H
Myricetin	OH	OH
Isorhamnetin	OCH₃	H

FIGURE 2.9 Structures of flavonols modified. (From Tsao, R., *Nutrients*, 2, 1231–1246, 2010. With permission.)

and luteolin. These components are part of a wide variety of edible plants, fruits, vegetables, and beverages such as tea, coffee, beer, and wine (Singh et al., 2014). These have anti-inflammatory, antioxidant, antiproliferative, antitumor, and antimicrobial estrogenic activities and are also used for cancer, cardiovascular diseases, and neurodegenerative diseases prevention (Martens and Mithöfer, 2005).

6. *Isoflavones:* Isoflavones are made of two phenolic rings attached to carbon triple bond. The most important are genistein, daidzein, and glycitein. These are found almost exclusively in legumes, soybeans, and soy products, and are considered the main source of input in our diet (Markovic et al., 2015). These compounds act as phytoestrogens providing health benefits when these are used in diseases that depend on hormonal behavior. Some of these benefits include chemoprevention of breast and prostate cancer, prevention of cardiovascular disease and cancer, and osteoporosis in older women relieves menopause symptoms (Vitale et al., 2013).

2.2.2 Vitamins

Vitamins are defined as a complex group of micronutrients of biological origin with different structures and functions, which are present in food in small amounts (traces), and they can be classified in water-soluble and fat-soluble vitamins (Figure 2.10). These compounds are necessary for normal functioning of our body because they are involved in metabolic reactions; therefore, vitamin deficiency causes a number of pathologies or even death.

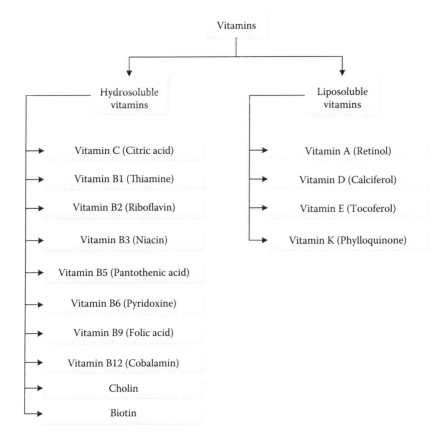

FIGURE 2.10 Classification of vitamins according to solubility.

2.2.2.1 Liposoluble Vitamins (A, D, E, and K)

Vitamin A or retinol is an essential vitamin. As it is not synthesized by the body, we can obtain it from vegetables and green, yellow, or orange fruits. It is also found in animal products such as organ meats, liver, other meats, eggs, and dairy products. This vitamin is necessary for normal vision, reproduction, embryonic development, cell and tissue differentiation, and immune function (Ross and Harrison, 2007). Vitamin A deficiency is the major cause of preventable blindness in malnourished children, particularly in developing countries (Hsu et al., 2015).

Vitamin D or calciferol is a vitamin widely distributed in vegetable food sources, where it can be found as vitamin D2 or ergocalciferol and in animal food sources as vitamin D3 or cholecalciferol. This vitamin is essential for life, because it is responsible for regulating the passage of calcium (Ca^{2+}) to the bones (Zempleni et al., 2007). Vitamin D deficiency is associated with rickets in children (Wagner and Greer, 2008) and osteomalacia in adults (Holick et al., 2011).

Vitamin E or tocopherol is a generic descriptor for all molecules with α-tocopherol activity. This acts as a lipid-soluble chain-breaking antioxidant that protects polyunsaturated fatty acids from lipid peroxidation, preserving the integrity of biological membranes, lipoproteins, and lipid stores against oxidation (Guzmán et al., 2015).

Vitamin K is present in plant foods as phylloquinone or vitamin K1, but it is also synthesized by bacteria such as *Escherichia coli*, producing menaquinone or vitamin K2 (Priyadarshia et al., 2009). This vitamin plays essential roles as a cofactor for enzymes in bone metabolism and blood coagulation, among others. No vitamin K deficiencies occur in people normally as it is synthesized by the intestinal bacteria (Ahmeda and Mahmoud, 2015).

2.2.2.2 Vitamin C

Vitamin C or ascorbic acid is present in fruits and vegetables, which can be easily degraded during processing by heat (Özdemir and Gökmen, 2015). It is a natural antioxidant, which decreases the risks of developing diseases such as cancer, cardiovascular disease, common cold, and asthma (Gabriel et al., 2015). Vitamin C reduces vascular oxidative stress, hypertension, coronary artery disease, and chronic ethanol and vascular relaxation (Hipólito et al., 2015). Vitamin deficiencies are associated with bruising, muscle weakness, gum, scarring disease, and scurvy (Du et al., 2012).

2.2.2.3 B-Complex

Vitamin B1 is an essential vitamin for humans, required for normal growth and functioning of the heart, nervous, and digestive systems, and it plays an important role as a coenzyme in the metabolism of carbohydrates, through addition of magnesium and ATP (Sofi et al., 2015). This vitamin is usually present in whole grains, legumes, and nuts. Lack of this vitamin causes cardiovascular and neurological damage that is clinically known as *beriberi* (Zhou et al., 2013).

Vitamin B2 or riboflavin is present in cells and tissues as flavin adenine dinucleotide and flavin mononucleotide adenine, which is involved in a series of redox reactions in primary metabolic pathways and processes of electron transfer (Lienhart et al., 2013). Riboflavin deficiency produces many effects on intermediary metabolism of lipids and proteins that result in damage to the skin, mouth, and eyes, and it can produce ulcers and anemia (Alam et al., 2015).

Vitamin B3 or niacin can be found in plants as nicotinic acid, and in animals as nicotinamide. This vitamin is important because it is a precursor of two important coenzymes: nicotinamide adenine dinucleotide (NAD) and its phosphorylated form (NADP), which are responsible for hydrogen transfer in many metabolic reactions of dehydrogenases acting on lipids, proteins, and carbohydrates. The niacin deficiency causes pellagra or the *three D disease*: dermatitis, diarrhea, and dementia (Zempleni et al., 2007).

Vitamin B5 or pantothenic acid is stable at neutral pH, but acids, alkalis, and heat destroy it quickly. It is fundamental as a preliminary step of coenzyme A biosynthesis in mammalian cells. Vitamin B5 is present in all kinds of plant and animal tissues, especially queen bee gelatin and liver (Schittl et al., 2007).

Vitamin B6, also known as pyridoxine, pyridoxal, and pyridoxamine, is responsible for generating antibodies, transmits signals in the brain, and maintains normal nerve function and normal blood sugar levels. Vitamin deficiency can lead to nervous system problems, affecting the skin, mucous membranes, and circulatory system (Koneswaran and Narayanaswamy, 2015).

Vitamin B9 is a group of essential biomolecules also called folate, which are essential for one-carbon units transfer reactions, the metabolism of individual amino acids, neurotransmitter synthesis, membrane phospholipids, and myelin for epigenetic control of gene expression (Araújo et al., 2015). Severe folate deficiency has been associated with megaloblastic anemia characterized by usually large and nucleated red blood cells that accumulate in the bone marrow. It also causes abnormalities in the formation of the fetus, atherosclerosis, cognitive decline, and neurodegenerative diseases such as Alzheimer's (Koike et al., 2012).

Vitamin B12 or cobalamin is essential for the metabolism of all cells as it acts as coenzyme in reactions of isomerization, dehydrogenation, and methylation and activation of folic acid. It is also important for DNA synthesis and for the energy extraction from protein and fat in mitochondrial citric acid cycle, maintaining the integrity of the nervous system (Wong et al., 2015). Deficiency of this vitamin can cause macrocytosis, anemia, or neurological abnormalities (Sharma and Biswas, 2012).

2.2.2.4 Choline

Choline is a precursor for the synthesis of the major phospholipid (phosphatidylcholine) of the neurotransmitter acetylcholine and betaine, a major donor of methyl groups. Choline also plays an important role in systemic lipid metabolism, hepatic and muscle function, and participates in the formation of all biological membranes (Schenkel et al., 2015). This vitamin is synthesized by the body, but not at the level that is required; hence, it is necessary to consume it from a variety of foods. Free choline is present in particularly high quantities in ready-to-eat cereals, wheat germ, and beef liver. Choline deficiency has several negative consequences; for example, steatohepatitis prevents proper neural tube during embryonic development (Da Silva et al., 2015).

2.2.2.5 Biotin

Biotin is synthesized by microorganisms and therefore is an additional food source that can partially be available for human organism. Biotin acts as a coenzyme in mammal's covalent binding of these four carboxylase: acetyl-CoA carboxylase, pyruvate carboxylase, propionyl-CoA carboxylase, and β-methylcrotonyl carboxylase (Hernández et al., 2012). The deficiency can be caused by poor intestinal absorption (short bowel syndrome) or in people who consume large quantities of raw egg white (Ochoa-Ruiza et al., 2015).

2.2.3 MINERALS

Minerals are micronutrients like vitamins, but some of these are considered essential because they are necessary for normal growth, reproduction, and good health. Several essential minerals can also be considered components of *functional foods* because of the expected preventive or therapeutic effects on chronic diseases (Phan-Thien et al., 2012).

TABLE 2.1
Examples of the Major Essential and Trace Elements and Their Function in the Body

Nutrients	RNI[a]/Day	Functions	Deficiency
Calcium (mg)	525	Healthy bones and teeth, regulation of muscle contraction and nerve conduction, blood clotting, enzyme activation, and hormone secretion	Rickets, irritability, jitteriness, tremors, and convulsions in newborn babies
Magnesium (mg)	75	Involved in glycolysis, replication of DNA and synthesis of RNA, regulatory effect on Ca and K level, parathyroid hormones secretion, vitamin D metabolism and subsequently bone function, normal neuromuscular function, and steady heart rate	Muscle spasm and weakness, sleep disorders, irritability, poor nail growth, and neuromuscular excitability
Iron (mg)	7.8	Major part of hemoglobin and role in cognitive development	Anemia, pale skin, tiredness, and motor and mental problems
Zinc (μg)	5	Growth and immune function, site-specific antioxidant, synthesis and activation of enzymes and proteins such as insulin, vitamin A, and nucleic acids	Growth retardation, immune deficiency, loss of appetite, and night blindness
Copper (mg)	0.3	Catalyst in mobilization of iron and connective tissue synthesis	Skeletal demineralization, decreased skin tone, and decreased plasma iron
Selenium (μg)	10	Antioxidant, maintenance of healthy immune system, and interaction with heavy metals	Asthma, vulnerability to infection

[a] UK-recommended nutrient intake for an infant 6–9 months of age (Zand et al., 2015).

2.2.3.1 Calcium

Calcium is the most abundant chemical element in the human body, which is present in the bones and teeth, and helps in the regulation of muscle contraction. It is involved in functions such as clotting of blood, functional contraction, enzyme activation, and nerve impulses transmission (Amalraj and Pius, 2015). The lack or loss of this mineral can cause diseases such as osteoporosis, rickets, hypertension, and colorectal cancer.

2.2.3.2 Magnesium

Magnesium is an essential mineral that plays an important role in skeletal development and keeps electric potential in both nerve and muscle membranes. It can also act as a cofactor for many enzymes that require ATP; therefore, magnesium is essential for the stabilization, energy production, membrane oxidative phosphorylation, glycolysis, DNA transcription, and protein synthesis. It is also involved in the active transport of calcium and potassium across cell membranes, which is essential for the conduction of nerve impulses and muscle contraction (Gröber et al., 2015).

Magnesium deficiencies include progressive muscle weakness, neuromuscular dysfunction, tachycardia, coma, and even death (Zand et al., 2015).

2.2.3.3 Iron

Iron is an essential nutrient for cellular oxygen transport, and is also a cofactor in more than 200 metalloenzymes; however, excess of iron is toxic. The World Health Organization estimates that about two billion people suffer from anemia and approximately half of these patients are related to iron-deficiency anemia (Zand et al., 2015).

2.2.3.4 Zinc

Zinc is an important mineral because it acts as a cofactor for a number of enzymes, acts as an antioxidant, and is anti-inflammatory, where zinc is an inhibitor of NADPH oxidase enzyme that produces reactive oxygen species. Zinc is also required for DNA synthesis, RNA transcription, division, and activation calls (Prasad, 2008). Zinc deficiency can cause growth retardation, poor appetite, delayed healing, and especially neurosensory problems (Zhao et al., 2015).

2.3 OHMIC HEATING

OH is a volumetric form of heating and a high-temperature short-time process in which thermal energy is generated by the passage of an alternating electrical current through a food material (Figure 2.11). The resistance it imposes causes the excitation of cells, where the friction of the vibration leads to the generation of heat inside the product (Lima et al., 2002; Ramaswamy et al., 2014).

In the food industry, OH is used as a method of preservation, including cooking, blanching, drying/concentration, pasteurization, and sterilization (Ramaswamy et al., 2014).

2.3.1 MINIMAL PROCESSING AND OH

Dietary guidelines emphasize the importance of a diet high in fruits and vegetables, as regular consumption of those natural foods protects against many diseases such as hypertension, stroke, coronary heart disease, some cancers, and diabetes

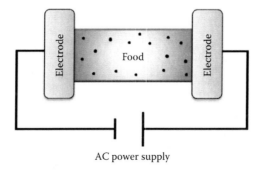

FIGURE 2.11 Schematic diagram of an ohmic heating system.

(Mytton et al., 2014). All these benefits of fruits and vegetables are attributed to their high content of vitamins, minerals, dietary fiber, and a host of beneficial non-nutrient substances, including plant sterols, flavonoids, and other antioxidants. Nevertheless, the consumption of fruits and vegetables remains below the recommended levels in many countries partly because these products are perishable. This deterioration can be caused by microorganisms and/or by a variety of physicochemical reactions that occur after harvesting; therefore, it is important to develop technologies to extend the shelf life of fruits and vegetables, so that they could be to extend the shelf life of fruits and vegetables.

To achieve this challenge, preservation techniques are applied to control the deterioration of food quality. As a result, new technologies to develop minimally processed fruits and vegetables have rapidly advanced, whose consumption is increasing due in part for a growing focus on health, and also the preference for convenience (ready-to-eat foods).

The purpose of minimally processed foods is to provide the consumer a fruit or vegetable product with similar characteristics to fresh food and long shelf life, ensuring safety of them, maintaining nutritional and sensory qualities (Wiley, 1997).

Methods to prevent alteration of these minimally processed foods are based on different processes (Figure 2.12), which can be applied in combination or individually.

However, in recent years, new conservation technologies have been developed, called *emerging technologies*. These can be classified into two groups: thermal and nonthermal technologies (Figure 2.13).

Within thermal technologies, OH has gained interest because products are of a superior quality compared to those processed by conventional technologies. For example, during conventional thermal processing in aseptic processing systems for particulate foods, significant product quality damage may occur due to slow conduction or convection heat transfer. Conversely, OH provides a rapid and uniform heating that reduces the treatment time and result in less thermal damage to vitamins, pigments, and other elements; therefore, this technology improves the quality with minimal structural, nutritional, or organoleptic changes. Many of these effects

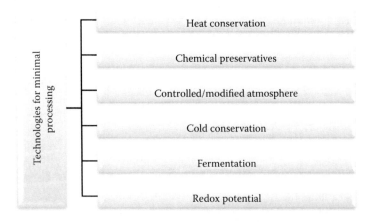

FIGURE 2.12 Conventional technologies for minimally processed fruits and vegetables.

Ohmic Heating and Bioactive Compounds

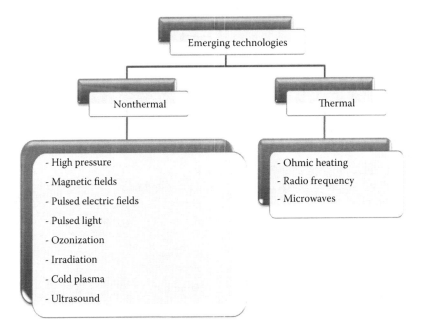

FIGURE 2.13 Emerging technologies in food processing.

are not only due to the heat produced by the application of OH, but also due to the association of an electric field, which has significant nonthermal effects on cells, known as *moderate electric field* (MEF) processing, paving the way to a wide range of improved food processes. The MEF application can cause changes in the physical properties and the permeability of vegetable cells, such as pores formation in cell membranes below the temperature at which they are normally permeable. This effect, known as *electropermeabilization or electroporation*, has received increasing attention because it can be used to manipulate cells and tissues, and it is beneficial for increasing the diffusion of materials.

The electroporation effect is originated by the application of MEF through the cell, which causes a redistribution of internal ions because cell membranes act as insulators. Ions move according to their charge in the electric field, but remain inside the cell accumulating at the poles in line with the electric field. This polar accumulation of ions into the cell creates an electrical potential across the cell membrane, and when the transmembrane voltage exceeds the threshold, conformational changes within the membrane induce a state of permeability, enabling the uptake of exogenous molecules found in the medium. Mass transfer enhancement is significant when the product initially possesses an intact cell structure. No enhancement is observed when a cell structure is either absent or previously permeabilized. Consequently, the mechanism of diffusion enhancement may be attributed to pore formation in cell membranes.

The permeabilized zone on each end of a cell will be asymmetrical, because the cell has a slight natural negative charge across its membrane, where it will be slightly larger on the anode (+) side than in the cathode (−) side (Figure 2.14).

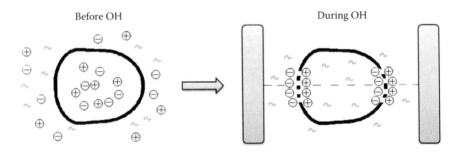

FIGURE 2.14 Electroporation produced in a cell by the electric field effect.

The plant membrane permeabilization process intensifies by increasing electric field and decreasing frequency. For an electrical field with hundreds of V/cm the electroporation becomes nonreversible. The electric field (E) strength can be represented by the following equation:

$$E = V/l \tag{2.1}$$

This can be controlled by changing either the applied voltage (V) or the distance between the electrodes (l).

The level of polarization is an important factor to control, and besides the frequency of the AC, it depends on the nature of the electrode surface. Usually, treated or coated electrodes are used, providing the frequency in the range that is normally used in commercial instruments (50–1000 Hz). It has been studied that the effect of different frequencies in heat generation on foodstuffs, and between the examined frequencies, 50 Hz gave the shortest time for raising the desired temperature of process (Ramaswamy et al., 2014).

The effect of frequency is particularly important for the electrical conductivities (ECs) in foods. EC is the ability of a material—such as food—to transport an electric charge, and it may depend on the composition, soluble salt percentage, electrolyte mobility, and temperature. By introducing an electrical voltage difference across a conducting material, the transferable charges will flow through the conductor that produces an electrical current (primarily alternating). The passage of an electric current through a foodstuff is the basis of the OH technique; therefore, EC can be considered the most important influencing parameter.

It has been found that the application of nonconventional technologies, such as OH combined with other technologies, can reduce the loss of some sensorial characteristics during the treatment and after storage in refrigerated conditions. The combination of different technologies may be used to potentiate the effect of each treatment and to increase the shelf life of minimal processing products.

2.3.2 OD/VI AND OH

OD is considered a minimal processing method that is widely used to remove a portion of the fruit's water content to obtain a product of intermediate moisture with the advantage of preserving fresh-like characteristics of fruits, such as color, firmness, and taste. This partial dehydration is achieved by immersion of the food in an aqueous hypertonic solution, where there are two major simultaneous countercurrent flows: A water flow from the food into the osmotic solution and another flow of solutes from the solution into the food. In a multiphase food system, mass transfer rates are attributed to the water and solute activity gradients across cell membranes until seek equilibrium.

Actually, if the osmotic solution possesses physiologically active compounds, it could be introduced into the solid food matrix to enhance its nutritional or functional characteristics. This technology has been recognized as suitable for formulating new products.

It is well known that the synergy of electricity and temperature makes this type of dehydration more efficient. As a result, many authors have been pursuing the application of OH to increase mass transfer into the food materials, where diffusivity is one of the most important factors in mass transfer. Diffusivity is affected for different process variables in OD treatment: structure and composition of plant tissue, characteristics of osmotic agent, concentration of the osmotic solution, temperature, and pressure.

In addition to reducing water content, OD prevents food contact with atmospheric oxygen and offers some advantages over conventional enzyme inactivation pretreatments, for example, sulfited or blanching; hence, it improves product quality and decreases energy consumption. OD can also be used as a pretreatment to hot air drying. However, a disadvantage of the OD process is the long time required to reduce the water activity. To enhance the efficiency of the mass transfer and the nutrient uptakes, VI process can be applied.

VI utilizes pressure gradients to accelerate the incorporation of a solution into the structural matrix of high-porosity food samples. This implies that the gas of intercellular spaces is exchanged for the external fluid due to the action of hydrodynamic mechanisms (HDM) promoted by pressure changes. The process consists of two principal stages: The phase of reduced pressure and the phase of atmospheric pressure, where impregnation of the material occurs because of two phenomena: HDM and deformation–relaxation phenomena (DRP), which lead to the filling of intercellular capillaries (Figure 2.15). After immersion of the material in solution (t_0), the inside pressure (p_i) and the exterior pressure (p_e) are equal to atmospheric pressure in the capillary ($p_i = p_e = p_{at}$). The initial volume of the capillary (Vg_0) is filled with gas (Figure 2.15, Step 0). In the first phase of the process, the pressure is reduced ($p_1 < p_{at}$) causing gas leakage, and origins the deformation and expansion of the capillary, which is the first part of the DRP, and the volume of the capillary is increased. This stage lasts until pressure equilibrium ($p_i = p_e$) is reached (Figure 2.15, Step 1A). Next, the capillary begins to be partially filled with liquid because of the HDM. The pressure inside the capillary increases slightly, while the free volume inside it decreases (Figure 2.1, Step 1B). In the second phase of VI, the pressure returns to the atmospheric value. This causes the transition of the DRP to the relaxation phase. The capillary shrinks more than it was at the beginning of the process. At the same time, as a result of the action of capillary pressure and decompression, an intensive inflow of liquid from the outside to the capillary is observed and the final volume of gas inside it decreases (Figure 2.15, Step 2). The relaxation phase is particularly

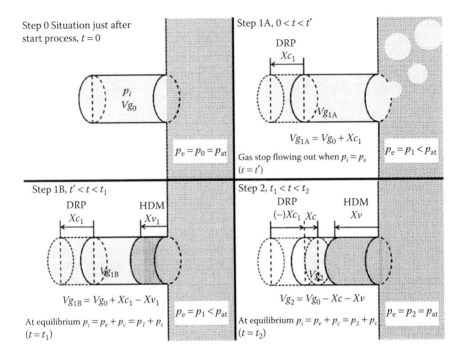

FIGURE 2.15 Hydrodynamic mechanism (HDM) and deformation–relaxation phenomena (DRP) contribute to the filling of ideal capillary with liquid during vacuum impregnation. (Adapted from Radziejewska-Kubzdela, E. et al., *Int. J. Mole. Sci.*, 15, 16577–16610, 2014. With permission.)

important from the practical point of view, since matrix impregnation occurs at this stage. Removal of vacuum should not be too rapid, since the excessively fast pressure equalization could close the capillary vessels and inhibit the HDM (Radziejewska-Kubzdela et al., 2014).

VI is an emerging technology that has grown significantly in popularity because it has been successfully used to incorporate vitamins (Cortés et al., 2007a), minerals (Barrera et al., 2004; Xie and Zhao, 2003), and probiotic microorganisms (Betoret et al., 2003) into fruits and vegetables matrix structure without substantially modifying their organoleptic properties. This treatment has the potential to protect those bioactive compounds within foods, even more when it is combined with OH, because it increases the permeability of fruits cells, where those components are introduced inside the fruit cells, in addition to accelerate mass transference and reduce process time. Therefore, VI application for a short period at the beginning of the OD and/or OH processes is considered a useful tool for development of new products with energy saving.

The applications of these three treatments in different products are found in some studies; for example, in strawberries (cv. Camarosa) where the combination of OH (13 V/cm), OD (65°B), and VI (5 KPa for 5 min) at 30°C, 40°C, or 50°C for 300 min had beneficial effects on the acceleration of mass transference. When the electric field is varied (9.2, 13, and 17 V/cm), it is concluded that VI/OH treatment at 13 V/cm

Ohmic Heating and Bioactive Compounds

is the best processing condition for dehydrating strawberries (Moreno et al., 2012). Similar results were obtained in pear cubes (cv. Packham's Triumph) at the same conditions, in which increases in the permeability of cell by OH explains the acceleration of mass transference and process time reduction (Moreno et al., 2011). Besides, the application of OH promotes the preservation of bioactive compounds, such as antioxidants by inactivating the function of polyphenoloxidase (PPO) enzyme in apple cubes, which it inhibited even during 4 weeks at refrigerated storage, concluding that VI/OH treatment at 50°C and stored at 5°C may be considered the better minimal processing that preserves the fresh-like properties of the apples (Moreno et al., 2013).

Allali et al. (2010) studied VI and OH pretreatments in osmodehydrated apple fruit, and concluded that those treatments led to profound changes in structure, which were evaluated by measurements of fruit firmness and electrical conductivity.

Several varieties of ohmic heaters exist, but most configurations consist of three modules: heater assembly, power supply, and control panel. The equipment used for most of the studies mentioned above is a batch laboratory unit. A schematic of a laboratory-scale OH setup is shown in Figure 2.16, which includes two concentric cylinder tanks made of stainless steel with a plastic bottom connected to a variable transformer of 60 Hz and an input of 220 V AC.

The temperature is measured by Teflon-coated thermocouple junctions, to prevent electrical interference due to the passage of current through the thermocouple wires, but these junctions tend to slow the temperature response. Signals from the thermocouples and the voltage and current transducers can be fed to a data logger and values can be recorded. Care should be taken while selecting the electrode material. They could be made of aluminum, stainless steel, rhodium-plated stainless steel, carbon (graphite), glassy carbon, platinum, titanium, and platinized titanium.

Besides this equipment in batch, there are continuous heaters. Figure 2.17 is a schematic of a continuous-flow OH process. These are used to viscous food product that contains particulates, which enters to the continuous-flow OH system via a feed pump hopper. The product then flows through a series of electrodes in the

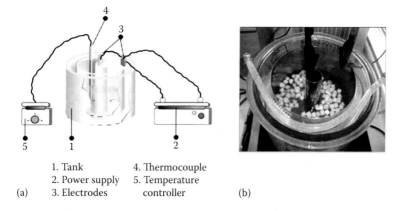

1. Tank
2. Power supply
3. Electrodes
4. Thermocouple
5. Temperature controller

FIGURE 2.16 Laboratory-scale ohmic heating equipment (a) schematic diagram and (b) top view of a real ohmic heating tank.

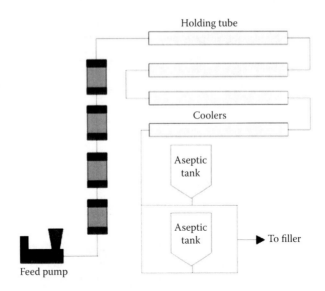

FIGURE 2.17 Schematic of a continuous-flow ohmic heating process. (From Ruan, R. et al., in: *Thermal Technologies in Food Processing: Ohmic Heating*, P. Richardson (ed.), CRC Press, Boca Raton, FL, 2001, pp. 243–244. With permission.)

ohmic column, where it is heated to process temperature for a fixed time to achieve commercial sterility. Next, the product flows through tubular coolers and into storage tanks, where it is stored until filling and packaging.

The purpose of the continuous system is to elevate the temperature in a rapid and uniform way to get the temperature of pasteurization or sterilization. However, in the batch system, the MEF application has a different objective, where in addition to heating, the effect of electroporation offers some benefits to mass diffusion. Moreover, it can be combined with other technologies such as OD and VI to be a good alternative to generate foods that retain or improve sensory and nutritional properties of raw materials.

2.4 ADVANTAGES AND LIMITATIONS OF OHMIC HEATING

In OH treatments, the MEF application offers an interesting addition for enhanced diffusion via electric field treatment, and the potential of low field strength and low frequencies for plant membrane permeabilization processes increases with increasing field strength and decreasing frequency. The enhancement of mass transfer is significant when the product initially possesses an intact cell structure (Moreno et al., 2011). In general, OH systems are advantageous due to an optimization of investment (increased efficiency), instant shutdown of the system, and reduced maintenance costs because of the lack of moving parts. In conventional heating, the time it takes to increase the temperature at this cold point may overprocess the remaining particles and the surrounding liquid. Heating food materials by internal heat generation without the limitation of conventional heat transfer and some of the nonuniformity commonly associated with microwave heating is due to limited dielectric penetration. Higher temperature in particulates than liquid can be achieved, which is impossible for conventional heating. These reduce risks of fouling on heat transfer surface and burning of the food products, resulting in minimal mechanical damage and better nutrients and vitamin retention. A high energy efficiency is achieved because 90% of the electrical energy is converted into heat. These help in ease of process control with instant switch on and shutdown reducing maintenance cost (no moving parts). These have ambient temperature storage and distribution when combined with an aseptic filling system. It is a quite environment-friendly system (Ruan et al., 2001). This overprocessing leads to nutrients destruction and decreased flavor. OH processes the particles and surrounding liquid simultaneously, preventing overcooking (Parrott, 1992).

A disadvantage of ohmic systems is its relationship with the costs of commercial OH systems, including installation, which can be in excess in comparison to commercial conventional pasteurizer's cost, including installation. Another disadvantage related to the type of food that can be processed lies in the presence of fat globules. A food that has fat globules can be troublesome to be effectively heated ohmically, as it is nonconductive due to lack of water and salt. If these globules are present in a highly electrical conductive region where currents can bypass them, they may heat slower due to lack of electrical conductivity. Any pathogenic bacteria that may be present in these globules may receive less heat treatment than the rest of the substance (Sastry and Palaniappan, 1992).

Another slight disadvantage relates to the electrical conductivity of a substance. As the temperature of a system rises, the electrical conductivity also increases due to the faster movement of electrons. An OH system that has not been cleaned thoroughly enough may result in electrical arcing due to protein deposits on the electrodes. However, by utilizing the knowledge of the issues mentioned above in designing an OH system, these disadvantages may be controlled more easily.

2.5 VEGETABLE MATRIX AS CARRIERS OF BIOACTIVE COMPOUNDS

2.5.1 ENRICHMENT WITH BIOACTIVE COMPOUNDS

The technologies used to develop functional food have changed considerably over the years. Although traditional processing operations generally have a negative effect on functionality, recent studies show that proper management of the processing technologies can reduce their impact, even improving the functional properties of the final product. A group of emerging technologies such as microencapsulation, edible films, VI, OH, and high pressure has grown significantly in popularity because of the growing interest in forming a structure that prevents the deterioration of physiologically active compounds (Betoret et al., 2015). In OH treatments, the MEF application offers an interesting addition for enhanced diffusion via electric field treatment, and the potential of low field strength and low frequencies for plant membrane permeabilization processes increases with increasing field strength and decreasing frequency. The enhancement of mass transfer is significant when the product initially possesses an intact cell structure (Moreno et al., 2011).

VI is a technique that is often used as a complement for impregnation with different substances or OD solutions. Betoret et al. (2015) studied the effects of the vacuum level on the structure and mechanical properties of the food products during VI operation, as well as the effects on the impregnation liquid on the resulting product structure and mechanical properties (Betoret et al., 2015). The VI operation allows the incorporation of technological and bioactive compounds into natural structures, taking advantage of both the protective effect of these ones and the synergistic effect of certain biocompounds. The possibility to consider in the structural matrix of a porous food bioactive compounds shows VI as an effective technology for new products design. There are several studies in which fruits and vegetables are impregnated with different isotonic or hypertonic solutions of one or more bioactive compounds, in order to achieve the recommended amount in the final product (Cortes et al., 2007b; Gras et al., 2003). This treatment has benefits for the kinetic process and quality of the product and can further reduce the costs (Moreno et al., 2013). In this treatment, air in the samples was replaced by solutions of the HDM (Atarés et al., 2009; Moraga et al., 2009), and the volume of liquid was similar to the total volume in the intracellular space where it was aired, promoting the loss of intracellular water (Schulze et al., 2012). VI is a mass transfer mechanism that takes place in a complex cellular and porous structure, and a homogeneous VI means that all components of a substrate are incorporated equally (Betoret et al., 2012).

Food-engineer processes such as OH and VI have been used for generating apple enriched with bioactive molecules such as vitamin E, calcium, selenium, folic acid, and arginine, which may offer functional foods with clinical applications L-arginine incorporation in apples. It might offer a suitable strategy for enhancing the compliance of oral administration of this amino acid, reducing intestine breakdown, and favoring L-arginine supply for its metabolic actions (Escudero et al., 2013).

Figure 2.18 shows the structural solids, such as the cell walls and membranes. A high degree of cell compartmentalization and small intracellular spaces in the fresh apples are observed. In the cells, a large vacuole occupies most of the protoplast, and the plasmalemma and tonoplast are close to the cellular wall (Figure 2.18d). Differences in the microstructural effects of the single OH treatments are shown in Figure 2.18b and e. The OH-treated samples showed a high degree of cell decompartmentalization, which was reflected by the lack of cell walls and membrane definition in the micrographs when compared to the micrographs of the well-preserved cells in the VI/OH-treated samples (Figure 2.18e and f). The protoplast appears in plasmolysis and is retracted to the center of the cell (Figure 2.18e and f); cellular breaking by electrothermal effect can be observed in Figure 2.18b, c, e, and f. These observations agree with the mass transfer behavior. Moreno et al. (2012) reported that the greatest solute gain, least firmness loss, and least color loss were obtained in the VI/OH treatment at 13 V/cm. The shelf life of strawberries treated with VI/OH at 13 V/cm and stored at 5°C was extended from 12 days to 25 days; similar phenomenon was observed in apples.

Folic acid (vitamin B9) can be efficiently incorporated into the structural matrix of fresh apple slices by using apple juice as the impregnation vehicle. VI and OH during impregnation treatment had a significant effect on the kinetics of folic acid gain in apple slices. The dehydration of the impregnated samples produced increases or decreases in the folic acid content, whereas the samples from apples impregnated

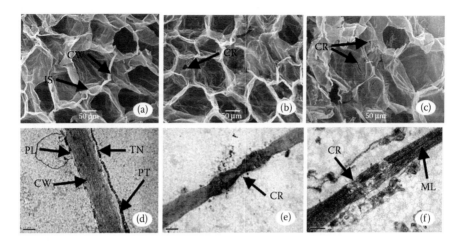

FIGURE 2.18 SEM and TEM micrographs of parenchyma tissue from fresh and treated apples with a 65°Brix sucrose solution at 50°C. (a, d) Fresh control; (b, e) OH; (c, f) VI/OH. IS, intercellular space; CW, cell wall; PL, plasmalemma; TN, tonoplast; PT, protoplast; CR, cell rupture; ML, middle lamella.

with the VI/OH treatment at 50°C that were dried at 60°C exhibited an increase in folic acid due to the electropermeabilization effect, which induced a retention of folic acid and an increase in the concentration by decreasing the water content. The VI/OH treatment at 50°C followed by drying at 60°C was determined to be the best process for creating dehydrated apple slices that are rich in folic acid (Moreno et al., 2016).

OH, VI, and dried operations, used individually or in combination, are examples of technologies that can be used to achieve this objective due to their high versatility, extended use, and structural effects. Some studies show the relationship among the application of these technologies, the structural characteristics of food, and their bioactive compounds together with the technologically and nutritionally achieved functionality. Applying OH leads, in some cases, to structural changes that improve the bioavailability of bioactive compounds.

2.5.2 Enzymatic Inactivation

Enzymatic browning caused by PPO activity in fruits and vegetables is a problem for the food industry and is one of the main causes of spoilage during processing and storage. The inhibition of browning is important in order to maintain the sensory attribute of color in fruit. PPO is an oxidoreductase enzyme that, in the presence of oxygen, catalyzes the oxidation of *o*-phenolic compounds to *o*-quinones, which are subsequently polymerized to dark-colored pigments (Van Loey, 2002). Inhibiting browning is important to maintain the sensory attribute of color in apples (Burdurlu and Karadeniz, 2003). Enzymatic browning can be inhibited by chemicals such as ascorbic acid or sulfites or by heat treatment (Kim et al., 2005). Studies of native enzymes in cloudy apple juice demonstrated enhanced inactivation of PPO with the increasing temperature, pressure, and time of continuous high-pressure carbon dioxide treatment

(Xu et al., 2011). Applying electrical fields via OH inactivated PPO and lipoxygenase more rapidly than did conventional heating. OH has been shown to be a useful alternative method for pasteurizing or sterilizing food products that may also enhance the rate of PPO inactivation in food materials (Jakób et al., 2010).

The combined effect of OH and OD with VI, on the PPO inactivation, physical properties, and microbial stability of apples stored at 5°C or 10°C have been studied. The treatments were performed using a 65% (w/w) sucrose solution and with OH at 13 V/cm at 30°C, 40°C, or 50°C for 90 min. The combination of OH and VI accelerated the mass transfer in the apple cube samples. The greatest color change and Browning index change during refrigerated storage were observed in the samples treated at atmospheric pressure at 30°C or 40°C (Figure 2.19). The enzymatic activity in this samples recovered, especially in the samples treated at atmospheric pressure (OD and OD/OH) at 30°C, in which the RAs reached 29% and 21%, respectively. Under the conditions studied, the VI/OH treatment at 50°C was the most effective treatment for inhibiting molds and yeasts, and mesophiles bacteria, particularly when the samples were stored at 5°C. The shelf life of the osmotically dehydrated apples was extended to more than 4 weeks. The VI/OH treatment at 50°C is the optimal process for dehydrating apples in a sucrose solution of 65°Brix at 13 V/cm. The greatest water loss resulted from the OD/OH and VI/OH treatments at 50°C, and the largest amount of solid gain was obtained with the VI/OH and OD/OH treatments at 50°C. PPO was completely inactivated with OD/OH and VI/OH treatments at 50°C (Moreno et al., 2013).

2.6 IMPACT OF OH IN PROTECTION OF BIOACTIVE COMPOUNDS

The OH technology has gained interest because the products are of better quality than those processed by conventional technologies, and because it is a thermal process that uses electrical energy to heat foods as a method of preservation, which can be used for microbial and enzymatic inactivation. The advantages of this technology

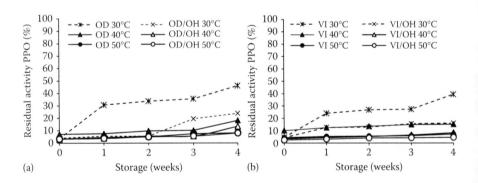

FIGURE 2.19 Browning index (BI) in treated apples and stored at 5°C. (a, b) Osmotic dehydration at atmospheric pressure, VI: osmotic dehydration with vacuum impregnation, OH: ohmic heating (OD/OH and VI/OH) at 13 V/cm.

combined with others include uniform heating and improvements in quality with minimal structural, nutritional, or organoleptic changes. The OH operation assists the incorporation of bioactive compounds into natural structures by taking advantage of the protective effect of these compounds and the synergistic effect of certain other compounds. On the other hand, the VI treatments allow the incorporation of bioactive compounds into natural structures by taking advantage of both the protective effect of these compounds and the synergistic effect of certain other compounds.

2.7 CONCLUSIONS AND FUTURE PERSPECTIVES

The OH technology offers a potential to processing applications in the development of functional food. This emergent technology's ability to produce food with enhanced bioactive compounds has grown meaningfully. Fruits and vegetables naturally represent the simplest form of functional foods and can be a source of functional compounds. Industrially, electrotechnologies provide opportunities and advantages not found in other technologies. Food processes help obtaining foods with nutritional and health benefits. Technologies need to be used in ways to increase functional properties of foods. Many studies show the relationship between the application of these technologies, the structural characteristics of food, and their bioactive compounds together with the technologically and nutritionally achieved functionality. Applying OH induces structural changes that improve the bioaccessibility of bioactive compounds. The adverse effects related to the use of high temperatures in dried process can be reduced by incorporating components that generate protective structures. The knowledge of the relations structure properties process is an element necessary to achieve this objective. By following this strategy, it is possible to optimize processing procedures in order to obtain final products, tailored based on consumer needs. Therefore, the OH, VI, and dried operations, used individually or in combination, are examples of technologies that can be used to achieve this objective to improve food functionality due to their high versatility.

ACKNOWLEDGMENTS

Author Jorge J. Moreno is grateful for the financial support provided by CONICYT through the FONDECYT project 1103453, and the project GI 152222/C "Group of Emergent Technology and Bioactive Components of Food (TECBAL)" of Universidad del Bío-Bío.

REFERENCES

Adlercreutz H., Lignans and human health, *Critical Reviews in Clinical Laboratory Sciences*, 44, (2007), 483–525.

Ahmeda S. and Mahmoud A., A novel salting-out assisted extraction coupled with HPLC-fluorescence detection for trace determination of vitamin K homologues in human plasma, *Talanta*, 144, (2015), 480–487.

Alam M., Iqbal S., and Naseem I., Ameliorative effect of riboflavin on hyperglycemia, oxidative stress and DNA damage in type-2 diabetic mice: Mechanistic and therapeutic strategies, *Archives of Biochemistry and Biophysics*, 584, (2015), 10–19.

Ali H., Almagribi W., and Al-Rashidi M., Antiradical and reductant activities of anthocyanidins and anthocyanins, structure–activity relationship and synthesis, *Food Chemistry*, 194, (2015), 1275–1282.

Allali H., Marchal L., and Vorobiev E., Effects of vacuum impregnation and ohmic heating with citric acid on the behaviour of osmotic dehydration and structural changes of apple fruit. *Biosystems Engineering*, 106(1), (2010), 6–13.

Amalraj A. and Pius A., Bioavailability of calcium and its absorption inhibitors in raw and cooked green leafy vegetables commonly consumed in India—An in vitro study, *Food Chemistry*, 170, (2015), 430–436.

Atarés L., Chiralt C., and Gonzaléz-Martínes C., Effect of solute on osmotic dehydration and rehydration of vacuum impregnated Apple cylinders (cv. Granny Smith), *Journal of Food Engineering*, 89 (2009), 49–56.

Araújo J., Martel F., Borges N., Araújo J., and Keating E., Folates and aging: Role in mild cognitive impairment, dementia and depression, *Ageing Research Reviews*, 22, (2015), 9–19.

Aron P. and Kennedy J., Flavan-3-ols: Nature, occurrence and biological activity, *Molecular Nutrition and Food Research*, 52, (2008), 79–104.

Barrera C., Betoret N., and Fito P., Ca^{2+} and Fe^{2+} influence on the osmotic dehydration kinetics of apple slices (var. Granny Smith), *Journal of Food Engineering*, 65(1), (2004), 9–14.

Betoret N., Puente L., Díaz M. J., Pagán M. J., García M. J., Gras M. L. et al. Development of probiotic-enriched dried fruits by vacuum impregnation, *Journal of Food Engineering*, 56(2–3), (2003), 273–277.

Betoret E., Sentandreu E., Betoret N., Codoñer-Franch P., Valls-Belles V., and Fito P., Technological development and functional properties of an apple snack rich in flavonoid from mandarin juice, *Innovative Food Science and Emerging Technologies*, 16, (2012), 298–304.

Betoret E., Betoret N., Rocculi P., and Dalla-Rosa M., Strategies to improve food functionality: Structureeproperty relationships on high pressures homogenization, vacuum impregnation and drying technologies, *Trends in Food Science & Technology*, 46, (2015), 1–12.

Burdurlu H. S. and Karadeniz F., Effect of storage on nonenzymatic browning of apple of apple juice concentrates, *Food Chemistry*, 80, (2003), 1438–1441.

Chanet A., Milenkovic D., Manach C., Mazur A., and Morand C., Citrus flavanones: What is their role in cardiovascular protection? *Journal Agricultural and Food Chemistry*, 60, (2012), 8809–8822.

Chang J., Rimando A., Pallas M., Camins A., Porquet D., Reeves J., Hale B., Smith M., Joseph J., and Casadesus G., Low-dose pterostilbene, but not resveratrol, is a potent neuromodulator in aging and Alzheimer's disease, *Neurobiology of Aging*, 33, (2012), 2062–2071.

Cortés M., Osorio A., and García E. Manzana deshidratada fortificada con vitamina E utilizando la ingeniería de matrices, *Vitae*, 14, (2007a), 17–26.

Cortes M., Osorio A., and Garcia E. Air dried apple fortified with vitamin E using matrix engineering, *VITAE. Revista de la Facultad de Quimica Farmaceutica*, 14(2), (2007b), 17–26.

Da Silva R., Kelly K., Lewis E., Leonard K., Goruk S., Curtis J., Parra D., Proctor S., Campo C., and Jacobs R., Choline deficiency impairs intestinal lipid metabolism in the lactating rat, *The Journal of Nutritional Biochemistry*, 26, (2015), 1077–1083.

Donovan J. and Waterhouse A., Bioavailability of flavanol monomers, *Flavonoids in Health and Disease*, (2003), 413–440.

Du J., Cullena J., and Buettner G., Ascorbic acid: Chemistry, biology and the treatment of cancer, *Biochimica et Biophysica Acta (BBA)—Reviews on Cancer*, 1826, (2012), 443–457.

Durst R. and Werolstad R., Separation and characterization of anthocyanins by HPLC. In: *Handbook of Food Analytical Chemistry*, R. E. Wrolstad et al. (eds.), John Wiley & Sons, New Jersey, (2001), pp. 33–45.

Ensafia A., Heydari-Soureshjania E., Jafari-Asla M., Rezaeia B., Ghasemib J., and Aghaeeb E., Experimental and theoretical investigation effect of flavonols antioxidants on DNA damage, *Analytica Chimica Acta*, 887, (2015), 82–91.

Escudero A., Petzold G., Moreno J., Gonzalez M., Junod J., Aguayo C., Acurio J., and Escudero C., Supplementation with apple enriched with L-arginine may improve metabolic control and survival rate in alloxan-induced diabetic rats, *Biofactor*, 39, (2013), 564–574.

Gabriel A., Usero J., Rodríguez K., Díaz A., and Tiangson C., Estimation of ascorbic acid reduction in heated simulated fruit juice systems using predictive model equations, *LWT—Food Science and Technology*, 64, (2015), 1163–1170.

Guzmán R., Prieto A., Cameán A., and Cameán A., Beneficial effects of vitamin E supplementation against the oxidative stress on Cylindrospermopsin-exposed tilapia (Oreochromis niloticus), *Toxicon*, 104, (2015), 34–42.

Gras M. L., Vidal D., Betoret N., Chiralt A., and Fito P., Calcium fortification of vegetables by vacuum impregnation, *Journal of Food Engineering*, 56(2–3), (2003), 279–284.

Gröber U., Schmidt J., and Klaus K., Magnesium in prevention and therapy, *Nutrients*, 7, (2015), 8199–8226.

Heima K., Tagliaferro A., and Bobilya D. Flavonoid antioxidants: Chemistry, metabolism and structure-activity relationships, *The Journal of Nutritional Biochemistry*, 13, (2002), 572–584.

Hernández A., Ochoa E., Ibarra I., Ortega D., Salvador A., and Velázquez A., Temporal development of genetic and metabolic effects of biotin deprivation. A search for the optimum time to study a vitamin deficiency, *Molecular Genetics and Metabolism*, 107, (2012), 345–351.

Hipólito U., Callera G., Simplicio J., De Martinis B. S., Touyz R. M., and Tirapellid C. R., Vitamin C prevents the endothelial dysfunction induced by acute ethanol intake, *Life Sciences*, 141, (2015), 99–107.

Holick M., Binkley N., Bischoff-Ferrari H., Gordon C., Hanley D., Heaney R., Murad M., and Weaver C., Evaluation, treatment, and prevention of vitamin D deficiency: An endocrine society clinical practice guideline, *Journal Clinical Endocrinol Metabolic*, 96, (2011), 1911–1930.

Hsu H., Tsai I., Kuo L., Tsai C., Liou S., and Woung L., Herpetic keratouveitis mixed with bilateral Pseudomonas corneal ulcers in vitamin A deficiency, *Journal of the Formosan Medical Association*, 114, (2015), 184–187.

Jakób A., Bryjak J., Wójtowics H., Illoevá V., Annus J., and Polakovic M., Inactivation kinetics of food enzymes during ohmic heating, *Food Chemistry*, 123, (2010), 369–376.

Kirkland J. B., Niacin, In: *Handbook of Vitamins*, 4th edition, J. Zempleni, R. Rucker, D. McCormick, and J. Suttie (eds.), CRC Press, Boca Raton, FL, (2007b), pp. 192–223.

Koike H., Hama T., Kawagashira Y., Hashimoto R., Tomita M., Iijima M., and Sobue G., The significance of folate deficiency in alcoholic and nutritional neuropathies: Analysis of a case, *Nutrition*, 28, (2012), 821–824.

Koneswaran M. and Narayanaswamy R., Ultrasensitive detection of vitamin B6 using functionalised CdS/ZnS core–shell quantum dots, *Sensors and Actuators B: Chemical*, 210, (2015), 811–816.

Landete J. M., Plant and mammalian lignans: A review of source, intake, metabolism, intestinal bacteria and health, *Food Research International*, 46, (2012), 410–424.

Lima M., Zhong T., and Lakkakula N. R., Ohmic heating: A value-added food processing tool, *Louisiana Agriculture*, 45(4), (2002), 16–17.

Lienhart W., Gudipati V., and Macheroux P., The human flavoproteome, *Archives of Biochemistry and Biophysics*, 535, (2013), 150–162.

Markovic R., Báltico M., Pavlovic M., Glisic M., Radulovic E., Djordjevic V., and Sefer D., Isoflavones—From biotechnology to functional foods, *Procedia Food Science*, 5, (2015), 176–179.

Martensa S. and Mithöfer A., Flavones and flavone synthases, *Phytochemistry*, 66, (2005), 2399–2407.

Moraga M. J., Moraga G., Fito P. J., and Martínez-Navarrete N., Effect of vacuum impregnation whit calcium lactate on the osmotic dehydration kinetics and quality of osmodehydrated grape fruit, *Journal of Food Engineering*, 90, (2009), 372–379.

Moreno J., Simpson R., Baeza A., Morales J., Muñoz C., Sastry S. et al. Effect of ohmic heating and vacuum impregnation on the osmodehydration kinetics and microstructure of strawberries (cv. Camarosa), *LWT—Food Science and Technology*, 45(2), (2012), 148–154.

Moreno J., Simpson R., Pizarro N., Pavez C., Dorvil F., Petzold G. et al. Influence of ohmic heating/osmotic dehydration treatments on polyphenoloxidase inactivation, physical properties and microbial stability of apples (cv. Granny Smith), *Innovative Food Science & Emerging Technologies*, 20, (2013), 198–207.

Moreno J., Simpson R., Sayas M., Segura I., Aldana O., and Almonacid S., Influence of ohmic heating and vacuum impregnation on the osmotic dehydration kinetics and microstructure of pears (cv. Packham's Triumph), *Journal of Food Engineering*, 104(4), (2011), 621–627.

Moreno J., Espinoza C., Simpson R., Petzold G., Nuñez H., and Gianelli M. P., Application of ohmic heating/vacuum impregnation treatments and air drying to develop an apple snack enriched in folic acid, *Innovative Food Science & Emerging Technologies*, (2016), doi:10.1016/j.ifset.2015.12.014.

Mytton O. T., Nnoaham K., Eyles H., Scarborough P., and Mhurchu C. N., Systematic review and meta-analysis of the effect of increased vegetable and fruit consumption on body weight and energy intake, *BMC Public Health*, 14(1), (2014), 886.

Norman A. W., and Henry H. L., Vitamin D, In: *Handbook of Vitamins*, 4th edition, J. Zempleni, R. Rucker, D. McCormick, and J. Suttie (eds.), CRC Press, Boca Raton, FL, (2007a), pp. 41–109.

Ochoa-Ruiza E., Díaz-Ruiza R., Hernández-Vázqueza A., Ibarra-Gonzáleza I., Ortiz-Plata A., Rembao D. et al. Biotin deprivation impairs mitochondrial structure and function and has implications for inherited metabolic disorders, *Molecular Genetics and Metabolism*, 116, (2015), 204–214.

Osakabe N., Hoshi J., Kudo N., and Shibata M., The flavan-3-ol fraction of cocoa powder suppressed changes associated with early-stage metabolic syndrome in high-fat diet-fed rats, *Life Sciences*, 114, (2014), 51–56.

Özdemir K. and Gökmen V., Effect of microencapsulation on the reactivity of ascorbic acid, sodium chloride and vanillin during heating, *Journal of Food Engineering*, 167, (2015), 204–209.

Parisi O., Puoci F., Restuccia D., Farina G., Iemma F., and Picci N., Polyphenols and their formulations: Different strategies to overcome the drawbacks associated with their poor stability and bioavailability, *Polyphenols in Human Health and Disease*, 4, (2014), 29–45.

Parrot D. L., Use of ohmic heating for aseptic processing of food particulates, *Food Technology*, 46, (1992), 68–72.

Phan K., Wright G., and Lee N., Inductively coupled plasma-mass spectrometry (ICP-MS) and -optical emission spectroscopy (ICP–OES) for determination of essential minerals in closed acid digestates of peanuts (*Arachis hypogaea* L.), *Food Chemistry*, 134, (2012), 453–460.

Prasad, A. S., Clinical, immunological, anti-inflammatory and antioxidant roles of zinc, *Experimental Gerontology*, 43, (2008), 370–377.

Priyadarshia A., Kima E., and Hwang K., Structural and functional analysis of Vitamin K2 synthesis protein MenD, *Biochemical and Biophysical Research Communications*, 388, (2009), 748–751.

Radziejewska-Kubzdela E., Bieganska-Marecik R., and Kidon M., Applicability of vacuum impregnation to modify physico-chemical, sensory and nutritive characteristics of plant origin products—A review, *International Journal of Molecular Sciences*, 15(9), (2014), 16577–16610.

Ramaswamy H. S., Marcotte M., Sastry S., and Abdelrahim K., *Ohmic Heating in Food Processing*, (2014), CRC Press, Boca Raton, FL.

Reinisalo M., Kårlund A., Koskela A., Kaarniranta K., and Karjalainen R., Polyphenol stilbenes: Molecular mechanisms of defence against oxidative stress and aging-related diseases, *Oxidative Medicine and Cellular Longevity*, Article ID 340520, (2015), 1–24.

Rentzsch M., Wilkens A., and Winterhalter P., Non-flavonoid phenolic compounds, In: *Wine Chemistry and Biochemistry*, C. Polo and M. V. Moreno-Arribas (eds.), Springer, New York, (2009), pp. 509–527.

Ross C. and Harrison E., Vitamin A: Nutritional aspects of retinoids and carotenoids, In: *Handbook of Vitamins*, 4th edition, J. Zempleni, R. B. Rucker, and D. B. McCormick (eds.), CRC Press, Boca Raton, FL, (2007), pp. 25–30.

Ruan R., Ye X., Chen P., Doona C. J., and Taub I., Ohmic heating. In: *Thermal Technologies in Food Processing: Ohmic Heating*, P. Richardson (ed.), CRC Press, Boca Raton, FL, (2001), pp. 243–244.

Ruiza A., Bustamante L., Vergaraa C., von Baera D., Hermosín-Gutiérrez I., Obandod L., and Mardones C., Hydroxycinnamic acids and flavonols in native edible berries of South Patagonia, *Food Chemistry*, 167, (2015), 84–90.

Sastry S. K. and Palaniappan S., Influence of particle orientation on the effective electrical resistance and Ohmic heating rate of a liquid–particle mixture, *Journal Food Process Engineering*, 15, (1992), 213–227.

Schittl H., Quint R., and Getoff N., Products of aqueous vitamin B5 (pantothenic acid) formed by free radical reactions, *Radiation Physics and Chemistry*, 76, (2007), 1594–1599.

Schenkel L., Sivanesan S., Zhang J., Wuyts B., Taylor A., Verbrugghe A., and Bakovic M., Choline supplementation restores substrate balance and alleviates complications of Pcyt2 deficiency, *The Journal of Nutritional Biochemistry*, 26, (2015), 1221–1234.

Schulze B., Peth S., Hubbermann E. M., and Schwarz K. The influence of vacuum impregnation on the fortification of apple parenchyma with quercetin derivatives in combination with pore structure X-ray analysis, *Journal of Food Engineering*, 109, (2012), 380–387.

Sharma V. and Biswas D., Cobalamin deficiency presenting as obsessive compulsive disorder: case report, *General Hospital Psychiatry*, 34, (2012), 578.e7–578.e8.

Singh M., Kaur M., and Silakari O., Flavones: An important scaffold for medicinal chemistry, *European Journal of Medicinal Chemistry*, 84, (2014), 206–239.

Sofi N., Raja W., Dar I., Kasana B., Latief M., Arshad F., Hussain M., Irfan H., Parray M., and Iqbal K., Role of thiamine supplementation in patients with heart failure—An Indian perspective, *Journal of Indian College of Cardiology*, 5, (2015), 291–296.

Tsao R., Chemistry and biochemistry of dietary polyphenols, *Nutrients*, 2, (2010), 1231–1246. doi:10.3390/nu2121231.

Van Loey A., Verachtert B., and Hendrickx M., Effect of high electric field pulses on enzymes, *Trends in Food Science and Technology*, 12, (2002), 94–102.

Vermerris W. and Nicholson R., *Phenolic Compound Biochemistry*, Springer, Dordrecht, the Netherlands, (2006), pp. 1–34.

Vitale D., Piazza C., Filippo B., and Salomone S. Isoflavones: Estrogenic activity, biological effect and bioavailability, *European Journal of Drug Metabolism and Pharmacokinetics*, 38, (2013), 15–25.

Wagner C. and Greer F., Prevention of rickets and vitamin D deficiency in infants, children, and adolescents, *American Academy of Pediatrics*, 122, (2008), 1142–1152.

Wiley R. C. *Frutas y Hortalizas Mínimamente Procesadas y Refrigeradas*, Acribia, Zaragoza, Spain, (1997), 362p.

Wolfe K. L. and Liu R. H., Structure–activity relationships of flavonoids in the cellular antioxidant activity assay, *Journal of Agricultural and Food Chemistry*, 56, (2008), 8404–8411.

Wong C., Ip C., Leung C., and C. Siu, Vitamin B12 deficiency in the institutionalized elderly: A regional study, *Experimental Gerontology*, 69, (2015), 221–225.

Xie J. and Zhao Y., Nutritional enrichment of fresh apple (Royal Gala) by vacuum impregnation. *International Journal of Food Sciences and Nutrition*, 54(5), (2003), 387–398.

Xu Z., Zhang L., Wang Y., Bi X., Buckow R., and Liao X. Effects of high pressure CO_2 treatments on microflora, enzymes and some quality attributes of apple juice, *Journal of Food Engineering*, 104, (2011), 577–584.

Zand N., Christides T., and Loughrill E., Dietary intake of minerals, In: *Handbook of Mineral Elements in Food*, M. de la Guardia and S. Garrigues (eds.), John Wiley & Sons, Hoboken, NJ, (2015), pp. 23–39.

Zhao J., Han J., Jiang J., Shi S., Ma X., Liu X. et al. The downregulation of Wnt/β-catenin signaling pathway is associated with zinc deficiency-induced proliferative deficit of C17.2 neural stem cells, *Brain Research*, 1615, (2015), 61–70.

Zhou Z., Tan C., Zheng Y., and Wang Q., Electrochemical signal response for vitamin B1 using terbium luminescent nanoscale building blocks as optical sensors, *Sensors and Actuators B: Chemical*, 188, (2013), 1176–1182.

3 Encapsulation as a Carrier System to Enrich Foods with Antioxidants

Aline Schneider Teixeira and Lorena Deladino

CONTENTS

3.1 Encapsulation of Antioxidants: Why to Encapsulate?.. 61
3.2 Encapsulation of Antioxidants: Considerations about the Antioxidant Source.... 62
    3.2.1 Synthetic Antioxidants 62
    3.2.2 Natural Antioxidants 64
3.3 Encapsulation of Antioxidants: Considerations about the Role of the Antioxidant in the Food System 66
    3.3.1 Antioxidants as Food Additives... 66
    3.3.2 Antioxidants as a Health Supplement 67
3.4 Encapsulation Techniques: Advantages and Disadvantages in Relation to the Type of Antioxidant Protected.. 68
3.5 Application of Encapsulated Antioxidants into Food Systems: Case Studies 71
3.6 Future Trends... 75
References... 75

3.1 ENCAPSULATION OF ANTIOXIDANTS: WHY TO ENCAPSULATE?

Antioxidants can be defined as substances that when present in low concentration with respect to a susceptible substrate, in the food or in the human body, significantly delay or inhibit the substrate oxidation. In the case of foods, it is one of the major causes of chemical spoilage, resulting in rancidity and/or deterioration of the nutritional quality, color, flavor, texture, and safety of foods. It is estimated that half of the world's fruit and vegetable crops are lost due to postharvest deteriorative reactions (Antolovich et al. 2002).

Lipids occur in almost all foodstuffs, and most of them (more than 90%) are in the form of triacylglycerols, which are esters of fatty acids and glycerol. Two major components involved in lipid oxidation are unsaturated fatty acids and oxygen. Oxidative degradation of lipids may be initiated by active oxygen and related species which are more active than triplet oxygen molecules present in air, as well as by exogenous agents (UV, ionization radiation, and heat) (Yanishlieva-Maslarova 2001).

In many food products, it is suitable to add mixtures of antioxidants that usually have higher activities than single compounds, and which guarantee that the limits for single compounds have not been exceeded. Physical mixture of the selected antioxidants is not always possible, so encapsulation of the compounds, together or in a separate form, could be the solution to solve these practical situations.

On the other hand, there is a trend toward a healthier way of living, which includes a growing awareness among consumers about what they eat and what benefits certain ingredients provide to maintain good health. Preventing illness by diet is a unique opportunity for the so-called innovative functional foods. Existing and new ingredients need to be incorporated into food systems, considering the fact that over time these ingredients slowly degrade and lose their activity, or become hazardous, by oxidation reactions. Ingredients can also react with components present in the food system, which may limit bioavailability, or change the color or taste of a product. In many cases, microencapsulation can be used to overcome these challenges (Schrooyen et al. 2001).

In this sense, research on the application of polyphenols has recently attracted great interest in the functional foods, nutraceutical, and pharmaceutical industries, because of their potential health benefits to humans. However, the effectiveness of polyphenols depends on preserving the stability, bioactivity, and bioavailability of the active ingredients. The unpleasant taste of most phenolic compounds also limits their application. The utilization of encapsulated polyphenols, instead of free compounds, can effectively alleviate these deficiencies (Fang and Bhandari 2010).

For many nutritional supplements, formulation and route of administration have a great effect on both these parameters. Food nanotechnology, involving the utilization of nanocarrier systems to stabilize the bioactive molecules against a range of environmental and chemical changes as well as to improve their bioavailability, presents exciting opportunities for nutritional supplement industries (Tavano et al. 2014).

Microencapsulation has been defined as the technology of packaging solid, liquid, and gaseous materials into small capsules, which release their contents at controlled rates over prolonged periods of time (Champagne and Fustier 2007). There are various types of encapsulation technologies that can be employed in the food industry, including spray drying, coacervation, entrapment, inclusion complexation, cocrystallization, freeze drying, yeast encapsulation, emulsion surfactant micelles, nanospheres, nanoparticles, nanoemulsions, liposomes, and niosomes. However, specific studies on the interaction of encapsulating material with the antioxidant, as well as the measurement of the stability of the encapsulated antioxidant are mandatory for each case study, so, generalization should be avoided.

3.2 ENCAPSULATION OF ANTIOXIDANTS: CONSIDERATIONS ABOUT THE ANTIOXIDANT SOURCE

3.2.1 Synthetic Antioxidants

The application of antioxidants in foods is governed by the U.S. Food and Drug Administration (FDA) regulations. The regulations require that the ingredient labels of products declare the antioxidants used and their carriers followed by an explanation of their intended purpose (Dziezak 1986). Some of the more popular synthetic antioxidants used in food industry are phenolic compounds such as butylated hydroxyanisol (BHA), butylated hydroxytoluene (BHT), tertiary butylhydroquinone (TBHQ), and esters of gallic acid, for example, propyl gallate (PG) (Figure 3.1). The synthetic phenolic antioxidants are always substituted by alkyls to improve their solubility in fats and oils. The four major synthetic antioxidants in use are subjected to a *good manufacturing practice* limit of 0.02% of the fat or oil content of the food. The most suitable antioxidant for vegetable oils is TBHQ, BHA, and BHT, which are fairly heat stable and are often used for stabilization of fats in baked and fried products. Antioxidants that are heat stable have the property referred to as *carry-through*. The disadvantage of gallates lies in their tendency to form dark precipitates with the iron ions and their heat sensitivity.

The process of autoxidation of polyunsaturated lipids in food involves a free radical chain reaction that is generally initiated by exposure of lipids to light, heat, ionizing radiation, metal ions, or metalloprotein catalysts. The enzyme lipoxygenase can also initiate oxidation. The classical route of autoxidation includes initiation (production of lipid free radicals), propagation, and termination (production of nonradical products) reactions. Antioxidants, therefore, according to their mode of action, have been classified as the compounds that terminate the free radical chain in lipid oxidation by donating electrons or hydrogen to fat containing a free radical and to the formation of a complex between the fat chain and a free radical. Antioxidants stop the free radical chain of oxidative reactions by contributing hydrogen from the phenolic hydroxyl groups, themselves forming stable free

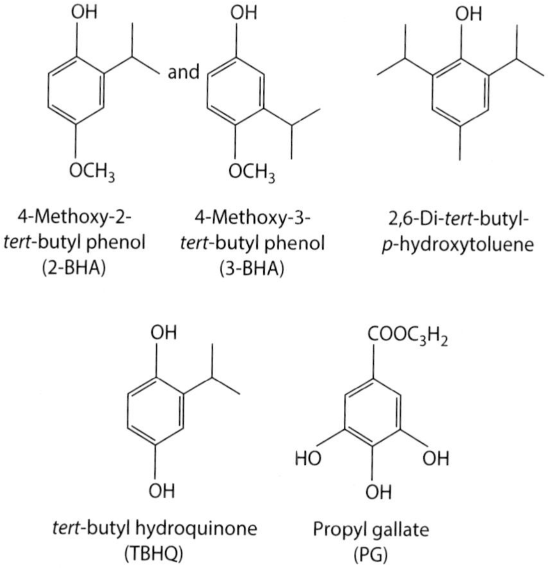

FIGURE 3.1 Chemical structure of the synthetic phenolic antioxidants, butylated hydroxyanisol (BHA), butylated hydroxytoluene (BHT), tertiary butylhydroquinone (TBHQ), and propyl gallate (PG).

radicals that do not initiate or propagate further oxidation of lipids (free-radical terminators) (Kaur and Kapoor 2001).

Not all antioxidant activity is conferred by free-radical interceptors. Reducing agents that function by transferring hydrogen atoms have also been categorized as oxygen scavengers. Some of these are ascorbyl palmitate, sulfites, ascorbic acid, glucose oxidase, and erythorbic acid. To be effective in foods these must be added during manufacturing or to the finished products. No single antioxidant offers a panacea to oxidative deterioration for all food products. The selection of an appropriate antioxidant appears to be determined by compatibility between the effect and the food-related properties of the product (Kaur and Kapoor 2001).

The synthetic antioxidants have been very thoroughly tested for their toxicological behaviors, but some of them are coming, after a long period of use, under heavy pressure as new toxicological data impose some caution in their use. In this context, natural products appear healthier and safer than synthetic antioxidants (Shahidi et al. 1992).

3.2.2 Natural Antioxidants

The natural antioxidants consumed in diet include phenolic and polyphenolic compounds, being the most important groups of natural antioxidants, the tocopherols, flavonoids, and phenolic acids (Figure 3.2). The mechanism by which antioxidants act can vary depending on food composition, including minor components. The beneficial effects of consuming fruits and vegetables have been attributed to the

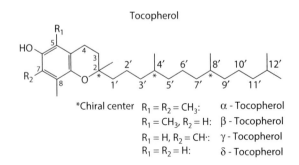

FIGURE 3.2 Chemical structures of the natural phenolic and polyphenolic antioxidants, tocopherol, flavonoid, and phenolic acids as well as their derivates.

presence of phenolic compounds, associated with the lowering of the risk of cardiovascular disease, cancer, waterfalls, and a great number of degenerative diseases. Antioxidants can prevent these diseases by reducing the reactive oxygen species (ROS) and thus limiting cell damage by ROS as well as reactive nitrogen species and reactive chlorine species.

It is a paradox that oxygen, which is considered as essential for life, is also reported to be toxic. Its toxicity is because of the process that unleashes the free radicals. The term free radical seems to appear a lot lately in everything, from vitamin brochures to cosmetic advertisements. Free radicals are unstable, highly reactive, and energized molecules having unpaired electrons. Examples of oxygen-derived free radicals include super oxide (O_2), hydroxyl (OH), hydroperoxyl (HOO), peroxyl (ROO), and alkoxyl (RO) radicals. Other common ROS produced in the body include nitric oxide (NO) and the peroxynitrite anion (ONOO). The free radicals react quickly with other compounds, trying to capture the electrons needed to gain stability. Generally, free radicals attack the nearest stable molecules, *stealing* its electrons. When the molecule that has been attacked loses its electron, it becomes a free radical itself, beginning a chain reaction. Once the process is started, it can cascade, initiating lipid peroxidation which results in destabilization and disintegration of the cell membranes or oxidation of other cellular components like proteins and DNA, generally resulting in the disruption of cells. Oxidation caused by free radicals sets reduced capabilities to combat aging and serious illnesses, including cancer, kidney damage, atherosclerosis, and heart diseases. Some free radicals arise normally during metabolism. Sometimes the body's immune system's cells purposefully create them to neutralize viruses and bacteria. However, environmental factors such as pollution, radiation, cigarette smoke, and herbicides can also generate free radicals. Thus, free radicals can not only produce beneficial effects but also induce harmful oxidation and cause serious cellular damage, if generated in excess (Kaur and Kapoor 2001).

The replacement of synthetic antioxidants by natural ones may have benefits due to health implications and functionality such as solubility in both oil and water, of interest for emulsions, in food systems. However, some of them such as those from spices and herbs (oregano, thyme, dittany, marjoram, lavender, and rosemary) have limited applications in spite of their high antioxidant activity, as they impart a characteristic herb flavor to the food, and deodorization steps are required (Moure et al. 2001).

The most commonly employed mean of consuming beneficial herbs is preparing a traditional medicine by making an aqueous infusion or decoction from the valuable parts of plants and herbs such as flowers, leafs, and roots. Some of them as tea (*Camellia sinensis*) and yerba mate (*Ilex paraguariensis*) are commonly drunk more than once a day as part of diet in many countries. Since most polyphenols are hydrophilic compounds that require an extraction process in order to be exhausted from the source material, their direct use in food matrix is limited. This problem can be solved by applying the microencapsulation technology, which provides the required technique for conversion of the liquid extract to an effective functional ingredient. The intake of natural bioactive compounds, especially polyphenols through food is of great interest, but difficulties associated with the susceptibility of those compounds to adverse external effects, or detrimental food processing conditions and their chemical instability have provided many efforts to improve oral bioavailability, among them microencapsulation, which represents a promising concept. The application of microencapsulated bioactive compounds as functional ingredients in various food and beverage applications exhibits a significant potential, since it could enable the enrichment of various food products with natural antioxidants (Diplock et al. 1999).

3.3 ENCAPSULATION OF ANTIOXIDANTS: CONSIDERATIONS ABOUT THE ROLE OF THE ANTIOXIDANT IN THE FOOD SYSTEM

Taking into account all the exposed above, the incorporation of antioxidants in foods may have two different purposes: (1) to act as a food additive, preventing or delaying food deterioration or (2) to function as a healthy/nutritional supplement.

In their review about encapsulation in food industry, Gibbs et al. (1999) highlighted that an important requirement is that the encapsulation system has to protect the bioactive component from chemical degradation (e.g., oxidation or hydrolysis) to keep the bioactive component fully functional. In the case of antioxidants, this point is crucial despite the function of the antioxidant in the foodstuff. However, this first requirement is not only a suggestion that we mainly have to overcome this chemical degradation in the gastrointestinal tract (pH, enzymes, presence of other nutrients), but also the deleterious circumstances during storage of the product that serves as vehicle for the bioactive components. Many food components may interfere with the bioactivity of the added bioactive food components. This is a big challenge, as only a small proportion of the molecules remain available following oral administration, due to insufficient gastric residence time, low permeability, and/or solubility within the gut. It is therefore mandatory that the encapsulation procedure protects the bioactive component during the whole period of food processing and storage (temperature, oxygen, and light).

3.3.1 ANTIOXIDANTS AS FOOD ADDITIVES

Fats, oils, and lipid-based foods deteriorate through several degradation reactions, both on heating and on long-term storage. The main deterioration processes are oxidation reactions and the decomposition of oxidation products which result in decreased nutritional value and sensory quality. The retardation of these oxidation processes is important for the food producer and, indeed, for all persons involved in the entire food chain from the factory to the consumer (Pokorny et al. 2001).

The antioxidants can inhibit or delay oxidation in two ways: either by scavenging free radicals (in which case the compound is described as a primary antioxidant), or by a mechanism that does not involve direct scavenging of free radicals (in which case the compound is a secondary antioxidant). Primary antioxidants include phenolic compounds such as vitamin E (α-tocopherol). These components are consumed during the induction period. Secondary antioxidants operate by a variety of mechanisms including binding of metal ions, scavenging oxygen, converting hydroperoxides to nonradical species, absorbing UV radiation, or deactivating singlet oxygen. Normally, secondary antioxidants only show antioxidant activity when a second minor component is present. This can be seen in the case of sequestering agents such as citric acid, which are effective only in the presence of metal ions, and reducing agents such as ascorbic acid, which are effective in the presence of tocopherols or other primary antioxidants.

3.3.2 Antioxidants as a Health Supplement

The polyphenolic compounds are used in numerous sectors of the food-processing industry as natural additives (natural coloring agents, conservative agents, natural antioxidants, nutritional additives). However, it is probably in the field of human health that the economic implication of polyphenols is most important. Actually, many plant extracts rich in phenolic molecules of interest are used as food complements or can be integrated into cosmetic or pharmaceutical formulations (Munin and Edwards-Lévy 2011). Figure 3.3 summarizes the main effects of antioxidants on health.

A rapidly growing body of research works day by day to show antioxidants is important in maintaining health and reducing the risk of disease. Increasingly, companies want to make claims concerning antioxidants in their products. These claims can fall into any of three claim categories: nutrient content claims, structure/function claims, or health claims. Another, more specific category of claim is a dietary guidance claim. In its Food Labeling Guide 2013, FDA stated the following requirements to use the term *antioxidant*:

For claims characterizing the level of antioxidant in a food,

1. An RDI must be established for each of the nutrients that are the subject of the claim.
2. Each nutrient must have existing scientific evidence of antioxidant activity.

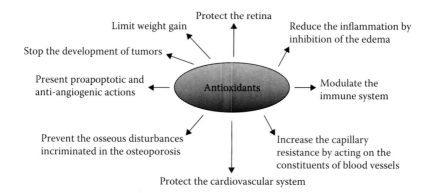

FIGURE 3.3 Properties of antioxidants on health.

3. The level of each nutrient must be sufficient to meet the definition for *high, good source,* or *more*; β-carotene may be the subject of an antioxidant claim when the level of vitamin A present as β-carotene in the food is sufficient to qualify for the claim.
4. Name(s) of nutrient(s) that is (are) the subject of the claim is (are) included as part of the claim (e.g., high in antioxidant vitamins C & E).

As an example, in Appendix D from U.S. FDA regulations: Qualified Health Claims, figures a claim for "Antioxidants, Vitamins, and Cancer," for dietary supplements containing vitamin E and/or vitamin C. The disclaimer (i.e., evidence is limited and not conclusive) is placed immediately adjacent to and below the claim, with no intervening material, in the same size, typeface, and contrast as the claim itself: "Some scientific evidence suggests that consumption of antioxidant vitamins may reduce the risk of certain forms of cancer. However, the FDA has determined that this evidence is limited and not conclusive" (FDA 2013).

In this context, long-term studies evaluating the effect of polyphenols on humans are mandatory. Although no information on causality can be obtained, epidemiologic studies are useful for the human health effects of long-term exposure to physiologic concentrations of polyphenols. Reliable data on polyphenol contents of foods are needed for studies of the potential role of dietary polyphenols in cancer and cardiovascular disease prevention. Accordingly, Arts and Hollman (2005) resumed an overview of epidemiologic studies on the health effects of flavonols, flavones, catechins, and lignans. They concluded that there is a need for more research on some specific diseases and with data becoming available for other polyphenols, as most studies have included flavonols and flavones only.

3.4 ENCAPSULATION TECHNIQUES: ADVANTAGES AND DISADVANTAGES IN RELATION TO THE TYPE OF ANTIOXIDANT PROTECTED

Depending on the encapsulation process used, several types of microcapsules, agglomerates, or capsules can be obtained. Figure 3.4 shows a schematic representation of the most common systems: Reservoirs consist of a core material constituted by the active compound and a coating called shell, wall material, coating barrier,

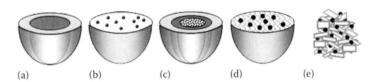

FIGURE 3.4 Morphologies of microparticles: (a) mononuclear core and homogeneous shell microcapsule, also called core–shell microcapsule, (b) polynuclear core and homogeneous shell microcapsules, (c) mononuclear core and multishell microcapsule, (d) polymer matrix, also called microsphere, where active is homogeneously or heterogeneously dispersed, and (e) agglomerates containing the active compound.

carrier, or encapsulant (Figure 3.4a). The release rate from these systems will depend on the thickness, the area, and the permeability of the barrier. Matrix systems consist of a homogeneous mixture of the active agent and the carrier (Figure 3.4b). A part of the active agent may be at the surface of the capsule, being unprotected. The release in these systems depends on the type of carrier, the system shape, and the amount of active agent. Whether the active agent is dissolved or dispersed in the polymer will affect its release rate well. Several methods allow obtaining mixed systems by adding an extra layer over the mentioned systems or adding a crosslinking agent to increase protection of the active agent or to delay its release (Figure 3.4c and d). Agglomerates offer extended surface areas to contain active agents, facilitate handling, and enhance their storage conditions (Figure 3.4e). Different shapes and sizes depending on the active agent, barrier material, technology used, and food structure can be found (Deladino et al. 2007b).

Encapsulation of most plant-derived antioxidant compounds such as lycopene, olive leaf extract, *Amaranthus* betacyanin extracts, β-carotene, D-limonene, and procyanidins were achieved using spray-drying method. Generally, the common excipient materials are limited to maltodextrins, gum arabic, and gelatin. The produced particles are of matrix type, which implies that some bioactive components may be exposed and generally they release the whole of the content rapidly after being incorporated in the foodstuff. Even though the process does not lead to a perfect encapsulate, the properties obtained are often sufficient to achieve the desired delayed release of the ingredient in the actual application and commercial examples. The relative ease and also the low cost are the main reasons for the broad application of spray drying in industrial settings. Despite the popularity, spray drying may trigger degradation of active compounds (Kosaraju et al. 2006; Zhang and Kosaraju 2007). The susceptibility of these compounds to high operating temperature during spray drying is the major reason of degradation. As an example, Bakowska-Barczak and Kolodziejczyk (2011) encapsulated black currant polyphenols by spray drying using maltodextrins with different dextrose equivalent and inulin as wall material. They found that although process yield was high (close to 80% for one of the maltodextrines) higher air inlet temperatures (>180°C) caused more polyphenol and anthocyanin losses. Therefore, a mild encapsulation process is needed to encapsulate these heat-sensitive compounds (Yim et al. 2010).

It is a relative gentle methodology in terms of application of solvents and matrix molecules. Only water-based dispersions are applied in spray drying. Therefore, the matrix should have a high solubility in water. In most instances hydrophilic carbohydrate molecules are applied. These carbohydrates undergo a transition to a so-called glass (i.e., an amorphous solid) when the dispersion is rapidly evaporated. Usually, the product is very stable and allows for a significant increase in shelf life (Augustin and Hemar 2009). The fact that the matrix is hydrophilic does not suggest that the procedure is only applicable for hydrophilic molecules. Both hydrophilic and hydrophobic bioactive molecules can be used as the core materials. However, hydrophobic molecules are usually first dissolved in an oil phase after which an oil-in-water emulsion is formed prior to drying (de Vos et al. 2010).

Concentrating in a natural antioxidant extract as case study, yerba mate liquid or lyophilized extracts has been encapsulated by different techniques (Deladino et al.

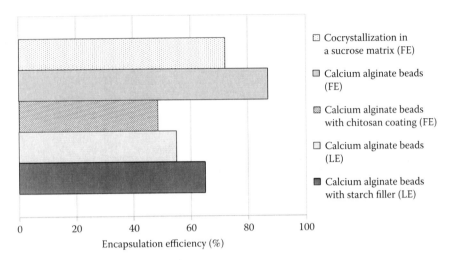

FIGURE 3.5 Encapsulation efficiency of different systems of encapsulation of liquid (LE) and freeze-dried (FE) yerba mate extracts.

2007a, 2008, 2014; Córdoba et al. 2013). In yerba mate aqueous extracts, polyphenol levels are higher than those of green tea and parallel to those of red wines (Gugliucci and Markowicz Bastos 2009; Gugliucci et al. 2009). Encapsulation efficiency of each system is shown in Figure 3.5; between systems prepared with the freeze-dried extracts, calcium alginate microcapsules incorporated more polyphenols from yerba mate, however their release was too fast in simulated gastric medium. Agglomerates of sucrose carrying lyophilized extract were obtained through cocrystallization showed very high efficiency, adding to the advantages of its powdered form, highly stable at high ambient humidity, and with very good flowing properties. When adding chitosan as external layer in calcium alginate microcapsules, the encapsulation efficiency decreased by half in comparison with the same capsule without chitosan layer, but more polyphenols reached the simulated intestinal medium so the chitosan barrier protected against the acidic conditions of the stomach. When the liquid extracts where encapsulated directly after its preparation, encapsulation efficiency decreased with respect to the freeze-dried extract, but incorporation of the starch filler to alginate matrix increased the entrapment capacity of yerba mate polyphenols, modulated the antioxidants release rate, and diminished the contribution of matrix erosion to the whole release mechanism. At least, encapsulation in corn starch partially gelatinized with high hydrostatic pressure was developed as yerba mate polyphenols carrier (Deladino et al. 2014).

As can be seen, several materials and techniques are suitable for the encapsulating of the same yerba mate extract, depending on the final use. In contrast to spray drying, it is worth to mention that none of these techniques reached 100°C during its preparation.

As mentioned earlier, solvent affinity is sometimes a limitation. Tavano et al.'s (2014) study overcame the limitation by encapsulating gallic and ascorbic acid, curcumin, and quercetin in tween 60 niosomes, as single components or in combination.

Co-encapsulation of two different compounds (one lipid soluble and the other water soluble) in the same niosomes formulation, gave in some cases an increase of the antioxidants amount encapsulated and the possibility to regulate and promote the compound release rate compared to that of vesicles containing only single compounds. Moreover, these formulations could represent a strategy to improve the solubility of curcumin and quercetin as molecules were poorly absorbed from the gastrointestinal tract after oral administration. Finally, enhanced antiradical properties, due to a synergic action between the two compounds (gallic acid/curcumin and ascorbic acid/quercetin) were achieved.

Chitosan has gained interest since 2005 due to its mucous adhesiveness, nontoxicity, biocompatibility, and biodegradability. Zhang and Kosaraju (2007) developed a chitosan-based delivery system using the sodium tripolyphosphate ionic crosslinking technique for the controlled release of polyphenolic antioxidants such as catechin. There are several advantages of using this method including the use of aqueous solutions, formation of smaller particles, and manipulation of particle charge by the variation in pH among others. There are some potential drawbacks in this method such as the relatively weak interactions between chitosan and sodium tripolyphosphate under low and high pH conditions. This may cause disintegration of particles under such conditions. Entrapment efficiency of the particles ranged between 27% and 40%. *In vitro* release studies indicated that the release of catechin in simulated gastric and intestinal fluids was between 15% and 40%, depending on the structural interactions between catechin and the chitosan matrix. The released amounts in simulated gastric and intestinal fluids were not more than 50%, which means there was still some catechin available for further release. Since chitosan particles are both mucoadhesive and also degradable in the colonic environment, catechin could be further released in large intestine for prolonged periods.

Tea polyphenols were encapsulated in chitosan nanoparticles by ionic gelation method using carboxymethyl chitosan and chitosan hydrochloride. These polymers are two different water-soluble chitosan with anionic and cationic groups, respectively. They can form nanoparticles through ionic gelation between the carboxyl groups of carboxymethyl chitosan and the amine groups of chitosan hydrochloride in aqueous solution. The encapsulant prepared as a novel nano-scale carrier has biocompatible and biodegradable characteristics, and also can limit the release of encapsulated materials more effectively. It is also thought to contribute to longer *in vivo* circulation times and allow encapsulation of water-soluble biomolecules. *In vitro* studies showed that tea polyphenols released in a sustained and controlled way, and the effect on cancer cell apoptosis by the released polyphenols was verified by an *in vitro* toxicity assay (Liang et al. 2011).

3.5 APPLICATION OF ENCAPSULATED ANTIOXIDANTS INTO FOOD SYSTEMS: CASE STUDIES

Many encapsulation procedures have been proposed but none of them can be considered as a universally applicable procedure for bioactive food components. This is caused by the fact that individual bioactive food components have their own characteristic molecular structure. They demonstrate extreme differences in molecular weight, polarity, solubility, and so on, which implies that different encapsulation approaches have to be applied in order to meet the specific physicochemical and molecular requirements for a specific bioactive component (Augustin and Hemar 2009). However, compatibility with the bioactives is not the only requirement an encapsulation procedure has to meet. It also should have specific characteristics to withstand influences from the environment. Besides, an ideal food-grade antioxidant should be safe, not impart color, odor, or flavor; be effective at low concentrations; be easy to incorporate; survive after processing; and be stable in the finished product as well as available at a low cost. Actually, few cases of application of encapsulated antioxidants in foods have been reported in scientific literature and they will be discussed here (Table 3.1).

Polyphenols from yerba mate encapsulated in calcium alginate with and without chitosan external layer were added to commercial instant soups; however, the product was a low fat powder containing pieces of dehydrated vegetables (Deladino et al. 2013). In this case the encapsulated extract was added to obtain a product enriched with antioxidants. Experimental results indicated that the extra amount of 0.1 mg of polyphenols incorporated by capsules in enriched soups was not modified by the other soup components. Besides, after 7 months of storage no significant differences were detected in polyphenol content for all formulations of enriched soups. Sensory analysis revealed that none of the enriched soups were significantly different from the commercial one. It is worth noting that yerba mate has a particular flavor and bitter notes. Thus, after the sensory results, it could be stated that encapsulation helped mask this typical taste not associated with these kinds of soups.

An eugenol-rich clove extract obtained from clove buds by supercritical carbon dioxide extraction, was encapsulated in maltodextrin and gum arabic matrices using spray drying by Woranuch and Yoksan (2013). A microencapsulated powder with maximum encapsulation efficiency of 65% was obtained. They applied the microcapsules in soybean oil, using the encapsulated clove powder as a source of natural antioxidant. They evaluated the antioxidant efficacy of encapsulated clove extract, unencapsulated clove extract, and commercial antioxidant BHT, individually administered in soybean oil, in frying of potato wedges. From a 30-day storage and frying stability study of soybean oil, it was found that there is no significant difference between the antioxidant activities of unencapsulated and encapsulated clove extracts in soybean oil. Further, encapsulated clove powder allowed controlled release of the antioxidant. No pro-oxidative activity at the initial stage of storage was observed contrary to that obtained with unencapsulated clove extract. The panel could not sensorially distinguish between potato wedges fried in soybean oil containing unencapsulated and encapsulated clove extracts both at the beginning (0 day) and at the end of the storage period (30 days). All potato wedges had pleasant taste with no disagreeable aftertaste. However, the potato wedges fried in BHT administered fresh soybean oil had unpleasant sour taste; also the potato wedges fried in 10-day-old stored oil had rancid odor, with total unacceptability at the end of the storage period. Therefore, from organoleptic evaluation, it could be concluded that the soybean oil with unencapsulated and encapsulated clove extracts conferred increased shelf stability to potato wedges, compared to that by the oil containing the commercial antioxidant (BHT).

TABLE 3.1
Encapsulated Antioxidants: Type of Antioxidant, Method of Encapsulation, and Bibliography

Antioxidant (Single Compounds or Extracts)	Encapsulation Technique	Encapsulating Material	Reference
Axtaxanthin	Solvent displacement	Poly(ethylene oxide)-4-methoxy-cinnamoylphthaloylchitosan, poly(vinylalcohol-*co*-vinyl-4-methoxycinnamate) and ethylcellulose	Tachaprutinun et al. (2009)
Apple and olive leaf extracts, grapeseed	Spray drying	Lecithin and sodium caseinate	Kosaraju et al. (2008)
Ascorbic acid, curcumin, gallic acid, quercetin	Multilamellar niosomal vesicles	Tween 60	Tavano et al. (2014)
Astaxanthin	Solvent displacement	Poly(ethylene oxide)-4-methoxy-cinnamoylphthaloylchitosan (PCPLC)	Tachaprutinun et al. (2009)
β-Carotene	Freeze drying	Mannitol	Sutter et al. (2007)
Catequin	Ionic crosslinking	Chitosan and sodium tripolyphosphate (STP)	Zhang and Kosaraju (2007)
Curcumin	Micelle	Beta casein	Esmaili et al. (2011)
Eugenol	Oil-in-water emulsion and ionic gelation	Chitosan	Woranuch and Yoksan (2013)
Eugenol-rich clove extract	Spray drying	Maltodextrin and gum arabic	Chatterjee and Bhattacharjee (2013)
Gallic acid	Electrospinning	Zein	Neo et al. (2013)
Ground ivy, hawthorn, nettle, olive and yarrow leaf extracts and raspberry leaf	Electrostatic extrusion	Alginate and chitosan	Belščak-Cvitanović et al. (2011)
Ilex paraguariensis	Ionic gelation	Alginate chitosan	Deladino et al. (2008)
	Ionic gelation	Chitosan hydrochloride and sodium tripolyphosphate	Harris et al. (2011)
Quercetin and ferulic acid	Electrospinning	Hybrid amaranth protein isolate (API): pullulan ultrathin fibers	Aceituno-Medina et al. (2015)
Rosemary	Co-precipitation by supercritical antisolvent	Poloxamers	Visentin et al. (2012)
Rutin	Ionic gellation	Methoxil pectin	Jantrawut et al. (2015)
St. John's wort (*Hypericum perforatum*)	Inclusion complex	β-Cyclodextrin	Kalogeropoulos et al. (2010)
Tea polyphenols	Crosslinking	Carboxymethyl chitosan, chitosan hydrochloride	Liang et al. (2011)

A novel application of encapsulated antioxidants was carried out by Barrett et al. (2011), where fish and flaxseed oils were improved by encapsulation and antioxidants. The oils, containing various levels of vitamin E, rosmarinic acid, ethylenediaminetetraacetic acid, and citric acid, plus carnosic acid and ascorbyl palmitate for fish oil were encapsulated by emulsification with lecithin and spray drying. Physical encapsulation markedly improved the stability of flaxseed oil and moderately improved the stability of fish oil. The oxidative stability of encapsulated fish oil was improved by all the antioxidants tested, being the most effective carnosic acid. Sensory results showed that bars containing 20% encapsulated fish or flaxseed oils were not significantly different from the control (unsupplemented) formulation for assessment of aroma quality, off-odor intensity, and flavor quality; however, both fat powders produced significantly higher scores for off-flavor intensity (with assessments for fish oil higher than those for flaxseed oil).

A promising application of encapsulated antioxidants is their incorporation in food packaging materials. Chatterjee and Bhattacharjee (2013) prepared eugenol-loaded chitosan comprising oil-in-water emulsion and ionic gelation with pentasodium tripolyphosphate. They incorporated them in a thermoplastic flour (TPF) active packaging prepared through extrusion technology at temperatures up to 155°C and 170°C to obtain products in the form of pellets and sheets, respectively. TPF containing encapsulated eugenol exhibited superior radical scavenging activity (~1.3–3.3-folds) compared with TPF containing naked eugenol.

Abdalla and Roozen (2001) encapsulated sage and oregano extracts in liposomes by ultrasonification or microfluidization. The antioxidant effect of these preparations was evaluated in salad dressings during storage in the dark at ambient temperature, at 40°C and at 60°C. The oxidation process was followed by measuring the formation of conjugated diene hydroperoxides as primary and hexanal as secondary oxidation products, as well as changes in the compositions of fatty acids and tocopherols. Oregano and sage extracts homogenized with olive oil as a carrier showed higher antioxidative effects than these extracts dissolved in ethanol during storage in the dark at ambient temperature and at 40°C. Exposure of salad dressings to light changed the antioxidative effect of plant extracts into a pro-oxidative effect. The preparation of liposomes by microfluidization showed higher encapsulation efficiency and more homogeneous vesicles than liposomes prepared by ultrasonification. Sage liposomes prepared by microfluidization showed high antioxidative effects similar to BHT liposomes in salad dressings during storage in the dark at ambient temperature and at 40°C.

Rashidinejad et al. (2014) encapsulated green tea catechin and epigallocatechin gallate (EGCG) in soy lecithin liposomes and they included cheese. High efficiency (>70%) and yield (80%) were achieved from the incorporation of catechin or EGCG inside the liposome structure. The nanocapsules containing these antioxidants were effectively retained within a low-fat hard cheese, presenting a simple and effective delivery vesicle for antioxidants. The encapsulation method of this study is a promising method to deliver antioxidants such as catechin and EGCG in a cheese matrix.

A novel approach for incorporating encapsulated antioxidants into peanut kernel was developed by Girardi et al. (2015). These authors encapsulated the synthetic antioxidants, BHA and BHT, by complex coacervation using a gelatin–gum arabic system as encapsulating agent with or without formaldehyde or glutaraldehyde as crosslinking agent. Significant differences were observed between crosslinking agents, where microcapsules reticulated with glutaraldehyde were 19% and 21% higher than microcapsules with formaldehyde and without crosslinking agents. A noticeable result is that reticulation variables assayed did not significantly affect the encapsulation efficiency; however, the permanence of antioxidant compounds for both, the formulated and in microcapsules added to a peanut, the antioxidant levels inside the microcapsule were highest in that reticulated with formaldehyde. The reticulation among molecules would lead to major stiffness of the microcapsules and therefore obstructing the compound release. The permanence experiment carried out on microcapsules added to peanut revealed that stability and storage period of BHA and BHT were greatly enhanced and a durable controlled released effect and long residual action were achieved by microencapsulation. In this regard, after 40 days of storage, both antioxidants still remained inside the microcapsules that were reticulated with formaldehyde in high levels.

3.6 FUTURE TRENDS

In this domain, the progress should accelerate the reasoned use of natural polyphenolic compounds. The challenge of converting the most powerful polyphenols into usable compounds has then been resolved through innovative formulations. Many companies and research institutes are looking for adding them to food systems will often require technological innovations. Encapsulation will certainly play an important role in this process, although cost increase and bioavailability should always be considered carefully.

REFERENCES

Abdalla, A. E. and J. P. Roozen. The effects of stabilised extracts of sage and oregano on the oxidation of salad dressings. *European Food Research and Technology* 212(5): (2001) 551–560.

Aceituno-Medina, M., S. Mendoza, B. A. Rodríguez, J. M. Lagaron, and A. López-Rubio. Improved antioxidant capacity of quercetin and ferulic acid during in-vitro digestion through encapsulation within food-grade electrospun fibers. *Journal of Functional Foods* 12: (2015) 332–341.

Antolovich, M., P. D. Prenzler, E. Patsalides, S. McDonald, and K. Robards. Methods for testing antioxidant activity. *Analyst* 127(1): (2002) 183–198.

Arts, I. C. W. and P. C. H. Hollman. Polyphenols and disease risk in epidemiologic studies. *The American Journal of Clinical Nutrition* 81(1): (2005) 317S–325S.

Augustin, M. A. and Y. Hemar. Nano- and micro-structured assemblies for encapsulation of food ingredients. *Chemical Society Reviews* 38(4): (2009) 902–912.

Bakowska-Barczak, A. M. and P. P. Kolodziejczyk. Black currant polyphenols: Their storage stability and microencapsulation. *Industrial Crops and Products* 34(2): (2011) 1301–1309.

Barrett, A. H., W. L. Porter, G. Marando, and P. Chinachoti. Effect of various antioxidants, antioxidants levels, and encapsulation on the stability of fish and flaxseed oils: Assessment by fluorometric analysis. *Journal of Food Processing and Preservation* 35(3): (2011) 349–358.

Belščak-Cvitanović, A., R. Stojanović, V. Manojlović et al. Encapsulation of polyphenolic antioxidants from medicinal plant extracts in alginate—Chitosan system enhanced with ascorbic acid by electrostatic extrusion. *Food Research International* 44(4): (2011) 1094–1101.

Champagne, C. P. and P. Fustier. Microencapsulation for the improved delivery of bioactive compounds into foods. *Current Opinion in Biotechnology* 18(2): (2007) 184–190.

Chatterjee, D. and P. Bhattacharjee. Comparative evaluation of the antioxidant efficacy of encapsulated and un-encapsulated eugenol-rich clove extracts in soybean oil: Shelf-life and frying stability of soybean oil. *Journal of Food Engineering* 117(4): (2013) 545–550.

Córdoba, A. L., L. Deladino, and M. Martino. Effect of starch filler on calcium-alginate hydrogels loaded with yerba mate antioxidants. *Carbohydrate Polymers* 95(1): (2013) 315–323.

de Vos, P., M. M. Faas, M. Spasojevic, and J. Sikkema. Encapsulation for preservation of functionality and targeted delivery of bioactive food components. *International Dairy Journal* 20(4): (2010) 292–302.

Deladino, L., P. S. Anbinder, A. S. Navarro, and M. N. Martino. Co-crystallization of yerba mate extract (Ilex paraguariensis) and mineral salts within a sucrose matrix. *Journal of Food Engineering* 80(2): (2007a) 573–580.

Deladino, L., P. S. Anbinder, A. S. Navarro, and M. N. Martino. Encapsulation of natural antioxidants extracted from Ilex paraguariensis. *Carbohydrate Polymers* 71(1): (2008) 126–134.

Deladino, L., A. Navarro, and M. N. Martino. Encapsulation of active compounds: Ionic gelation and cocrystallization as case studies. In *Functional Properties of Food Components*, edited by C. E. Lupano (2007b) 125–157.

Deladino, L., A. S. Navarro, and M. N. Martino. Carrier systems for yerba mate extract (Ilex paraguariensis) to enrich instant soups. Release mechanisms under different pH conditions. *LWT–Food Science and Technology* 53(1): (2013) 163–169.

Deladino, L., A. S. Teixeira, A. S. Navarro, I. Alvarez, A. D. Molina-García, and M. Martino. Corn starch systems as carriers for yerba mate (Ilex paraguariensis) antioxidants. *Food and Bioproducts Processing* 94: (2014) 463–472.

Diplock, A., T. Aggett, P. J. Ashwell, M. Bornet, F. Fern, E. B. Roberfroid, and M. Prelims. *British Journal of Nutrition* 81(01): (1999) 1–27.

Dziezak, J. D. Preservatives: Antioxidants. *Food Technology* 9(94): (1986) 94–102.

Esmaili, M., S. M. Ghaffari, Z. Moosavi-Movahedi et al. Beta casein-micelle as a nano vehicle for solubility enhancement of curcumin; food industry application. *LWT–Food Science and Technology* 44(10): (2011) 2166–2172.

Fang, Z. and B. Bhandari. Encapsulation of polyphenols: A review. *Trends in Food Science & Technology* 21(10): (2010) 510–523.

FDA (2013). A Food Labelling Guide. US Department of Health and Human Services. Food and Drug Administration. Center for Food Safety and Human Nutrition, Silver Spring, MD, January 2013.

Gibbs, B.F., S. Kermasha, A. Inteaz, and C. N. Mulligan. Encapsulation in the food industry: A review. *International Journal of Food Sciences and Nutrition* 50(3): (1999) 213–224.

Girardi, N. S., D. Garcia, A. Nesci, M. A. Passone, and M. Etcheverry. Stability of food grade antioxidants formulation to use as preservatives on stored peanut. *LWT—Food Science and Technology* 62(2): (2015) 1019–1026.

Gugliucci, A. and D. H. Markowicz Bastos. Chlorogenic acid protects paraoxonase 1 activity in high density lipoprotein from inactivation caused by physiological concentrations of hypochlorite. *Fitoterapia* 80(2): (2009) 138–142.

Gugliucci, A., D. H. Markowicz Bastos, J. Schulze, and M. F. Ferreira Souza. Caffeic and chlorogenic acids in Ilex paraguariensis extracts are the main inhibitors of AGE generation by methylglyoxal in model proteins. *Fitoterapia* 80(6): (2009) 339–344.

Harris, R., E. Lecumberri, I. Mateos-Aparicio, M. Mengíbar, and A. Heras. Chitosan nanoparticles and microspheres for the encapsulation of natural antioxidants extracted from Ilex paraguariensis. *Carbohydrate Polymers* 84(2): (2011) 803–806.

Jantrawut, P., O. Chambin, and W. Ruksiriwanich. Scavenging activity of rutin encapsulated in low methoxyl pectin beads. *Cellulose Chemistry and Technology* 49(1): (2015) 51–54.

Kalogeropoulos, N., K. Yannakopoulou, A. Gioxari, A. Chiou, and D. P. Makris. Polyphenol characterization and encapsulation in β-cyclodextrin of a flavonoid-rich Hypericum perforatum (St John's wort) extract. *LWT—Food Science and Technology* 43(6): (2010) 882–889.

Kaur, C. and H. C. Kapoor. Antioxidants in fruits and vegetables—The millennium's health. *International Journal of Food Science & Technology* 36(7): (2001) 703–725.

Kosaraju, S. L., L. D'Ath, and A. Lawrence. Preparation and characterisation of chitosan microspheres for antioxidant delivery. *Carbohydrate Polymers* 64(2): (2006) 163–167.

Kosaraju, S. L., D. Labbett, M. Emin, I. Konczak, and L. Lundin. Delivering polyphenols for healthy ageing. *Nutrition & Dietetics* 65: (2008) S48–S52.

Liang, J., F. Li, Y. Fang et al. Synthesis, characterization and cytotoxicity studies of chitosan-coated tea polyphenols nanoparticles. *Colloids and Surfaces B: Biointerfaces* 82(2): (2011) 297–301.

Moure, A., J. M. Cruz, D. Franco et al. Natural antioxidants from residual sources. *Food Chemistry* 72(2): (2001) 145–171.

Munin, A. and F. Edwards-Lévy. Encapsulation of natural polyphenolic compounds; A review. *Pharmaceutics* 3(4): (2011) 793–829.

Neo, Y. P., S. Ray, J. Jin et al. Encapsulation of food grade antioxidant in natural biopolymer by electrospinning technique: A physicochemical study based on zein–gallic acid system. *Food Chemistry* 136(2): (2013) 1013–1021.

Pokorny, J., N. Yanishlieva, and M. Gordon. *Antioxidants in Food—Practical Applications*. Boca Raton, FL: Woodhead Publishing (2001).

Rashidinejad, A., E. John Birch, D. Sun-Waterhouse, and D. W. Everett. Delivery of green tea catechin and epigallocatechin gallate in liposomes incorporated into low-fat hard cheese. *Food Chemistry* 156: (2014) 176–183.

Schrooyen, P. M. M., R. van der Meer, and C. G. De Kruif. Microencapsulation: Its application in nutrition. *Proceedings of the Nutrition Society* 60(04): (2001) 475–479.

Shahidi, F., P. K. Janitha, and P. D. Wanasundara. Phenolic antioxidants. *Critical Reviews in Food Science and Nutrition* 32(1): (1992) 67–103.

Sutter, S. C., M. P. Buera, and B. E. Elizalde. β-Carotene encapsulation in a mannitol matrix as affected by divalent cations and phosphate anion. *International Journal of Pharmaceutics* 332(1–2): (2007) 45–54.

Tachaprutinun, A., T. Udomsup, C. Luadthong, and S. Wanichwecharungruang. Preventing the thermal degradation of astaxanthin through nanoencapsulation. *International Journal of Pharmaceutics* 374(1–2): (2009) 119–124.

Tavano, L., R. Muzzalupo, N. Picci, and B. de Cindio. Co-encapsulation of antioxidants into niosomal carriers: Gastrointestinal release studies for nutraceutical applications. *Colloids and Surfaces B: Biointerfaces* 114: (2014) 82–88.

Visentin, A., S. Rodríguez-Rojo, A. Navarrete, D. Maestri, and M. J. Cocero. Precipitation and encapsulation of rosemary antioxidants by supercritical antisolvent process. *Journal of Food Engineering* 109(1): (2012) 9–15.

Woranuch, S. and R. Yoksan. Eugenol-loaded chitosan nanoparticles: II. Application in bio-based plastics for active packaging. *Carbohydrate Polymers* 96(2): (2013) 586–592.

Yanishlieva-Maslarova, N. V. Inhibiting oxidation. In: *Antioxidants in Foods: Practical Applications*, edited by J. Pokorny, N. Yanishlieva, and M. Gordon. Cambridge, England: Woodhead Publishing (2001) 22–57.

Yim, Z.H., C.B. Tiong, R.F. Mansa, P. Ravindra, and E.S. Chan. Release kinetics of encapsulated herbal antioxidants during gelation process. *Journal of Applied Sciences* 10: (2010) 2668–2672.

Zhang, L. and S.L. Kosaraju. Biopolymeric delivery system for controlled release of polyphenolic antioxidants. *European Polymer Journal* 43(7): (2007) 2956–2966.

4 Electric Field Treatment in the Retention of Vitamins in Fruit Pulps and Beverages Processing

Giovana D. Mercali and Júlia R. Sarkis

CONTENTS

4.1 Introduction 79
4.2 Vitamins—Structure and Properties 82
    4.2.1 Fat-Soluble Vitamins 83
    4.2.2 Water-Soluble Vitamins 84
4.3 Vitamin Degradation Mechanisms during Electric Field Treatments 85
4.4 Vitamins Stability during Electric Field Treatments Applied in Fruits Pulp 89
    4.4.1 Ohmic Heating 90
        4.4.1.1 Effect of the Electric Field Strength 90
        4.4.1.2 Effect of the Electric Field Frequency 92
        4.4.1.3 Effect on Shelf Life 93
    4.4.2 Pulsed Electric Fields 94
        4.4.2.1 Effect of the Electric Field Strength and the Treatment Time 94
        4.4.2.2 Effect on Shelf Life 96
4.5 Concluding Remarks and Future Trends 98
References 99

4.1 INTRODUCTION

Although many forms of food processing are designed to maximize vitamin retention, degradation still happens to some extent during manufacturing. Vitamin loss occurs mostly in water-soluble vitamins due to heat degradation, leaching into the liquid medium, and chemical reactions, such as oxidation and reaction with other food constituents (Kong and Singh 2012). Preservation of nutritional compounds during processing represents a major challenge for traditional techniques of food production because they generally involve preliminary and heat treatments that can reduce nutritional and sensory properties of the final product.

The use of conventional thermal processing techniques (e.g., water immersion, hot air drying, retorting, and heat transfer in heat exchangers, among others) has been favored by important technological developments over the past few years along with the easier handling of the equipment available. Nevertheless, heat processing under severe conditions may induce several chemical and physical changes that reduce the nutritional and organoleptic quality of the product.

Over the years, food industry and academia have increased the exploration of innovative processing technologies with minimal heat treatment to reduce nutrient degradation. Some of these technologies include ohmic heating (OH), microwave heating, and infrared heating. These technologies are an alternative for thermal processing with the advantage of rapid heating up to the desire holding temperature; this reduces the treatment time, which is critical to avoid excessive thermal damage of labile substances, such as vitamins and pigments (Palaniappan and Sastry 1991, FDA 2000, Castro et al. 2004).

OH is characterized as a process where electric current passes through foods to heat them by generating internal energy, without involving any heating medium or heat transfer surface. Therefore, this technology does not have the limitations of conventional high-temperature short-time (HTST) processes, which depend on heat transfer mechanisms (FDA 2000, Ruan et al. 2001). As OH can simultaneously generate heat in solid and liquid phases and does not need to transfer heat either within a solid or through a solid–liquid interface, it becomes particularly interesting for viscous products and foods containing particulates, such as soups, purées, and fruit pulps (de Alwis and Fryer 1990, Sastry and Palaniappan 1992). Thus, this technology can be applied uniformly to homogeneous and heterogeneous products comparable to HTST treatments with a significant process time reduction. Moreover, OH is a high-energy efficiency process. It consists of the application of alternating current (AC) voltage to electrodes in direct contact with the product. As shown in Figure 4.1, a bath system contains an ohmic cell, a power supply unit, temperature sensors, a data logger, and a computer. The potential applications of this technique in food industry are very wide, including blanching, evaporation, dehydration, fermentation, pasteurization, and sterilization (FDA 2000, Sarang et al. 2008).

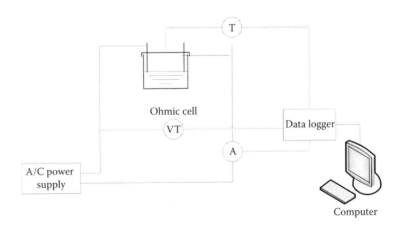

FIGURE 4.1 Diagram representing a OH batch equipment (V = voltage measurement; A = current transformer; T = temperature sensor).

For many decades, it was presumed that the effects of OH were purely thermal and electric field application had no influence on microorganisms and vegetable cells. However, recently it was observed that the nonthermal effects are significant within the range of OH voltages (Kulshrestha and Sastry 2003, Lakkakula et al. 2004, Mezadri et al. 2005, Loghavi et al. 2008, 2009, Moreno et al. 2012, Somavat et al. 2012). Although the main emphasis of OH is its use in pasteurization and sterilization processes, due to the electricity effects, a number of new applications have been identified in recent years. These processes, which depend upon either nonthermal effects of electricity or combined effects of electric and thermal phenomena, are named as moderate electric field (MEF) processing. The MEF processes involve the application of electric fields under 1000 V cm^{-1}, with or without heating, to achieve the specific goals (Sastry 2008).

Some other emerging technologies, such as PEFs, high pressure (HP), and power ultrasound, use lower temperatures combined with nonthermal effects to inactive enzymes and microorganisms, preserving the organoleptic characteristics of the product. PEF consists of the application of high-voltage short pulses (1–100 µs) to food products placed between two electrodes in a treatment chamber structure (Torregrosa et al. 2006). Figure 4.2 shows a typical system for the application of PEF technology in pumpable products. The equipment consist of a PEF generation unit, which is composed of a high voltage generator and a pulse generator, a treatment chamber, a suitable product handling system, and a set of monitoring and controlling devices (Soliva-Fortuny et al. 2009).

Even though PEF is a nonthermal processing technology, an increase in temperature can occur during PEF treatment depending on sample composition and processing parameters (Wouters et al. 2001). A high electric field strength within the food is generated due to a large flux of electric current that flows through the treatment chamber for a very short time (Sharma et al. 1998). The electric field affects the cell membranes and causes changes in enzyme structure and alteration in transport of ions (Wouters et al. 2001, Torregrosa et al. 2006). During PEF treatments, cell membrane permeabilization, also called electroporation, can occur. Depending on cell size and shape and the electric field intensity, the permeabilization may be reversible

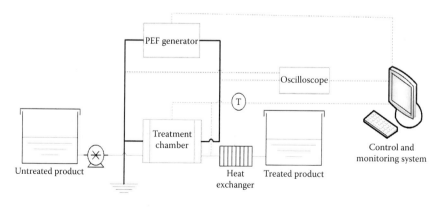

FIGURE 4.2 Diagram representing a PEF treatment unit for liquids (T = temperature sensor).

or irreversible. According to Knorr et al. (2011), electric field intensities in the range of 0.1–1 kV cm^{-1} cause reversible permeabilization for stress induction in plant cells; electric fields between 0.5 and 3 kV cm^{-1} cause irreversible permeabilization of plant and animal tissue; and electric fields ranging from 15 to 40 kV cm^{-1} cause irreversible permeabilization of microbial cells. The PEF technology is known as nonthermal cell disintegration or preservation process because the mechanism of electroporation is based essentially on a mechanical electrocompressive force affecting the cell membrane (Knorr et al. 2011).

PEF technology has been receiving considerable attention from scientists and the food industry. This process can be applied for the conservation of liquid foods, such as fruit juices (Akin and Evrendilek 2009, Odriozola-Serrano et al. 2013, Chen et al. 2014), milk (Riener et al. 2009, Zulueta et al. 2010a, 2013), and beer (Milani et al. 2015), among others. The main advantages of PEF are the minimal changes in fresh foods and the inactivation of a wide range of microorganisms and enzymes, leading to an increase of shelf life and food safety maintenance (Barbosa-Cánovas and Zhang 2001).

4.2 VITAMINS—STRUCTURE AND PROPERTIES

Vitamins are a heterogeneous group of organic nutrients, essential to health, required in small quantities for a variety of biochemical functions. They are not produced by the body, with the exception of vitamin D, and must, consequently, be obtained from the diet (Ottaway 2010, Bender 2012). Humans obtain vitamins from a diverse diet of fruits, vegetables, cereals, and meat. They are present in food products in different amounts and its concentration can be considerably affected by processing and storage conditions (Ottaway 2010).

Vitamins have diverse functions that may be described in one of four general classes: membrane stabilizers, hydrogen (H$^+$) and electron donors and acceptors, hormones, and coenzymes (Gallagher 2005). The B-complex vitamins have the highest level of biological similarity, since they play a vital role in enzyme reactions that are necessary for protein, carbohydrate, and fat metabolism (Marks 1993). They are usually divided into two categories: fat-soluble vitamins, such as vitamins A, D, E, and K; and water-soluble vitamins, such as vitamin C and all members of the vitamin B-complex. Table 4.1 summarizes vitamins according to their solubility.

Even though vitamins are divided into two categories, they have distinct attributes regarding its chemistry, function, daily requirements, and stability during processing. In terms of chemical structure, there are no similarities between substances. Some are single substances (such as biotin), while others (such as vitamin E) are groups of compounds all with vitamin activity. Regarding the daily requirements, some vitamins are required in tens of milligrams a day (vitamins C, E, and niacin), while others are only needed in microgram amounts per day—vitamins D and B12 (Ottaway 2010). This heterogeneity between vitamins also applies to their mechanisms of degradation and stability. While some vitamins are very sensitive to heat, light, and oxygen, such as vitamin C and folate, others are relatively stable, such as niacin and pantothenic acid.

TABLE 4.1
Vitamins Classification According to Their Solubility

Vitamins	
Water-Soluble	**Fat-Soluble**
Vitamin C	Vitamin A
Thiamine (vitamin B1)	Vitamin D
Riboflavin (vitamin B2)	Vitamin E
Vitamin B6	Vitamin K
Vitamin B12	
Biotin	
Niacin	
Folate	
Pantothenic acid	

4.2.1 Fat-Soluble Vitamins

The lipid-soluble vitamins (A, D, E, and K) are hydrophobic compounds that are absorbed passively along with dietary fats. They are transported in the blood by lipoproteins or attached to specific binding proteins and are not quickly eliminated from the body, being stored in the liver and fatty tissues (Gallagher 2005, Bender 2012).

Vitamin A can be obtained from two sources: retinoids, which include retinol, retinaldehyde, and retinoic acid, found only in animal tissue; and carotenoids, found in plants, known as provitamin A, as they can be cleaved to yield retinaldehyde and, therefore, retinol and retinoic acid (Gallagher 2005, Ottaway 2010). Vitamin A has essential roles in vision and several systemic functions, such as normal cell differentiation and cell surface function, growth and development, immune functions, and reproduction (Gallagher 2005, Ottaway 2010).

Vitamin D is produced in the skin of humans and animals by activation of sterols (cholesterol and ergosterol) by ultraviolet light from the sun (Potter and Hotchkiss 1995). It helps absorption and metabolism of calcium and phosphate, and cell differentiation (Bender 2012).

Vitamin E can be found naturally as a number of substances, including α, β, γ, and δ tocopherols and α tocotrienols. It has a fundamental role in protecting the body against the damaging effects of reactive oxygen species that are formed metabolically or encountered in the environment. Vitamin E works as a chain-breaking, free radical trapping antioxidant in cell membranes and plasma lipoproteins (Gallagher 2005, Bender 2012).

Vitamin K comprises the phylloquinones (the vitamin K1 series), which are synthesized by green vegetables, and the menaquinones (the vitamin K2 series), which are synthesized by intestinal bacteria, with differing lengths of side chain (Gallagher 2005). Menadione, menadiol, and menadiol diacetate are synthetic compounds that can be metabolized to phylloquinone. These substances play an essential role in blood clotting, bone formation, and regulation of multiple enzyme systems (Gallagher 2005, Bender 2012).

The dietary inadequacy of lipophilic vitamins can lead to deficiency syndromes, including night blindness and xerophthalmia (vitamin A); rickets in young children and osteomalacia in adults (vitamin D); neurologic disorders and hemolytic anemia of the newborn (vitamin E); and hemorrhagic disease of the newborn (vitamin K). On the other hand, excessive intake of vitamins A and D can cause toxicity (Bender 2012).

4.2.2 Water-Soluble Vitamins

Vitamin C and the B-complex vitamins are known as water-soluble. These vitamins are absorbed by passive and active mechanisms for being transported by carriers; they are not stored in appreciable amounts in the body and are excreted in the urine, making their regular consumption a necessity. These vitamins are distributed in the aqueous phases of the cells (cytoplasm and mitochondrial matrix space) and are essential cofactors (Gallagher 2005).

Thiamine, also known as vitamin B1, has a central role in energy-yielding metabolism, especially in the metabolism of carbohydrates. It is also important in the metabolism of fat and alcohol. Thiamine diphosphate and thiamine triphosphate are the active forms of the vitamin, which are required as cofactors in an ample diversity of reactions but especially those involving oxidative decarboxylation to form activated aldehyde species (Marks 1993, Bender 2012).

Riboflavin (vitamin B2) is vital for the metabolism of carbohydrates, amino acids, and lipids and supports antioxidant protection. It carries out these functions as the coenzymes flavin adenine dinucleotide (FAD) and flavin adenine mononucleotide (FMN) (Gallagher 2005). Niacin (vitamin B3) is the general term for nicotinamide and nicotinic acid. Its metabolic function is as the nicotinamide ring of the coenzymes NAD (nicotinamide adenine dinucleotide) and NADP (nicotinamide adenine dinucleotide phosphate) in oxidation–reduction reactions. They are crucial in all cells for energy production and metabolism and can be synthesized in the body from the essential amino acid tryptophan (Gallagher 2005, Bender 2012).

Pantothenic acid is broadly distributed in foods; cases of clinical deficiency are unusual. It is an integral part of coenzyme A, which is essential in energy production from the macronutrients. It links carbohydrate, fat, and amino acid metabolisms and forms the common normal pathway for these substances to enter the citric acid cycle. It is also essential for all acetylation processes and for steroid hormones formation via cholesterol (Marks 1993, Gallagher 2005).

Vitamin B6 is the name given to the closely related substances pyridoxine, pyridoxal, and pyridoxamine. All three compounds are converted to the metabolically active coenzyme form pyridoxal phosphate (PLP), which is primarily involved in the metabolism of amino acids, especially in transamination and decarboxylation. Altogether, the number of enzymes which are known to require PLP is higher than 60 (Marks 1993, Gallagher 2005, Bender 2012).

Biotin is a coenzyme for a substantial number of enzymes involved in carboxylation. Biotin is widely distributed in animal and plant tissues but in small concentrations. It can occur both in the free state (milk, fruit, and some vegetables) and in a form bound to a protein (animal tissues and yeast) (Gallagher 2005, Ottaway 2010).

Folates are a group of chemical substances with similar biochemical properties to a parent compound, pteroyl glutamic acid (folic acid). They function as an enzyme cosubstrate in many synthesis reactions in the metabolism of amino acids and nucleotides by donating or accepting single carbon units (Gallagher 2005).

There are several related chemical structures consisting of a corrin nucleus surrounding a cobalt atom, which show vitamin B12 activity. Collectively, they can be described as the cobalamins. Called the anti-pernicious anemia factor, this vitamin is also important for nucleic acid formation and for fat and carbohydrate metabolism (Potter and Hotchkiss 1995).

Vitamin C (ascorbic acid and dehydroascorbic acid) is widely distributed in nature and can occur at relatively high levels in some fruits and vegetables. Since ascorbic acid easily loses electrons and is reversibly converted to dehydroascorbic acid, it serves as a biochemical redox system involved in many electron transport reactions. By reacting with potentially toxic reactive oxygen species, this vitamin can prevent oxidative damage. It is required for the normal formation of the protein collagen, which is an important constituent of skin and connective tissues. Similar to vitamin E, vitamin C supports iron absorption. This vitamin is easily destroyed by oxidation, especially at high temperatures, and is easily lost during food processing, storage, and cooking (Potter and Hotchkiss 1995, Gallagher 2005, Ottaway 2010).

Overall, deficiency of B-vitamins and vitamin C is rare, with exceptions for alcoholics, those on a low calorie diet, and the elderly. However, specific syndromes are characteristic of deficiencies of individual vitamins, such as beriberi (thiamine); cheilosis, glossitis, and seborrhea (riboflavin); pellagra (niacin); megaloblastic anemia, methylmalonic aciduria, and pernicious anemia (vitamin B12); megaloblastic anemia (folic acid); and scurvy (vitamin C) (Bender 2012).

4.3 VITAMIN DEGRADATION MECHANISMS DURING ELECTRIC FIELD TREATMENTS

Many efforts have been made to evaluate the stability of vitamins during food processing; however, this task is difficult since there is a noticeable variation in stability among the various forms of vitamins (Gregory 2007). Moreover, degradation of vitamins is rather complex and entirely dependent on the type of processing (Rawson et al. 2011). The main factors that influence vitamin stability are temperature, moisture, oxygen, pH, light, presence of metallic ions, presence of oxidizing or reducing agents, and presence of other food components, such as proteins, sulfur dioxide, among others (Ottaway 2010).

Fat-soluble vitamins are generally more stable than water-soluble vitamins. Among the water-soluble vitamins, vitamin C, thiamine, and folate are the most unstable vitamins, being easily degraded by high levels of heat, light, and/or oxygen. Thiamine is hypersensitive to heat, and riboflavin is extremely light sensitive. B-group vitamins are normally sensitive to neutral and alkaline pH. Vitamins C and A are oxygen-sensitive vitamins (Kong and Singh 2012). Some vitamins, such as vitamin K, vitamin D, niacin, biotin, and pantothenic acid, are relatively stable during processing and storage.

Oxidation is one of the most important causes of foods nutritional changes. It occurs when food is exposed to air or because of the oxidative enzymes activity (such as peroxidase or lipoxygenase). In fruit products, losses of vitamin C and carotene due to oxidation are minimized by deaeration prior to thermal treatment (Fellows 2009, Rawson et al. 2011). It is important to point out that the degradation extent depends on the food's chemical nature, the extrinsic factors of the environment, the stabilities of the individual forms of vitamins in a particular foodstuff, and the opportunity for leaching (Gregory 2007).

Heat treatments, such as canning, blanching, pasteurization, baking, frying, extrusion, streaming, and sterilization, can cause loss of heat-labile vitamins (Gregory 2007). For this reason, thermal processing is a major cause of changes in nutritional properties of foods (Fellows 2009, Rawson et al. 2011). The conventional heating technologies depend essentially on the generation of heat outside the product to be heated (by fuel combustion or by an electric resistive heater) and its transference into the product by conduction and convection mechanisms (Rawson et al. 2011). The high temperature of the thermal processing accelerates reactions that would occur more slowly at ambient temperature (Gregory 2007). Degradation becomes more evident in processes in which high temperatures are maintained for a relatively long time, such as in canning. The use of high-temperature short-time (HTST) conditions improves vitamin retention. For example, in milk processing the lower-temperature longer-time process (LTLT) operating at 63°C for 30 min causes greater loss of vitamins than HTST processing at 71.8°C for 15 s (Fellows 2009).

Some of the new technologies, such as OH and PEF treatments, have emerged to reduce or eliminate the food exposure to heat. OH technology is an alternative to conventional thermal treatments that allows HTST processing (de Alwis and Fryer 1992). It can provide a fast and uniform heating with the ability to start/stop thermal processing instantaneously (Bansal and Chen 2006). The PEF technology uses lower temperatures and nonthermal effect of electricity for microbial and enzyme inactivation and avoids or greatly reduces detrimental changes in the nutritional, sensory, and physical properties of foods.

Studies to assess the effects of these new emerging technologies on the nutritional value of foods are of great importance to food technologists and consumers. While a considerable amount of studies have been published evaluating the microbial aspects of food preservation by OH and PEF, little information is available about the effect of these technologies on food constituents. Most of the studies performed to investigate the effect of electric field treatments on bioactive components evaluated vitamin C degradation. This vitamin is used as a quality index in fruits and vegetables because it is very sensitive to degradation during processing and storage. It can be used as a valid criterion to predict changes in organoleptic properties and/or nutritional components. For this reason, from now on, the discussion about the effect of the electric field on vitamins will be focused on vitamin C behavior and stability. There are still a number of unanswered questions regarding the stability of other vitamins during electric field processing. Systematic studies to better understand the effects of the electric field on the kinetic reactions of these nutritional compounds are still necessary.

According to Assiry et al. (2003), ascorbic acid degradation under electrical treatment conditions may occur due to chemical oxidation, chemical degradation via the anaerobic pathway, and electrochemical degradation by reactions at the electrodes. The first two types of degradation commonly occur in food processing, but, in the presence of oxygen, the chemical oxidation mechanism is dominant (Gregory 2007). Through chemical oxidation, ascorbic acid is converted into dehydroascorbic acid, which also has biological activity. Further oxidation generates diketogulonic acid, which has no biological function (Lee and Kader 2000, Furusawa 2001, Gregory 2007). After diketogulonic acid hydrolysis, additional oxidation, dehydration, polymerization, and reaction with amino acids and proteins generate various nutritionally inactive products (Lee and Kader 2000, Gregory 2007).

Vitamin C degradation rates are influenced by a number of factors, such as pH, acidity, metal ions, light, humidity, water activity, temperature, presence of amino acids, carbohydrates, lipids and enzymes, and oxygen solubility in the reaction temperature, among others (Assiry et al. 2003). The presence of metal ions, such as Fe^{3+} and Cu^{2+}, catalyzes the aerobic reaction. In the anaerobic pathway, ascorbic acid undergoes ketonization to form the intermediate keto-tautomer, which is in equilibrium with its anion. These compounds undergo further delactonization to form diketogulonic acid (Pátkai et al. 2002, Gregory 2007, Rawson et al. 2011).

Under electric field treatments, the electrochemical degradation becomes important due to the presence of an external power supply and the required contact between the electrodes and the processed food products to transfer the current uniformly into the medium (Samaranayake and Sastry 2005). In OH devices, AC voltage is applied to the electrodes at both ends of the product body; the electrical power, supplied for heating, can result in undesirable electrochemical reactions at the interface between the electrode and liquid phase (Bansal and Chen 2006). In PEF equipment, the large field intensities are achieved through storing a large amount of energy in a capacitor bank (a series of capacitors) from a direct current (DC) power supply, which is then discharged in the form of high-voltage pulses (FDA 2011). These large currents have to pass the electrode–liquid interface and can also cause electrochemical reactions.

According to Assiry et al. (2003), the electrochemical reactions that can occur at the electrodes are electrolysis of water and electrode corrosion. Electrolysis of water releases hydrogen and oxygen in the medium, which may cause additional oxidation of the ascorbic acid. For PEF processing, in which the capacitors are charged by a DC power source, electrolysis of water will yield hydrogen at the cathode and oxygen at the anode, as shown in Equations 4.1 through 4.4. As OH uses alternating voltage, both products may be released at each electrode.

$$\text{Anode:} \quad 2H_2O_{(l)} \rightarrow O_{2(g)} + 4H^+_{(aq)} + 4e^- \quad (4.1)$$

$$\text{Cathode:} \quad 2H^+_{(aq)} + 2e^- \rightarrow H_{2(g)} \quad (4.2)$$

$$2H_2O_{(l)} + 2e^- \rightarrow H_{2(g)} + 2OH^-_{(aq)} \quad (4.3)$$

$$\text{Overall:} \quad 2H_2O_{(l)} \rightarrow 2H_{2(g)} + O_{2(g)} \quad (4.4)$$

Electrode corrosion, either by direct metal oxidation or by electrochemical generation of corroding chemicals, releases metal ions in the medium. In the absence of agitation (convective transport), these reactions may be limited to the electrode vicinity. However, if the product is stirred inside the reactor, some of the products of the reactions may be dispersed into the medium (Assiry et al. 2003). The released electrode material may cause color changes, catalyze undesirable reactions in food products, and reduce the electrode lifetime (Kim and Zhang 2011). The products that are formed during electrode corrosion can react with other compounds in the liquid. For example, if stainless steel electrodes are used, ions such as Fe^{2+} and Fe^{3+} released from the electrodes' surface would catalyze the oxidation of ascorbic acid.

According to Ibrahim (1999), the damage caused by the use of a DC is higher than the damage caused by an AC, and is usually greater for lower frequency than for higher frequency currents. Thus, the use of high frequency is a good strategy to control the corrosion process. The modifications necessary in the pulse generator systems in PEF devices to reduce the amount of electrochemical reactions are still a challenge. The application of shorter pulses or switching systems without leak current could help to avoid electrochemical reactions (Martín-Belloso and Soliva-Fortuny 2011). In addition, an alternative way to avoid the occurrence of electrochemical reactions is by choosing an appropriate electrode material. Various materials (aluminum, carbon, platinum, platinized titanium, stainless steel, titanium, among others) have been used as electrodes in different OH and PEF studies and applications (Samaranayake and Sastry 2005). Each of these materials has different physical and chemical properties that influence the electrochemical reaction processes; therefore, there is an important relation between the electrode material and the reactions that degrade bioactive compounds. Samaranayake and Sastry (2005) investigated the behavior of various electrode materials at different pH values in OH process. They noticed corrosion of electrodes and apparent (partial) electrolysis of the heating medium with most types of electrodes at low-frequency ACs. The results showed that stainless steel, widely used in food industry, was the most electrochemically active electrode material during OH at all the pH values. On the other hand, platinized titanium showed to be the best material for electrode manufacturing because of the relatively inert electrochemical behavior.

Kim and Zhang (2011) studied different electrode materials (titanium, platinized titanium, stainless steel 316, and boron carbide) during PEF treatment aiming to reduce electrode corrosion and material migration into the food. Among the materials analyzed, titanium was the most resistant to corrosion and, for this reason, is considered to be toxicologically safe and to have a long lifetime, being recommended as an electrode material in PEF systems. Platinized titanium electrodes showed better corrosion resistance when compared to stainless steel 316 and boron carbide electrodes; however, it was not as good as titanium. Stainless steel 316 electrodes exhibit severe pitting corrosion in scanning electron microscope analysis.

Traditionally, the degradation of vitamin C in foods during their thermal processing and storage has been described in terms of first-order kinetics and the Arrhenius equation (Polydera et al. 2003, Serpen and Gökmen 2007, Lima et al. 2010).

Assiry et al. (2003) compared the degradation kinetics of ascorbic acid in a 3.5 pH buffer solution under conventional heating and OH, using a stirred ohmic heater with stainless steel electrodes. According to these authors, the degradation kinetics was adequately described by a first-order model, but electrode reactions and electrolysis products influenced the kinetics parameters. Moreover, the reaction temperature dependence for some ohmic treatments could not be represented by the Arrhenius relation, indicating that other factors beyond temperature may have taken effect during OH. As reported by the authors, the results highlight the importance of using inert electrode coatings or high-frequency power to control electrochemical reactions.

Mercali et al. (2012) evaluated vitamin C degradation in acerola pulp during thermal treatment by ohmic and conventional heating. Despite using platinum electrodes, electrolysis was observed at a low intensity. The authors reported that the presence of stainless steel temperature sensors may have contributed to the occurrence of these reactions. Qihua et al. (1993) observed bubble formation (as a result of electrolysis) during OH process probably because of some electrochemical reactions, especially when the orange juice temperature reached 50°C. Some studies have shown that electrochemical reactions are accelerated by the use of high voltages and low frequencies (Palaniappan and Sastry 1991, Qihua et al. 1993, Assiry et al. 2003, Mercali et al. 2012, 2014).

Roodenburg et al. (2005) investigated the metal release in a stainless steel PEF system during treatment of orange juice. The experiments showed that, due to PEF treatment, dissolved metals are present in the juice. The metal concentrations found did not exceed the legislation values for fruit juices and the EU Drinking Water Directive for human consumption.

While the microbial inactivation and engineering aspect of OH and PEF have been profoundly investigated, the effect of electrochemical reaction or electrode corrosion on vitamin degradation had little attention. From studies performed for vitamin C degradation, it can be inferred that other vitamins may suffer degradation through similar pathways. Studies showed electrochemical degradation of anthocyanins, which are, as vitamin C, molecules prone to react with free radicals (Sarkis et al. 2013). In the work of Bendicho et al. (2002), ascorbic acid degradation in milk and simulated skim milk ultrafiltrated was compared to other water-soluble vitamins (thiamine, riboflavin) and to fat-soluble vitamins (cholecalciferol and tocopherol). No changes in vitamin content were observed after PEF or thermal treatments except for ascorbic acid, which suffer degradation after both process. These results indicate that vitamin C is more sensitive than other vitamins and, for this reason, can be considered a suitable parameter to evaluate the sensibility of fruit pulps and beverages to degradation.

4.4 VITAMINS STABILITY DURING ELECTRIC FIELD TREATMENTS APPLIED IN FRUITS PULP

The application of electrical technologies for food processing has recently gained renewed attention in an effort to improve product quality. The research involving OH began to intensify in the 1980s, and currently there are several studies being performed involving PEF processes. Most researches have addressed the basic engineering and heat transfer aspects of the OH and PEF technologies, however the influence of these processes on reaction kinetics within food systems has been restricted to a few studies.

Due to lack of work evaluating the retention of some types of vitamins in the presence of an electric field, it is not possible to predict the exact effect of this variable on their stability. Some studies evaluated the influence of electrical heating techniques on vitamin C degradation and compared these technologies with conventional heating. This section summarizes the recent studies of emerging processing technologies applied to fruit pulps and their products and discusses the impact of processing conditions on the stability of vitamins, with a special focus on vitamin C.

4.4.1 Ohmic Heating

There are two kinds of studies evaluating the effects of the OH technology on bioactive compounds. The first kind of study is focused on the thermal effects, which is the principal mechanisms of microbial and enzyme inactivation in OH. These studies evaluated the thermal and nonthermal effects together by simply comparison with other conventional technologies, without eliminating the temperature as a variable during the experiments. One example of this is the study carried out by Vikram et al. (2005) that analyzed thermal degradation kinetics of nutrients in orange juice heated by electromagnetic (infrared, ohmic, and microwave heating) and conventional methods. The research evaluated vitamin C degradation and also changes in visual color as an index of carotenoids. The destruction of vitamin C was influenced by the heating method and the temperature of processing. Out of the four methods studied, OH gave the best result, with better vitamin C retention at all temperatures.

The second kind of study is focused on the nonthermal effects caused by the electricity. Recent, researches have eliminated thermal differences to determine these effects and evaluate its influence on the degradation of nutritional compounds. One of the techniques to evaluate the nonthermal effects during the heating process consists of matching the time–temperature histories during both conventional and OH processes. In this case, the thermal histories of both heating technologies are the same, and any difference in the results can be associated with the electrical effects.

There are several important process parameters in OH, such as applied voltage gradient, frequency, electrical conductivity, temperature, among others. The studies related to vitamins and OH generally focus on the effects that electric field strength and frequency have on the degradation kinetics during processing and, in some cases, on the product shelf life. Therefore, this section will be divided to approach these different items.

4.4.1.1 Effect of the Electric Field Strength

The electric field strength applied during OH is a fundamental parameter and can be varied by adjusting the electrode gap or the applied voltage, being expressed in volts (V) per centimeter (cm). The electric field strengths applied in OH and MEF are typically under 1000 V cm^{-1}, with or without heating (Sastry 2008).

Some studies have been performed to evaluate the effect of electrical field strength on food properties and vitamin stability. Due to the small amount of work performed evaluating fruit pulps and beverages, this section will cover also some fruit and vegetable products. Mercali et al. (2012) evaluated the influence of solids content (2–8 g/100 g) and heating voltage (120–200 V) on vitamin C degradation in acerola pulp during thermal treatment by OH. As the rate of OH is directly proportional to the square of the electric field strength and the electrical conductivity, samples were heated up to 85°C (and maintained at this temperature for 3 min) with different temperature profiles. The researchers found that ascorbic acid degradation was significantly influenced by both variables evaluated. OH, when performed with low-voltage gradients, exhibited vitamin C degradation similar to conventional heating. However, high-voltage gradients increased the degradation of vitamin C. This behavior may be explained by the increase of electrochemical reactions when using high-voltage gradients, which can adversely affect the ascorbic acid and catalyze the degradation pathways in the presence of oxygen. It is important to point out that in this study stainless steel temperature sensors were used, which might be the cause of increased electrochemical reactions, showing that this material may not be the most appropriate for OH of foods.

On the other hand, Lima and Sastry (1999) verified that the presence of an electric field had no significant effect on ascorbic acid degradation in orange juice. Although there was electrolysis and metal corrosion when stainless steel electrodes were used, these phenomena did not affect the final concentration of ascorbic acid. Castro et al. (2004) studied the effect of the electric field on the degradation rate of ascorbic acid in strawberry industrial pulp. The temperature range used was between 60°C and 97°C for both heating processes (conventional and ohmic). Power source was turned on and electric field was varied in order to simulate the conventional thermal history of the samples. During the holding phase of the experiments, electric field strength was kept lower than 20 V cm^{-1}. As reported by the authors, it is very important to have coincidence of the heating phase of both processes to ensure that any differences observed between both technologies would be solely due to the electric field. The obtained kinetic parameters were identical for the two types of heating processes, leading to the conclusion that the presence of an electric field (<20 V cm^{-1}) did not affect the ascorbic acid degradation of strawberry pulp.

Pereira et al. (2007) characterized cloudberry jam after OH and conventional heating, comparing the effects of both processes in terms of food's chemical properties, such as ascorbic acid content, anthocyanin content, and phenolic content. For the ohmic processing, it was used as a continuous ohmic heater comprising one teflon heating section (240 mm long and 90 mm wide) with two electrodes constituting side walls of the heating section. Inside the heating section, voltages of 40, 140, and 240 V of AC were applied across the two electrodes. Overall, the results of chemical analyses indicate that the OH technology provides products with chemical properties similar to those of the products obtained by conventional treatment.

Yildiz et al. (2010) studied the effect of electric field strength (10–40 V cm^{-1}) and temperature (60°C, 70°C, 80°C, or 90°C) of OH on β-carotene content, chlorophyll content, and color changes in spinach purée. The time necessary to reach the desired holding temperature was different in each experiment. The effect of voltage gradient applied was not statistically significant ($p < 0.05$) on chlorophyll content, β-carotene content, and color values.

Tumpanuvatr et al. (2012) compared quality parameters of 10 kinds of botanical beverages, concentrated juices, and purées of orange and pineapple ohmically and conventionally heated with the same temperature profile. The results showed that color, flavor, and vitamin C content of ohmically heated samples were not inferior to conventionally heated products.

Louarme and Billaud (2012) investigated the effects of conventional and OH treatments on the degradation of sugar and ascorbic acid in chunky fruit desserts prepared with apple purée and chunky peach pieces. According to the authors, OH reduced the levels of 5-HMF and furfural (products resulting from ascorbic acid degradation) by 89% and 53% when compared to the conventional counterpart, respectively. The results showed that ascorbic acid degradation during OH process mainly depended on oxidative reactions pathway.

In general, studies evaluating the electric field strength effects show that, in most cases, vitamin C degradation during OH was similar to conventional heating; however special attention is needed when choosing the electrodes and sensors materials. Results showing that nutrient retention is similar in both, conventional and ohmic treatments, can be considered as positive results, since they show there was not a significant negative effect of the electric field. Moreover, it must be considered that, in industrial applications involving OH, the heating step of the process will, in most cases, be faster and more homogeneous when compared to conventional heating.

4.4.1.2 Effect of the Electric Field Frequency

Another important parameter affecting vitamin degradation during OH is the electric field frequency. Since OH consists of the application of AC to electrodes in contact with the product, it is important to consider the frequency values. Bioactive components, such as vitamins, are subjected to an electric field distortion when the OH technology is applied, known as polarization. This phenomenon is influenced by the frequency alternating voltage and it can affect the rates and mechanism of the reaction associated with the degradation of bioactive compounds (Mercali et al. 2014).

Studies have been performed to evaluate the electric field frequency effects during MEF treatments on food properties and process parameters (Imai et al. 1995, Lima et al. 1999, Bansal and Chen 2006, Kulshrestha and Sastry 2006, Shynkaryk et al. 2010) on diffusion and extraction of certain constituents of foodstuffs (Lima and Sastry 1999, Wang et al. 2002, Kulshrestha and Sastry 2003, Lakkakula et al. 2004) on the stimulation of microbial growth (Loghavi et al. 2008, 2009) and on inactivation kinetics of spores (Somavat et al. 2012). However, little is known about the effect of the electric field frequency on nutrient stability during OH. Mercali et al. (2014) investigated the effects of the electric field frequency on ascorbic acid degradation and color changes in acerola pulp during OH and this technology was compared with the conventional heating process. The experiments were conducted matching the time-temperature histories of both technologies. Figure 4.3 shows ascorbic acid degradation for electric field frequencies ranging between 10 and 10^5 Hz. The authors reported that the use of low electric field frequency (10 Hz) led to greater ascorbic

Electric Field Treatment in the Retention of Vitamins

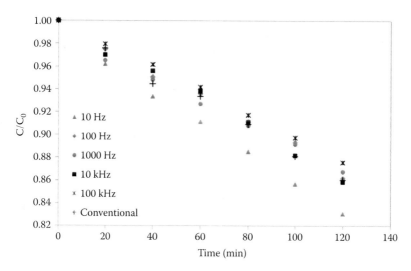

FIGURE 4.3 Ascorbic acid degradation during ohmic and conventional heating processing of acerola pulp. (Reproduced from Mercali, G.D. et al., *J. Food Eng.*, 123, 1–7, 2014. With permission.)

acid degradation and higher color changes, probably due to the occurrence of electrochemical reactions. Above 100 Hz, these reactions were minimized and, as a result, ohmic and conventional heating processes showed similar degradation rates of ascorbic acid and similar color changes. The use of high electric field frequency did not affect the degradation kinetics of ascorbic acid and pigment compounds. This indicates that the ascorbic acid molecule predisposition for hydrogen donation in redox reactions was not affected by the rapidly varying electric field.

4.4.1.3 Effect on Shelf Life

One of the main reasons of food processing is to extend shelf life; hence, when a novel technology viability is evaluated, this is an important aspect to consider. In the period from 1992 to 1994, a consortium of 25 partners from various industries (food processors, equipment manufacturers, and ingredient suppliers), academia (food science, engineering, microbiology, and economics), and government in the United States was formed. This group developed products and evaluated the capabilities of an OH system. The equipment used in these tests was a 5 kW pilot-scale continuous-flow ohmic system manufactured by APV Baker Ltd. (Crawley, UK). Several products were developed, including shelf-stable, low, and high-acid products, as well as refrigerated extended shelf-life products. Ohmically heated products were found to have texture, color, flavor, and nutrient retention that matched or exceeded those of traditional processing methods, such as freezing, retorting, and aseptic processing. Thus, the consortium concluded that the technology was viable (Ruan et al. 2001).

Pataro et al. (2011) investigated the effect of OH processing on the quality and shelf life of apricots in syrup. The liquid–solid mixture (pH < 4) was pasteurized at 90°C for about 113 s in a 30 kW aseptic continuous pilot ohmic system using a 25 kHz AC. The aseptically packaged samples were stored at 25°C and the ascorbic acid content was periodically examined over 1 year period. The ascorbic acid content was slightly reduced by the electrothermal treatment. However, it was well preserved for the entire storage time. According to the authors, the initial reduction in ascorbic acid content is congruent with that usually detected immediately after thermal treatments.

Leizerson and Shimoni (2005) examined the effects of OH on the stability of orange juice with comparison to conventional pasteurization. During storage at 4°C, degradation curves of ascorbic acid followed a linear decrease pattern in both ohmic and conventionally pasteurized orange juices. No significant difference was observed in ascorbic acid concentration between these two treatments. The results of the study suggest that a thermal treatment by continuous OH can be used to extend the shelf life of pasteurized freshly squeezed orange juice.

Overall, the most interesting aspect of the continuous OH treatment is the lack of overheating due to homogeneous heat generation inside the product, which results in high retention of the sensorial attributes of fresh juices and fruit pulps.

4.4.2 Pulsed Electric Fields

In Section 4.1, PEF technology is a nonthermal process and, for this reason, it is expected to maintain food quality attributes. Studies conducted using PEF indicated that the technology preserves the nutritional components of the food. However, the effects of pulses of high electric field strength on chemical and nutritional compounds, such as vitamins, must be better understood (FDA 2011).

The main process parameters that determine PEF treatments are electric field strength, shape and width of the pulse, treatment time, frequency, specific energy density, and temperature. Among all, electric field intensity and treatment time (number of pulses) are the basic control parameters affecting energy density applied during PEF processing (Altunakar and Barbosa-Cánovas 2011). Therefore, these factors, as well as the shelf life of the foods treated by PEF, will be discussed in the following sections.

4.4.2.1 Effect of the Electric Field Strength and the Treatment Time

Like OH, the electric field strength applied during PEF is one of the factors determining the treatment intensity and it can be adjusted by changes in the electrode gap or the applied voltage. The other important factor is the treatment time, which is described as the sum of the duration of each applied pulse during the whole treatment. The mechanisms that mediate the effects of PEF on food constituents are not yet completely elucidated. However, evidences have shown that PEF treatments induce fewer changes in vitamin content than conventional thermal processing treatments.

Marselles-Fontanet et al. (2013) compared unprocessed, PEF processed, and thermally processed grape juices belonging to different grape varieties. They evaluated the treatment effects on microbiology, physical, chemical, and nutritional properties, including vitamin C. Their results showed that the treatment effect on vitamin C content of the juices varied with the grape variety analyzed. For Mazuelo grapes, which showed a higher initial vitamin C content, the PEF treatment preserved vitamins more than the thermal treatment. However, for the Moscatel and Parellada varieties, which had significantly lower initial vitamin C content, the difference between treatments was not clearly observed. Furthermore, the reduction of radical scavenging activity of the juices (for all varieties) was higher for thermally processed grape juices than for PEF-treated juices.

Vitamin C degradation using different parameters during PEF application have showed that the lower the electric field strength, the treatment time, and the pulse frequency, the higher the vitamin C retention in orange, tomato, and strawberry juices (Elez-Martínez and Olga Martín-Belloso 2007, Odriozola-Serrano et al. 2007, 2008a, b). This effect can be visualized in Figure 4.4, reproduced from Odriozola-Serrano et al. (2008b), which analyzed the effect on tomato juice of different electric field strengths (20, 25, 30, and 35 kV cm^{-1}) during 100, 300, 600, 1000, 1500, and 2000 μs of treatment. All experiments used squarewave bipolar pulses at 250 Hz. The temperature was controlled in the inlet and outlet of the PEF chambers and the samples never exceeded 40°C.

Some studies, analogous to the one presented on Figure 4.4, have been performed to better elucidate the kinetics of vitamin C degradation under different PEF processing electric field strengths. The results showed that this vitamin depletion can be approached through simple first-order models (Torregrosa et al. 2006); however, models based on the Weibull distribution have better fitted vitamin C degradation kinetics (Odriozola-Serrano et al. 2008a, b).

Additional to the studies involving juices, some researches evaluated beverages containing juice and milk or other protein sources. Sharma et al. (1998) compared PEF treatment and heat pasteurization to process a protein-fortified fruit-based beverage at 30°C. The PEF-treated beverage had less protein denaturation and lower vitamin C loss compared with that of the heat-treated beverage. Rivas et al. (2007) evaluated the effects of PEF technology (15–40 kV cm^{-1}; 0–700 μs) and thermal processing (84°C and 95°C, 15–120 s) on orange juice and milk mixed beverage

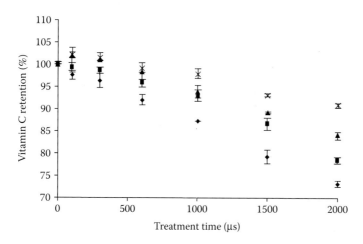

FIGURE 4.4 Effect of treatment time and electric field strength on the vitamin C retention of tomato juice (mean ± SD). Electric field strengths: (♦) 35 kV cm^{-1}, (■) 30 kV cm^{-1}, (▲) 25 kV cm^{-1}, and (x) 20 kV cm^{-1}. (Reproduced from Odriozola-Serrano, I., R. Soliva-Fortuny, V. Gimeno-Añó, and O. Martín-Belloso., *J. Food Eng.*, 89 (2008b): 210–216. With permission.)

fortified with water-soluble vitamins (biotin, folic acid, pantothenic acid, and riboflavin). After all PEF treatments (different field strengths or treatment times), no significant changes in vitamin content were observed. On the other hand, loss of vitamins after high heat treatments was observed, reaching 20% with the most intense treatment (95°C, 45 s).

Cortés et al. (2006) evaluated modifications of vitamin A and carotenoids in untreated orange juice, pasteurized orange juice (90°C, 20 s), and orange juice processed with high-intensity PEFs (30 kV cm^{-1}, 100 μs) during 7 weeks of storage at 2°C and 10°C. Less degradation of carotenoids and vitamin A was observed when PEF was used, in comparison with the pasteurized orange juice. Immediately after processing, vitamin A content decrease was 7.52% for PEF treated juice and 15.62% for the pasteurized juice.

A number of studies were performed comparing two emerging nonthermal technologies: HP and PEF. In a study conducted by Vervoort et al. (2011), the impact of thermal, HP, and PEF processing for mild pasteurization of orange juice was compared on a fair basis, using processing conditions leading to an equivalent degree of microbial inactivation. Examining the effect on specific chemical and biochemical quality parameters directly after treatment and during storage at 4°C revealed that there were significant differences only in residual enzyme activities. All other quality parameters investigated, including sugar profile, organic acid profile, bitter compounds, vitamin C concentration, carotenoid profile, and furfural and 5-hydroxymethylfurfural concentrations, experienced no significantly different impact from the three pasteurization techniques. Similar conclusions were found by Plaza et al. (2006), that analyzed the vitamin C content and radical scavenging activity of freshly squeezed orange juice processed by HP (400 MPa/40°C/1 min), PEF (35 kV cm^{-1} / 750 μs), and low pasteurization (70°C/30 s). The authors concluded that degradation after treatment was not significantly different and was around 8% for all technologies.

Overall, no additional degradation was observed during PEF treatments when compared to thermal processing and, in some cases, vitamin C showed higher retention when this electrotechnology was used. This indicates that, probably, there were no electrochemical reactions during processing. Thus, the higher retention of vitamins, when observed, is likely related to differences on the temperature reached during PEF and heat treatments. As ascorbic acid is heat-sensitive in the presence of oxygen, high temperatures during processing can greatly affect the rates of its degradation through an aerobic pathway (Odriozola-Serrano et al. 2013). PEF technology is, therefore, recommended for the treatment of fruit juices and beverages, especially if short treatment times and lower electric fields are applied. For beverages containing vegetable or milk proteins, this technology might also be able to lessen or avoid protein denaturation.

4.4.2.2 Effect on Shelf Life

Application of PEF technology is expected to lead to longer shelf life of fruit juices with minimal product quality loss and good retention of fresh-like flavor. Several studies have been performed on orange juice stability, evaluating changes in vitamin C content over time. Cortes et al. (2008) evaluated the effect of PEF on ascorbic acid content of orange juice in comparison with conventional pasteurization. In order to access modifications of ascorbic acid content, the juice was refrigerated at two temperatures (2°C and 10°C). The shelf life at 2°C, based on 50% degradation of the ascorbic acid, was 277 and 90 days for the PEF-treated (30 kV cm^{-1} and 100 µs) orange juice and the pasteurized juice, respectively. Zhang et al. (1997) evaluated the shelf life of reconstituted orange juice treated with an integrated PEF pilot plant system and compared the results with heat-treated juice. When stored at 4°C, both heat- and PEF-treated juices had a shelf life of more than 5 months. Vitamin C losses were lower and color was generally better preserved in PEF-treated juices compared to the heat treated ones up to 90 days (storage temperature of 4°C or 22°C) after processing. Higher vitamin C retention in orange juice was also found by Elez-Martínez et al. (2006). The researchers found that orange juice processed by PEF fitted the recommended daily intake standards of vitamin C throughout 56 days storage at 4°C, whereas heat-processed juice exhibited poor vitamin C retention beyond 14 days storage (25.2%–42.8% degradation).

Some other juices were also evaluated regarding vitamin C retention during PEF treatment. Odriozola-Serrano et al. (2008c) reported better retention of lycopene and vitamin C content on tomato juice when comparing PEF with thermally treated products during storage. Torregrosa et al. (2006) evaluated the shelf life of the orange–carrot juice treated by pulses at 25 kV cm^{-1} with two different treatment times (280 and 330 µs) and compared with a heat-treated juice (98°C, 21 s). PEF treatment at 25 kV cm^{-1} for 280 or 330 µs allowed the half-life of the juice to reach 50 days when stored at 2°C, whereas, when the preservation temperature is 10°C, the half-life was 19 days. After treatment, the ascorbic acid content in juice treated by PEF was greater than in juice pasteurized conventionally.

The previously mentioned work of Cortés et al. (2006) evaluated shelf life at 2°C and 10°C of orange juice treated by PEF and conventional pasteurization and showed that, during storage under refrigeration, total carotenoids and vitamin A were maintained for a longer time in the juice treated with PEF than in the juice conserved using conventional pasteurization treatments. Rivas et al. (2007) compared the effects of PEF technology (15–40 kV cm^{-1}; 0–700 µs) and thermal processing (84°C and 95°C, 15–120 s) on an orange juice and milk mixed beverage fortified with water-soluble vitamins (biotin, folic acid, pantothenic acid, and riboflavin) during storage for 60 days. Results showed that PEF technology affects water-soluble vitamin content (biotin, folic acid, pantothenic acid) at least in the same way as mild heat treatments (84°C) and better than high intensity thermal treatments (such as 95°C and 45 s). In the case of riboflavin, PEF treatments preserved its stability better than thermal treatments, when stored at 4°C for 60 days. The results confirmed the stability of the vitamins after the PEF treatment and during storage.

Different beverages were also evaluated for their shelf life after processing. Sharma et al. (1998) reported that a fruit-based beverage heat treated had a slightly higher apparent viscosity than the PEF-treated beverage and developed sedimentation in the container during storage. The PEF-processed beverage maintained its natural orange juice color better than the heat-treated product, which developed a slightly whitish color. However, the PEF-treated product was less microbiologically stable (5 months) at refrigeration temperature compared with the heat-treated beverage, which was stable for more than 12 months. Also, Zulueta et al. (2010b) reported that no significant differences of ascorbic acid degradation in a ready-to-drink orange juice–milk beverage were observed between heat pasteurization (90°C, 20 s) and PEF treatments (treatment (25 kV cm^{-1}, 280 µs) during storage at 4°C and 10°C.

4.5 CONCLUDING REMARKS AND FUTURE TRENDS

Vitamins are essential food constituents and very sensitive to the environmental conditions and to the presence of certain chemical compounds. During the application of electrical treatments, electrochemical reactions can occur, releasing ions and molecules that might influence vitamin stability. Therefore, combined studies involving OH or PEF and vitamin degradation are essential to the application of these emerging technologies.

Studies performed evaluating vitamin degradation demonstrated that OH can be viewed as a potential technology for heat treatment of foods rich in vitamins. These studies highlight the importance of using suitable materials for manufacturing electrodes and temperature sensors, as well as high electric field frequencies, to prevent the occurrence of electrochemical reactions. Studies evaluating the nonthermal effects associated passage of electricity through the foodstuff demonstrated that the nonthermal effects have no influence on vitamin degradation and comparative results of this technology with conventional heating were alike. It is important to point out that these results were obtained for similar temperature–time histories for both technologies, procedure necessary for the assessment of nonthermal effects of OH. In an industrial condition, OH has the advantage of faster heating which decreases the time exposure to heat and, consequently, decreases the degradation rates of heat-sensitive products.

Studies evaluating PEF technology suggest that it can be suitable to assure pasteurization without substantially affecting the vitamin content of foods, since they have been reported to cause less or similar change in these components content than conventional processing treatments. Thus, food applications of PEF can be a potential energy efficient nonthermal process to obtain products with enhanced content of bioactive compounds. It is important to point out that, as in OH, secure electrode materials need to be used and, since some heating can occur during PEF, temperature should be monitored. Furthermore, the treatment time and electric field strength should be as small as possible to maintain food quality.

Overall, the interaction of electric fields with food constituents is important, and needs to be deeply studied. Fundamental research is needed to characterize the interactions of electric fields with various food matrices at the molecular level and to study the chemistry and electrochemistry at/near the electrode area during the electrical treatments. By understanding these technologies on a more fundamental level, the process efficacy can be ensured and food quality maintained. In particular, further studies are needed to access the interaction of the electric field with other vitamins, once most of the woks in the area investigated only vitamin C stability. Also, additional work is required to better elucidate how vitamins behave when particulate or highly viscous foods are being processed, such as fruit purées and pulps containing fruit pieces. These studies will have significant economic consequences to industry because they will enable modeling and optimization to produce food products with higher nutritional and sensory characteristics.

REFERENCES

Akin, E. and G.A. Evrendilek. Effect of Pulsed Electric Fields on Physical, Chemical, and Microbiological Properties of Formulated Carrot Juice. *Food Science and Technology International* 15(3) (2009): 275–282.

Altunakar, B. and G.V. Barbosa-Cánovas. Engineering Aspects of Pulsed Electric Fields. In *Nonthermal Processing Technologies for Food*, edited by H.Q. Zhang, G.V. Barbosa-Cánovas, V.M. Balasubramaniam, C.P. Dunne, D.F. Farkas, and J.T.C. Yuan. Chichester, UK: John Wiley & Sons, (2011), 640p.

Assiry, A., S.K. Sastry, and C. Samaranayake. Degradation Kinetics of Ascorbic Acid during Ohmic Heating with Stainless Steel Electrodes. (In English). *Journal of Applied Electrochemistry* 33(2) (2003): 187–196.

Bansal, B. and X.D. Chen. Effect of Temperature and Power Frequency on Milk Fouling in an Ohmic Heater. *Food and Bioproducts Processing* 84(C4) (2006): 286–291.

Barbosa-Cánovas, G.V. and H.Q. Zhang. (eds.) In *Pulsed Electric Fields in Food Processing: Fundamental Aspects and Applications,* Food Preservation Technology. Lancaster, PA: Technomic Publishing Company, (2001).

Bender, D.A. Micronutrients: Vitamins & Minerals. In *Harper's Illustrated Biochemistry*, edited by R.K. Murray, K.M. Botham, P.J. Kennelly, V.W. Rodwell, and P.A. Weil. New York: McGraw-Hill Medical, (2012).

Bendicho, S., A. Espachs, J. Arántegui, and O. Martín. Effect of High Intensity Pulsed Electric Fields and Heat Treatments on Vitamins of Milk. *Journal of Dairy Research* 69(01) (2002): 113–123.

Castro, I., J.A. Teixeira, S. Salengke, S.K. Sastry, and A.A. Vicente. Ohmic Heating of Strawberry Products: Electrical Conductivity Measurements and Ascorbic Acid Degradation Kinetics. *Innovative Food Science & Emerging Technologies* 5(1) (2004): 27–36.

Chen, J., X.Y. Tao, A.D. Sun et al. Influence of Pulsed Electric Field and Thermal Treatments on the Quality of Blueberry Juice. *International Journal of Food Properties* 17(7) (2014): 1419–1427.

Cortes, C., M.J. Esteve, and A. Frigola. Effect of Refrigerated Storage on Ascorbic Acid Content of Orange Juice Treated by Pulsed Electric Fields and Thermal Pasteurization. *European Food Research and Technology* 227(2) (2008): 629–635.

Cortés, C., M.J. Esteve, D. Rodrigo, F. Torregrosa, and A. Frígola. Changes of Colour and Carotenoids Contents during High Intensity Pulsed Electric Field Treatment in Orange Juices. *Food and Chemical Toxicology* 44(11) (2006): 1932–1939.

de Alwis, A.A.P. and P.J. Fryer. A Finite-Element Analysis of Heat Generation and Transfer during Ohmic Heating of Food. *Chemical Engineering Science* 45(6) (1990): 1547–1559.

de Alwis, A.A.P. and P.J. Fryer. Operability of the Ohmic Heating Process: Electrical Conductivity Effects. *Journal of Food Engineering* 15(1) (1992): 21–48.

Elez-Martínez, P. and O. Martín-Belloso. Effects of High Intensity Pulsed Electric Field Processing Conditions on Vitamin C and Antioxidant Capacity of Orange Juice and Gazpacho: A Cold Vegetable Soup. *Food Chemistry* 102(1) (2007): 201–209.

Elez-Martínez, P., R. Soliva-Fortuny, and O. Martín-Belloso. Comparative Study on Shelf Life of Orange Juice Processed by High Intensity Pulsed Electric Fields or Heat Treatment. (In English). *European Food Research and Technology* 222(3–4) (2006): 321–329.

FDA. Kinetics of Microbial Inactivation for Alternative Food Processing Technologies: Ohmic and Inductive Heating. Department of Health and Human Services (2000). http://www.fda.gov/Food/FoodScienceResearch/SafePracticesforFoodProcesses/ucm101246.htm.

FDA. Kinetics of Microbial Inactivation for Alternative Food Processing Technologies—Pulsed Electric Fields. Department of Health and Human Services. (2011). http://www.fda.gov/Food/FoodScienceResearch/SafePracticesforFoodProcesses/ucm101662.htm.

Fellows, P.J. (ed.) Pasteurization. In *Food Processing Technology*. Oxford, UK: Woodhead Publishing, (2009a), pp. 381–395.

Fellows, P.J. (ed.) Properties of Food and Processing. In *Food Processing Technology*. Oxford, UK: Woodhead Publishing, (2009b), pp. 11–95.

Furusawa, N. Rapid High-Performance Liquid Chromatographic Identification/Quantification of Total Vitamin C in Fruit Drinks. *Food Control* 12 (2001): 27–29.

Gallagher, M. L. Vitaminas. Translated by Andréa Favano. In *Krause: Alimentos, Nutrição E Dietoterapia*, edited by L.K. Mahan and S. Escott-Stump. São Paulo, Brazil: Roca, (2005), pp. 57–89.

Gregory, J.F. Vitamins. In *Fennema's Food Chemistry*, edited by S. Damodaran, K.L. Parkin, and O.R. Fennema. New York: CRC Press, (2007), pp. 439–522.

Ibrahim, E.S. Corrosion Control in Electric Power Systems. *Electric Power Systems Research* 52 (1999): 9–17.

Imai, T., K. Uemura, N. Ishida, S. Yoshizaki, and A. Noguchi. Ohmic Heating of Japanese White Radish *Rhaphanus sativus* L. *International Journal of Food Science & Technology* 30(4) (1995): 461–472.

Kim, M. and H.Q. Zhang. Improving Electrode Durability of Pef Chamber by Selecting Suitable Material. In *Nonthermal Processing Technologies for Food*, edited by H.Q. Zhang, G.V. Barbosa-Cánovas, V.M. Balasubramaniam, C.P. Dunne, D.F. Farkas, and J.T.C. Yuan. Chichester, UK: John Wiley & Sons, (2011): 201–212.

Knorr, D., A. Froehling, H. Jaeger et al. Emerging Technologies in Food Processing. *Annual Review of Food Science and Technology* 2(1) (2011): 203–235.

Kong, F. and R.P. Singh. Effect of Processing on Nutrients in Food. In *Innovation in Healthy and Functional Foods*, edited by D. Ghosh, S. Das, D. Bagchi, and R.B. Smarta. Boca Raton, FL: CRC Press, (2012), 616p.

Kulshrestha, S. and S.K. Sastry. Frequency and Voltage Effects on Enhanced Diffusion during Moderate Electric Field (Mef) Treatment. *Innovative Food Science and Emerging Technologies* 4(2) (2003): 189–194.

Kulshrestha, S.A. and S.K. Sastry. Low-Frequency Dielectric Changes in Cellular Food Material from Ohmic Heating: Effect of End Point Temperature. *Innovative Food Science & Emerging Technologies* 7(4) (2006): 257–262.

Lakkakula, N.R., M. Lima, and T. Walker. Rice Bran Stabilization and Rice Bran Oil Extraction Using Ohmic Heating. *Bioresource Technology* 92(2) (2004): 157–161.

Lee, S.K. and A.A. Kader. Preharvest and Postharvest Factors Influencing Vitamin C Content of Horticultural Crops. *Postharvest Biology and Technology* 20 (2000): 207–220.

Leizerson, S. and E. Shimoni. Effect of Ultrahigh-Temperature Continuous Ohmic Heating Treatment on Fresh Orange Juice. *Journal of Agricultural and Food Chemistry* 53(9) (2005): 3519–3524.

Lima, J.R., N.J. Elizondo, and P. Bohuon. Kinetics of Ascorbic Acid Degradation and Colour Change in Ground Cashew Apples Treated at High Temperatures (100–180°C). *International Journal of Food Science and Technology* 45 (2010): 1724–1731.

Lima, M., B.F. Heskitt, and S.K. Sastry. The Effect of Frequency and Wave Form on the Electrical Conductivity-Temperature Profiles of Turnip Tissue. *Journal of Food Process Engineering* 22(1) (1999): 41–54.

Lima, M., and S.K. Sastry. The Effects of Ohmic Heating Frequency on Hot-Air Drying Rate and Juice Yield. *Journal of Food Engineering* 41 (1999): 115–119.

Loghavi, L., S.K. Sastry, and A.E. Yousef. Effect of Moderate Electric Field Frequency on Growth Kinetics and Metabolic Activity of *Lactobacillus acidophilus*. *Biotechnology Progress* 24 (2008): 148–153.

Loghavi, L., S.K. Sastry, and A.E. Yousef. Effects of Moderate Electric Field Frequency and Growth Stage on the Cell Membrane Permeability of *Lactobacillus acidophilus*. *Biotechnology Progress* 25(1) (2009): 85–94.

Louarme, L. and C. Billaud. Evaluation of Ascorbic Acid and Sugar Degradation Products during Fruit Dessert Processing under Conventional or Ohmic Heating Treatment. *LWT—Food Science and Technology* 49(2) (2012): 184–187.

Marks, J. Biological Functions of Vitamins. In *The Technology of Vitamins in Food*, edited by P.B. Ottaway. New York: Springer Science+Business Media, (1993), pp. 1–18.

Marselles-Fontanet, A.R., A. Puig-Pujol, P. Olmos, S. Minguez-Sanz, and O. Martin-Belloso. A Comparison of the Effects of Pulsed Electric Field and Thermal Treatments on Grape Juice. *Food and Bioprocess Technology* 6(4) (2013): 978–987.

Martín-Belloso, O. and R. Soliva-Fortuny. Pulsed Electric Fields Processing Basics. In *Nonthermal Processing Technologies for Food*, edited by H.Q. Zhang, G.V. Barbosa-Cánovas, V.M. Balasubramaniam, C.P. Dunne, D.F. Farkas, and J.T.C. Yuan. Chichester, UK: John Wiley & Sons, (2011) pp. 155–175.

Mercali, G.D., D.P. Jaeschke, I.C. Tessaro, and L.D.F. Marczak. Study of Vitamin C Degradation in Acerola Pulp during Ohmic and Conventional Heat Treatment. *LWT—Food Science and Technology* 47(1) (2012): 91–95.

Mercali, G.D., S. Schwartz, L.D.F. Marczak, I.C. Tessaro, and S. Sastry. Ascorbic Acid Degradation and Color Changes in Acerola Pulp during Ohmic Heating: Effect of Electric Field Frequency. *Journal of Food Engineering* 123(0) (2014): 1–7.

Mezadri, T., A. Pérez-Gálvez, and D. Hornero-Méndez. Carotenoid Pigments in Acerola Fruits (Malpighia Emarginata Dc.) and Derived Products. *European Food Research and Technology* 220(1) (2005): 63–69.

Milani, E.A., S. Alkhafaji, and F.V.M. Silva. Pulsed Electric Field Continuous Pasteurization of Different Types of Beers. *Food Control* 50 (2015): 223–229.

Moreno, J., R. Simpson, A. Baeza et al. Effect of Ohmic Heating and Vacuum Impregnation on the Osmodehydration Kinetics and Microstructure of Strawberries (Cv. Camarosa). *LWT—Food Science and Technology* 45(2) (2012): 148–154.

Odriozola-Serrano, I., I. Aguiló-Aguayo, R. Soliva-Fortuny, V. Gimeno-Añó, and O. Martín-Belloso. Lycopene, Vitamin C, and Antioxidant Capacity of Tomato Juice as Affected by High-Intensity Pulsed Electric Fields Critical Parameters. *Journal of Agricultural and Food Chemistry* 55(22) (2007): 9036–9042.

Odriozola-Serrano, I., I. Aguilo-Aguayo, R. Soliva-Fortuny, and O. Martin-Belloso. Pulsed Electric Fields Processing Effects on Quality and Health-Related Constituents of Plant-Based Foods. *Trends in Food Science & Technology* 29(2) (2013): 98–107.

Odriozola-Serrano, I., R. Soliva-Fortuny, V. Gimeno-Añó, and O. Martín-Belloso. Kinetic Study of Anthocyanins, Vitamin C, and Antioxidant Capacity in Strawberry Juices Treated by High-Intensity Pulsed Electric Fields. *Journal of Agricultural and Food Chemistry* 56(18) (2008a): 8387–8393.

Odriozola-Serrano, I., R. Soliva-Fortuny, V. Gimeno-Añó, and O. Martín-Belloso. Modeling Changes in Health-Related Compounds of Tomato Juice Treated by High-Intensity Pulsed Electric Fields. *Journal of Food Engineering* 89(2) (2008b): 210–216.

Odriozola-Serrano, I., R. Soliva-Fortuny, and O. Martín-Belloso. Changes of Health-Related Compounds throughout Cold Storage of Tomato Juice Stabilized by Thermal or High Intensity Pulsed Electric Field Treatments. *Innovative Food Science & Emerging Technologies* 9(3) (2008c): 272–279.

Ottaway, P.B. Stability of Vitamins during Food Processing and Storage. In *Chemical Deterioration and Physical Instability of Food and Beverages*, edited by L.H. Skibsted, J. Risbo, and M.L. Andersen. New York: CRC Press, (2010): 539–560.

Palaniappan, S. and S.K. Sastry. Electrical Conductivity of Selected Juices: Influences of Temperature, Solids Content, Applied Voltage, and Particle Size. *Journal of Food Process Engineering* 14(4) (1991): 247–60.

Pataro, G., G. Donsì, and G. Ferrari. Aseptic Processing of Apricots in Syrup by Means of a Continuous Pilot Scale Ohmic Unit. *LWT—Food Science and Technology* 44(6) (2011): 1546–1554.

Pátkai, G., I. Kormendy, and A. Kormendy-Domján. Vitamin C Decomposition Kinetics in Solutions, Modelling Citrus Juices. *Acta Alimentaria* 31 (2002): 125–147.

Pereira, R., M. Pereira, J.A. Teixeira, and A.A. Vicente. Comparison of Chemical Properties of Food Products Processed by Conventional and Ohmic Heating. (In English). *Chemical Papers* 61(1) (2007): 30–35.

Plaza, L., C. Sánchez-Moreno, P. Elez-Martínez et al. Effect of Refrigerated Storage on Vitamin C and Antioxidant Activity of Orange Juice Processed by High-Pressure or Pulsed Electric Fields with Regard to Low Pasteurization. (In English). *European Food Research and Technology* 223(4) (2006): 487–493.

Polydera, A.C., N.G. Stoforos, and P.S. Taoukis. Comparative Shelf Life Study and Vitamin C Loss Kinetics in Pasteurised and High Pressure Processed Reconstituted Orange Juice. *Journal of Food Engineering* 60 (2003): 21–29.

Potter, N.N. and J.H. Hotchkiss. (eds.) Nutritive Aspects of Food Constituents. In *Food Science*. New York: Springer, (1995): 46–68.

Qihua, T., V.K. Jindal, and J. van Winden. Design and Performance Evaluation of an Ohmic Heating Unit for Liquid Foods. *Computers and Electronics in Agriculture* 9(3) (1993): 243–253.

Rawson, A., A. Patras, B.K. Tiwari et al. Effect of Thermal and Non Thermal Processing Technologies on the Bioactive Content of Exotic Fruits and Their Products: Review of Recent Advances. *Food Research International* 44(7) (2011): 1875–1887.

Riener, J., F. Noci, D.A. Cronin, D.J. Morgan, and J.G. Lyng. Effect of High Intensity Pulsed Electric Fields on Enzymes and Vitamins in Bovine Raw Milk. *International Journal of Dairy Technology* 62(1) (2009): 1–6.

Rivas, A., D. Rodrigo, B. Company, F. Sampedro, and M. Rodrigo. Effects of Pulsed Electric Fields on Water-Soluble Vitamins and Ace Inhibitory Peptides Added to a Mixed Orange Juice and Milk Beverage. *Food Chemistry* 104(4) (2007): 1550–1559.

Roodenburg, B., J. Morren, H.E. Berg, and S.W.H. de Haan. Metal Release in a Stainless Steel Pulsed Electric Field (Pef) System: Part II. The Treatment of Orange Juice; Related to Legislation and Treatment Chamber Lifetime. *Innovative Food Science & Emerging Technologies* 6(3) (2005): 337–345.

Ruan, R., X. Ye, P. Chen, and C.J. Doona. Ohmic Heating. In *Thermal Technologies in Food Processing*, edited by P. Richardson. Londres, England: Woodhead Publishing, 2001: 242–265.

Samaranayake, C.P. and S.K. Sastry. Electrode and Ph Effects on Electrochemical Reactions during Ohmic Heating. *Journal of Electroanalytical Chemistry* 577(1) (2005): 125–135.

Sarang, S., S.K. Sastry, and L. Knipe. Electrical Conductivity of Fruits and Meats during Ohmic Heating. *Journal of Food Engineering* 87(3) (2008): 351–356.

Sarkis, J.R., D.P. Jaeschke, I.C. Tessaro, and L.D.F. Marczak. Effects of Ohmic and Conventional Heating on Anthocyanin Degradation during the Processing of Blueberry Pulp. *LWT—Food Science and Technology* 51(1) (2013): 79–85.

Sastry, S. Ohmic Heating and Moderate Electric Field Processing. *Food Science and Technology International* 14(5) (2008): 419–422.

Sastry, S. and S. Palaniappan. Mathematical Modeling and Experimental Studies on Ohmic Heating of Liquid-Particle Mixtures in a Static Heater. *Journal of Food Process Engineering* 15 (1992): 241–261.

Serpen, A. and V. Gökmen. Reversible Degradation Kinetics of Ascorbic Acid under Reducing and Oxidizing Conditions. *Food Chemistry* 104 (2007): 721–725.

Sharma, S.K., Q.H. Zhang, and G.W. Chism. Development of a Protein Fortified Fruit Beverage and Its Quality When Processed with Pulsed Electric Field Treatment. *Journal of Food Quality* 21(6) (1998): 459–473.

Shynkaryk, M.V., T. Ji, V.B. Alvarez, and S.K. Sastry. Ohmic Heating of Peaches in the Wide Range of Frequencies (50 Hz to 1 Mhz). *Journal of Food Science* 75(7) (2010): E493–E500.

Soliva-Fortuny, R., A. Balasa, D. Knorr, and O. Martín-Belloso. Effects of Pulsed Electric Fields on Bioactive Compounds in Foods: A Review. *Trends in Food Science & Technology* 20(11–12) (2009): 544–556.

Somavat, R., H.M.H. Mohamed, Y.-K. Chung, A.E. Yousef, and S.K. Sastry. Accelerated Inactivation of Geobacillus Stearothermophilus Spores by Ohmic Heating. *Journal of Food Engineering* 108(1) (2012): 69–76.

Torregrosa, F., M.J. Esteve, A. Frigola, and C. Cortes. Ascorbic Acid Stability During Refrigerated Storage of Orange-Carrot Juice Treated by High Pulsed Electric Field and Comparison with Pasteurized Juice. *Journal of Food Engineering* 73(4) (2006): 339–345.

Tumpanuvatr, T. and W. Jittanit. The Temperature Prediction of Some Botanical Beverages, Concentrated Juices and Purees of Orange and Pineapple during Ohmic Heating. *Journal of Food Engineering* 113(2) (2012): 226–233.

Vervoort, L., I. Van der Plancken, T. Grauwet et al. Comparing Equivalent Thermal, High Pressure and Pulsed Electric Field Processes for Mild Pasteurization of Orange Juice Part II: Impact on Specific Chemical and Biochemical Quality Parameters. *Innovative Food Science & Emerging Technologies* 12(4) (2011): 466–477.

Vikram, V.B., M.N. Ramesh, and S.G. Prapulla. Thermal Degradation Kinetics of Nutrients in Orange Juice Heated by Electromagnetic and Conventional Methods. *Journal of Food Engineering* 69 (2005): 31–40.

Wang, W.-C. and S.K. Sastry. Effects of Moderate Electrothermal Treatments on Juice Yield from Cellular Tissue. *Innovative Food Science & Emerging Technologies* 3(4) (2002): 371–377.

Wouters, P.C., I. Alvarez, and J. Raso. Critical Factors Determining Inactivation Kinetics by Pulsed Electric Field Food Processing. *Trends in Food Science & Technology* 12(3–4) (2001): 112–121.

Yildiz, H., F. Icier, and T. Baysal. Changes in B-Carotene, Chlorophyll and Color of Spinach Puree during Ohmic Heating. *Journal of Food Process Engineering* 33(4) (2010): 763–779.

Zhang, Q.H., X. Qiu, and S.K. Sharma. Recent Development in Pulsed Electric Field Processing. In *New Technologies Yearbook*, edited by D.I. Chandarana. Washington, DC: National Food Processors Association, (1997), 31–42.

Zulueta, A., F.J. Barba, M.J. Esteve, and A. Frigola. Effects on the Carotenoid Pattern and Vitamin A of a Pulsed Electric Field-Treated Orange Juice-Milk Beverage and Behavior during Storage. *European Food Research and Technology* 231(4) (2010a): 525–534.

Zulueta, A., F.J. Barba, M.J. Esteve, and A. Frigola. Changes in Quality and Nutritional Parameters during Refrigerated Storage of an Orange Juice-Milk Beverage Treated by Equivalent Thermal and Non-Thermal Processes for Mild Pasteurization. *Food and Bioprocess Technology* 6(8) (2013): 2018–2030.

Zulueta, A., M.J. Esteve, and A. Frigola. Ascorbic Acid in Orange Juice-Milk Beverage Treated by High Intensity Pulsed Electric Fields and Its Stability during Storage. *Innovative Food Science & Emerging Technologies* 11(1) (2010b): 84–90.

5 Effect of Combined High Pressure–Temperature Treatments on Bioactive Compounds in Fruit Purées

Snehasis Chakraborty, Nishant R. Swami Hulle, Kaunsar Jabeen, and Pavuluri Srinivasa Rao

CONTENTS

5.1 Introduction 105
5.2 Bioactive Compounds in Fruit Purées 106
5.3 Overview of High-Pressure Processing 107
5.4 Stability of Fruit Bioactives under High Pressure and Temperature 108
5.5 Effect of HPP on Bioactives in Fruit Purées 110
    5.5.1 Vitamins 111
    5.5.2 Flavonoids 114
    5.5.3 Carotenoids 115
    5.5.4 Total Phenolics 118
    5.5.5 Antioxidant Capacity 121
5.6 Storage Stability of Bioactives in HPP-Treated Purée 121
5.7 Kinetic Analyses for Degradation of Bioactives after HPP 124
5.8 Concluding Remarks and Future Research Needs 126
References 127

5.1 INTRODUCTION

Fruit products have their unique appearances and flavors, which distinguish them from various food products. These identities are due to their large content of micronutrients, bioactives, and other nutritional compounds. The fruit purée is normally produced by crushing the fleshy part of a fresh fruit. It serves as a basis for the preparation of juice, beverages, and also adds a fruit serving for dairy desserts, pastries, smoothies, breakfast bars, and others. Besides, it brings the color, texture, viscosity, and robust taste into many of its products or servings such as bakery, beverages, confections, dairy, snack foods, and sauces. Fruit purées are rich in polyphenols, essential vitamins, flavonoids, anthocyanins, and other metabolites which show different types of antioxidant capacity by scavenging free radicals. Overall, these are considered as bioactive compounds which offer a cumulative protection against degenerative diseases. Some of the epidemiologic studies also revealed that the consumption of fruit products in large amounts enhances both morbidity and mortality from these diseases harder (Pellegrini et al., 2003).

Main concern of the fruit purée processing industry is its limited stability throughout the post processing storage (up to 2–3 days under ambient conditions) (Buzrul et al., 2008). Besides microbial spoilage, quality degradation due to endogenous enzyme activity in squeezed purée is also a factor of concern (Hendrickx et al., 1998). Thermal treatment as a means of preservation of these products is the common process implemented so far. However, the application of heat renders a negative impact on the nutritional quality, appearance, and taste of these (Awuah et al., 2007). Today's consumer demand for minimally processed fruit products and *natural* fruit products tends to think of alternative to thermal processing.

5.2 BIOACTIVE COMPOUNDS IN FRUIT PURÉES

The balance between oxidants and antioxidants within the human body should be maintained properly for sustaining an optimal physiological condition. Imbalance or overproduction in oxidants leads to an oxidative damage to the large biomolecules such as lipids, DNA, and proteins. The bioactive compounds in the fruits have complementary and overlapping mechanisms of oxidative agents, stimulation of the immune system, regulation of gene expression in cell proliferation and apoptosis, hormone metabolism, and antibacterial and antiviral effects (Sun et al., 2002). In plants, these compounds occur as secondary metabolites having miscellaneous functions throughout its life cycle (Wu and Chappell, 2008). Unlike the primary metabolites, bioactive compounds are not needed for the growth and metabolism of the plant; rather these are involved in protection or signaling within the plant.

Depending upon the biosynthetic origins of plant physiology and developmental stages, these secondary metabolites can be classified into five groups: polyketides, isoprenoids, alkaloids, phenylpropanoids, and flavonoids (Oksman-Caldentey and Inze, 2004). On the other hand, based on the chemical groups, the bioactive compounds are classified as alkaloids, glycosides (cardiac glycosides, cyanogenic glycosides, glucosinolates, saponins, and anthraquinone glycosides), flavonoids, proanthocyanidins, tannins, terpenoids, and phenylpropanoids (Bernhoft, 2010). In case of fruit products, the major bioactive compounds are mainly represented by vitamin C and polyphenols such as anthocyanins, phenolic acids, flavonoids, stilbenes, carotenoids, and tannins. The different classes of bioactive compounds and their major fruit sources have been summarized in Table 5.1. Depending on the concentration and qualitative diversity, they act as natural antioxidants and scavenge the reactive oxygen species like hydroxyl, peroxide radicals, singlet oxygen, and other reactive forms of oxygen. Thus, they inhibit the oxidation process required for degenerative diseases by scalping the enzymes and metal ions. The purées from berry fruits are rich in phenolics like anthocyanins, phenolic acids, flavonols, and tannins; whereas, citrus fruit purées contribute sufficient amount of vitamin C and flavonoids (Szajdek and Borowska, 2008).

TABLE 5.1
Different Classes of Bioactive Compounds and Their Major Fruit Sources

Class	Specific Compounds	Major Fruit Sources
Carotenoids	Carotene (α and β), lutein, lycopene	Tomato, papaya, guava
Terpenes	Limonene, carvone	Citrus fruits, cherry
Phenolic acids	Hydroxycinnamic acids: caffeic, ferulic, chlorogenic	Blueberry, cherry, pears, apple, grape
	Hydroxybenzoic acid	Raspberries, strawberry
Lignans	Secoisalariciresinol, mataresinol	Berries
Saponins	Panaxadiol, panaxatriol	Tomato
Water-soluble pigments	Aglycones, anthocyanidin glycosides	Berries
Water-soluble vitamin	Ascorbic acid	Citrus fruits, berries

5.3 OVERVIEW OF HIGH-PRESSURE PROCESSING

High-pressure processing (HPP), a relatively new concept compared to conventional thermal processing, also sometimes known as high hydrostatic pressure (HHP), or ultrahigh pressure (UHP) processing, is a nonthermal food processing method that subjects liquid or solid foods, with or without packaging, to pressures ranging between 50 and 1000 MPa. HPP has already been established as an alternative to thermal processing. In food processing, it is of great interest due to its ability to inactivate food-borne microorganisms and enzymes, at low temperature, without the need for chemical preservatives and it is less detrimental than thermal processes to low-molecular-weight food compounds such as flavoring agents, pigments, vitamins, and so on as covalent bonds are not affected by pressure. Pressure-treated foods have sensory properties similar to fresh products, which is a major advantage in fruit purée processing as it matches consumer demand for healthy, nutritious, and *natural* products. HPP of fruit products offers the chance of producing food of high quality, greater safety, and increased shelf life (Chakraborty et al., 2014).

The basis for applying high pressure to foods is to compress the water surrounding the food. At room temperature, the volume of water decreases with increase in pressure. Owing to small volume change resulting from liquid compression, high-pressure vessels using water do not present the same operating hazards as vessels using compressed gases. High pressure can be generated by direct compression, indirect compression, and heating of the pressure medium. Indirect compression uses a high-pressure intensifier to pump a pressure medium from a reservoir into a closed high-pressure vessel until the desired pressure is reached.

In batch-wise HPP systems, the product is generally treated in its final primary package; commonly, the food and its package are treated together and so the entire pack remains a *secure unit* until the consumer opens it. Batch-type processing has been a preferred method for HPP of packaged fluid foods, thus eliminating any danger of cross contamination by the medium. The food is prepared and aseptically filled/sealed in flexible packaging, then placed in a pressure chamber for pressurizing

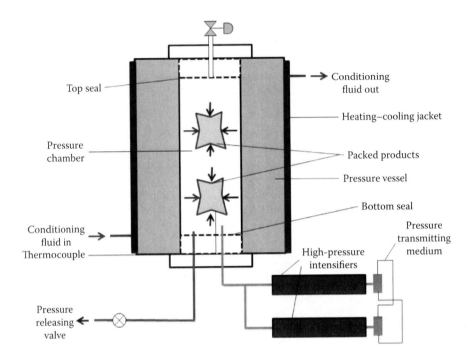

FIGURE 5.1 Schematic of batch type high-pressure processing (HPP) set up.

using a suitable pressure transmitting medium (PTM). Pressurization is done by pumping the medium either into the vessel or into the intensifier using high-pressure pumps (Figure 5.1).

Industrial application of high pressure to fruit purée restricts the pressure range within 400–600 MPa for optimum processing, as this range of pressure induces reversible and irreversible changes in several micro- or macromolecules in the medium (Knorr et al., 2006; Heinz and Buckow, 2009). When combined with pressure, the process temperature varies within 10°C–40°C for pasteurization (Deliza et al., 2005) and 60°C–90°C for sterilization of food products (Black et al., 2007; Barba et al., 2012). Being acidic in nature, high-pressure pasteurization of fruit purée may not require elevated temperature; instead it is achieved at ambient temperature and <300 MPa. However, the process temperature for high-pressure sterilization of fruit purée depends on the target yeast strain (vegetative or ascospores) typical to product (Guerrero-Beltrán et al., 2005).

5.4 STABILITY OF FRUIT BIOACTIVES UNDER HIGH PRESSURE AND TEMPERATURE

A single cycle of any high-pressure process consists of three steps: pressurization, pressure-hold period, and depressurization (Figure 5.2). The effects of all these three steps on the degradation of bioactives are dissimilar to each other. While the large structures like proteins, enzymes, nucleic acids, and polysaccharides can be

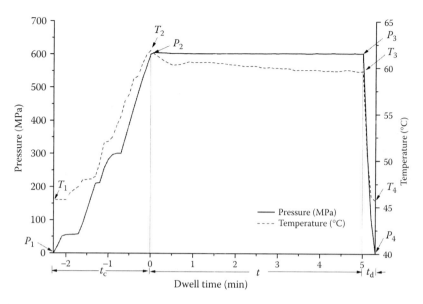

FIGURE 5.2 Pressure–temperature profile at the sample vicinity during a high-pressure treatment of 600 MPa/60°C/5 min.

altered by the application of high pressure; the effect of HPP on low-molecular-weight food components in the fruit purées such as phenolic compounds, ascorbic acid (AA), anthocyanin, carotenoids, and flavonoids is minimal. The reason is high pressure more likely affects the noncovalent bonds like hydrogen bond, hydrophobic interactions, and so on, where compaction in volume can take place. In general, the application of high-pressure helps in hydration as well as chelation in which the pro-oxidant metal ions get involved, thus making bioactives like AA unavailable for oxidation (Cheah and Ledward, 1997). The exclusion of oxygen and also inhibiting the action of peroxides during pressurization might be the factors that make the bioactive content stable in high-pressure-treated samples (Polydera et al., 2005).

On the other hand, covalent bonds which are the backbone of these low-molecular-weight compounds remain unaffected by the pressure (San Martin et al., 2002). Thus, one of the major advantages of using HPP in fruit purée processing is that the pressure-treated products mimic the sensory properties of fresh produce (Oey et al., 2008; Sanchez-Moreno et al., 2009). These compounds may be affected at higher pressures like more than 800 MPa which results in degradation of vitamins and fragmentation of larger flavoring compounds (Sumitani et al., 1994; Lambert et al., 1999).

The temperature associated with high pressure is the major contributing factor behind the thermal degradation of these bioactive components (Lambert et al., 1999; Rastogi et al., 2007). The possible mechanism describing the effects of high pressure and temperature on the bioactive compounds has been summarized in Figure 5.3. The degradation reactions like oxidation, nonenzymatic reaction, pigment destruction, and condensation are responsible for reducing the

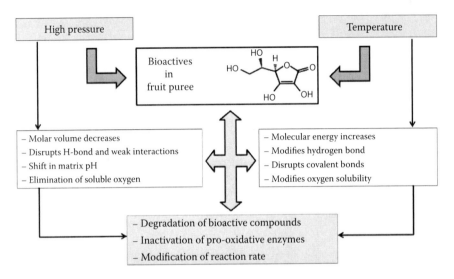

FIGURE 5.3 Proposed mechanism of individual and combined effects of high pressure and temperature on the stability of bioactive compounds in fruits.

bioavailability of these compounds in the product. At extreme pressure conditions, the changes in redox properties due to adiabatic heating during compression may induce some chemical reactions, finally resulting in loss of bioactive content in the purée (Fernández-García et al., 2000). It is easy to understand that increased molecular energy at an elevated temperature facilitates the chemical reaction in the system as the probability of molecular collision becomes high. Though the thermal resistance values (z-value) of these bioactive compounds are generally high compared to that of microbes and enzymes, the loss in nutritional quality due to thermal treatment in case of fruit products is significant (Holdsworth and Simpson, 2016). Vitamins are the most heat-sensitive constituents, among them vitamin C is the least resistant. Its degradation depends on the pH, oxygen concentration, pro-oxidant present, and enzymes. Several pigments like chlorophylls, anthocyanin, and carotenoids, among which some show antioxidant activity, are also likely to be degraded upon heating resulting in a discolored product. Hence, the post-processing nutrient quality of the purée becomes a yardstick before the application of combined high-pressure and thermal treatment.

5.5 EFFECT OF HPP ON BIOACTIVES IN FRUIT PURÉES

Since 2000, several literatures have described the effect of HPP combined with/without temperature on bioactive compounds in fruit purées (Fernández-Garcia et al., 2000; Oey et al., 2004; Sánchez-Moreno et al., 2005, 2006; Kaushik et al., 2014). The fruit sources include apple, avocado, blackberry, kiwifruit, mango, nectarine, pineapple, plum, raspberry, and strawberry. Most of these studies mainly focus on the retention of AA, phenolics, and antioxidant capacity of the fruit purée after the HPP treatment.

5.5.1 VITAMINS

AA content in a fruit sample is generally used as an indicator of oxidative stress exerted during the treatment so that stability of other water-soluble vitamins can be underlined (Barba et al., 2012). It is the only bioactive that is better preserved by HHP than by thermal treatment of a vegetable beverage. However, its stability in fruit purée depends upon various factors like type of fruit, cultivar, and so on. The change in vitamin C (one of the water-soluble vitamins) content in fruit purées under high-pressure environment has been studied frequently and the studies revealed that the stability of AA in high-pressure-treated samples varied in different ways as summarized in Table 5.2.

Landl et al. (2010) studied the effect of two commercially feasible HPP treatments (400 and 600 MPa/20°C/5 min) on the AA content of the Granny Smith apple purée. They reported that 93% and 78% retention in AA after treatments of 400 MPa/20°C/5 min and 600 MPa/20°C/5 min, respectively. The stability of AA during HPP was hypothesized as modulated by the presence of pro-oxidant metal ions and oxygen concentration in the purée. Fernandez-Sestelo et al. (2013) evaluated the stability of AA in HPP-treated (500 MPa/25°C/3 min) kiwifruit purée and obtained only 15% loss in its content. Participation of the AA molecule in nonenzymatic browning reaction was cited as the reason behind this.

In guava purée, AA was found to be sensitive to higher pressure (600 MPa) than 400 MPa (Yen and Lin, 1996). Reduction in AA of high-pressure processed products is due to the oxidation of L-AA to dehydro AA (DHAA), which is then further irreversibly converted into 2,3-diketogulonic acid (Figure 5.4). Patras et al. (2009b) reported a 53% reduction in vitamin C in tomato purée at 500 MPa followed by 400 MPa (39%), whereas <6% loss at 600 MPa. In another study, Sánchez-Moreno et al. (2006) found a loss of 30% and 39%, respectively, in vitamin C and AA in tomato purée at 400 MPa/25°C/15 min. However, pressurization (100–600 MPa) at ambient temperature (10°C–30°C) had been observed to have little effect on vitamin C in strawberry purée, though a slight reduction (5%–9%) occurred at 400–600 MPa (Patras et al., 2009a; Bodelon et al., 2013). Also, high pressure (100–400 MPa) in combination with temperature (20°C) had no effect on vitamin C in comparison to raw sample (control), whereas 15% loss was obtained at 50°C. At elevated temperatures like 50°C, the higher amount of AA loss was due to its increased thermolability at those temperatures toward its degradation.

The AA content was significantly reduced during pressurization (400–800 MPa) at high temperatures (50°C, 80°C, and 110°C) in strawberry and raspberry purée with a maximum loss of about 65% at 110°C (Verbeyst et al., 2012). Pressure (200–600 MPa/10–20 min) in combination with temperature (50°C–70°C) had a significant effect on AA in pineapple purée with a maximum reduction of 25% at 70°C followed by 20% at 60°C (Chakraborty et al., 2015). The loss in AA in pineapple purée as affected by pressure and temperature has been modeled and presented in Figure 5.5. The difference between temperature and pressure sensitivities of AA can be clearly seen. It can be visualized that temperature had major deterioration effect on AA stability than pressure. Hence, a decrease in vitamin C concentration could be distinguished as a function of temperature, but not driven

TABLE 5.2
Recent Studies Describing the Changes in Ascorbic Acid and Flavonoid Content in High-Pressure-Treated Fruit Purées

Purée	HPP Conditions	Response	Major Findings	Reference
Strawberry	100–400 MPa/ 20°C and 50°C/ 15 min	AA and anthocyanin content	• 15% loss in AA at 50°C as compared to 20°C • No significant change in anthocyanins	Bodelon et al. (2013)
	400–600 MPa/ 10°C–30°C/ 15 min	AA and anthocyanin content	• 9% loss in vitamin C content at 400 MPa • No changes in anthocyanin (Pg-3-glu) content except for 400 MPa (14% loss)	Patras et al. (2009a)
	400–800 MPa/ 20°C–110°C/ 20 min	AA and anthocyanin content	• Vitamin C was stable only at 20°C • Significant reduction (~65%) in anthocyanin content at 110°C	Verbeyst et al. (2012)
Tomato	400–600 MPa/ 20°C/15 min	Vitamin C	• Only 6% loss in vitamin C at 600 MPa • Maximum loss was at 500 MPa (53%) followed by 400 MPa (39%)	Petras et al. (2009b)
	400 MPa/ 25°C/15 min	Vitamin C and AA content	• Vitamin C and AA losses were about 30% and 39%, respectively • Dwell time had significant effect on ascorbic acid degradation	Sánchez-Moreno et al. (2006)
Guava	400–600 MPa/ 25°C/10–15 min	AA content	• Decreasing order of AA: 400 MPa/15 min < 600 MPa/15 min	Yen and Lin (1996)
Apple purée	400 and 600 MPa/ 20°C/5 min	Vitamin C and AA content	• 20% loss in vitamin C content at 600 MPa • Loss of 97% and 95% in AA at 400 and 600 MPa, respectively	Landl et al. (2010)
Pineapple	200–600 MPa/ 50°C–70°C/ 10–20 min	Flavonoids and AA content	• Maximum 25% loss in AA content at 500 MPa/70°C/10 min • Flavonoids were stable up to 50°C at all pressure levels	Chakraborty et al. (2015)
Kiwi fruit	500 MPa/25°C/ 3 min	AA content	• 8% loss in AA content after HPP	Fernandez-Sestelo et al. (2013)
Raspberry	400–800 MPa/ 20°C–110°C/ 20 min	AA and anthocyanin content	• Similar loss in vitamin C content at 50°C–110°C • 70% reduction in anthocyanin content at 110°C	Verbeyst et al. (2012)
	200–800 MPa/ 18°C–22°C/ 15 min	Anthocyanin content	• No significant loss in anthocyanin content during pressurization • Cy-3-glu was more stable than Cy-3-soph	Suthanthangja et al. (2005)

(Continued)

TABLE 5.2 (*Continued*)
Recent Studies Describing the Changes in Ascorbic Acid and Flavonoid Content in High-Pressure-Treated Fruit Purées

Purée	HPP Conditions	Response	Major Findings	Reference
Blackberry	400–600 MPa/ 10°C–30°C/ 15 min	Anthocyanin content	• No changes in anthocyanin (Cy-3-gly) content	Patras et al. (2009b)
Plum	300–900 MPa/ 60°C–80°C/ 1 min	Anthocyanin content	• Complete retention of Cy-3-glu and Cy-3-rut at 600MPa/70°C • Maximum 23% loss in anthocyanin content at 600 MPa/60°C	García-Parra et al. (2014)
	400–600 MPa/ 20°C–25°C/ 1 s–5 min	Anthocyanin content	• 88% anthocyanin was retained at 600 MPa/5 min • Max loss was obtained at 600 MPa/2.5 min	González-Cebrino et al. (2013)

HPP, high-pressure processing; AA, ascorbic acid.

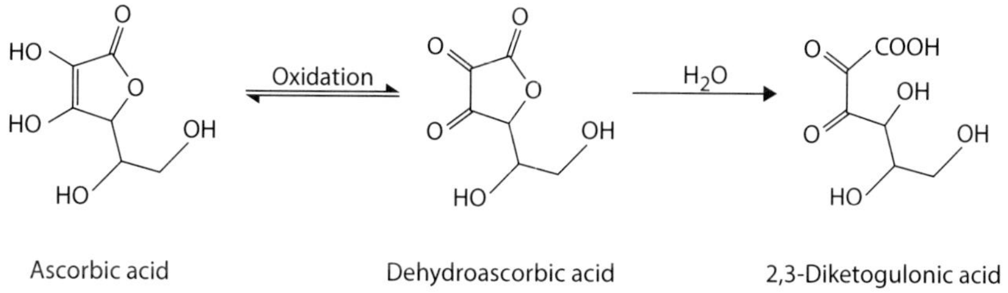

FIGURE 5.4 Aerobic oxidation pathway of L-ascorbic acid. (From Davey, M.W. et al., *J. S. Food Agric.*, 80, 825–860, 2000. With permission.)

by pressure. In addition, the stability of AA depends on various factors like pH, oxygen content, and metal ion.

Apart from AA, the number of studies on other water-soluble vitamins like folates in real fruit matrix is very scarce. Indrawati et al. (2004) studied the temperature and pressure stabilities of endogenous 5-methyltetrahydrofolic acid in kiwifruit purée at different pressure/temperature combinations (from 50 to 200 MPa/25°C and 500 MPa/60°C). They reported that folate in the purée was stable up to 60°C at all the pressure levels used which was hypothesized due to the presence of AA which acted as the folate protector in the sample. At 500 MPa, an increased amount of 5-CH$_3$-H$_4$ folate concentration in the sample was also reported. It might be due to the conversion from polyglutamates to monoglutamates by endogenous conjugate during HPP.

The high-pressure stability of water-insoluble vitamins in fruit purée has not been attempted frequently. Sánchez-Moreno et al. (2004) studied the effect of additives on

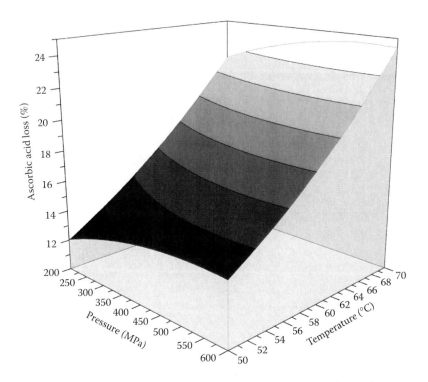

FIGURE 5.5 Response surface plot showing the pressure–temperature interaction on ascorbic acid loss (%) in pineapple purée for 15 min treatment. (From Chakraborty, S. et al., *Innov. Food Sci. Emerg. Technol., 28,* 10–21, 2015. With permission.)

stability of vitamin A in high-pressure-treated (50–400 MPa/25°C/15 min) tomato purée. The initial content of vitamin A in untreated purée was 58 retinol equivalents per 100 g sample and 24% increase in its content was reported at 400 MPa. An increased extractability was obtained at a pressure below 200 MPa when the purée was mixed with 1% citric acid and 0.4% NaCl. The compaction in tissue volume during pressurization was cited as the possible reason behind this.

5.5.2 Flavonoids

Flavonoids are the major parts of total phenolics in the fruit purée sample. This group includes several types of compounds like flavonols, flavonones, flavon-3-ols, flavones, and anthocyanin (Haminiuk et al., 2012). There is limited literature available on flavonoid content in high-pressure-treated fruit purées rather anthocyanin content has been extensively studied. The summary and findings of these studies have been presented in Table 5.2. Being phenolics and antioxidants, flavonoids are also susceptible to aerobic oxidation and temperature.

Anthocyanin in HPP-treated strawberry, raspberry, and blackberry purée was well retained at ambient temperature (Patras et al., 2009a; Suthanthangjai et al., 2005; Bodelon et al., 2013); whereas, a slight decrease in anthocyanin content was obtained in plum purée at 600 MPa/25°C/5 min (González-Cebrino et al., 2013). In case of strawberry purée, the major anthocyanins present in the purée were Pg-3-glu (15–17 mg/100 g), Pg-3-rut (2–2.5 mg/100 g), and Cy-3-glu (5–6 mg/100 g). No significant change in the anthocyanin content was reported within the HPP domain (100–400 MPa/15 min) at 20°C (Bodelon et al., 2013). Application of pressure of 100 MPa resulted in an extra amount (15%) of extracted anthocyanin in the sample at 20°C; however, at 50°C, due to thermal degradation, only 2% loss was reported at the same pressure–time level. Patras et al. (2009a) reported a similar stability of strawberry anthocyanins within the domain of 400–600 MPa/10°C–30°C/15 min; however, a 14% loss was obtained in Pg-3-glu content at 400 MPa. Verbeyst et al. (2012) registered that an increase in temperature up to 110°C at 400 MPa/20 min resulted in 65% and 70% loss in anthocyanin content in strawberry and raspberry purée, respectively. The anthocyanin in raspberry purée was stable within 200–800 MPa/20°C/15 min (Suthanthangjai et al., 2005) and 400–600 MPa/10°C–30°C/15 min (Patras et al., 2009b). However, the barostability of Cy-3-glu in the samples was more than Cy-3-soph. The anthocyanin in plum purée was relatively less stable within HPP conditions. García-Parra et al. (2014) reported that the content of Cy-3-glu and Cy-3-rut at 600 MPa/70°C/1 min was same as that of untreated; whereas, 23% loss in anthocyanin content was obtained at 300 MPa/60°C. For the same sample, González-Cebrino et al. (2013) revealed that 88% anthocyanin could be retained at 600 MPa/25°C/5 min; whereas, the maximum loss was obtained at 600 MPa/2.5 min. The resistance of polyphenoloxidase and other oxidative enzymes after HPP might be responsible for such reduction in anthocyanin content in the sample. Degradation of anthocyanin is primarily caused by oxidation, cleavage of covalent bonds, or enhanced oxidation reactions as presented in Figure 5.6. The general oxidation pathway of pelagonidin-3-glucoside is the conversion to pelargonidin through enzymatic deglycosylation followed by the formation of protocatechuic acid and/or phloroglucinaldehyde (Patras et al., 2010).

In another study, the total flavonoid content in pineapple was estimated by Chakraborty et al. (2015) within a high pressure domain of 200–600 MPa/30°C–70°C/0–20 min. The authors reported that flavonoids were stable up to 50°C at all the pressure levels; however, maximum 34% loss was obtained at 500 MPa/70°C/15 min. Degradation of flavonoids in the purée with an increase in pressure at the higher temperatures might be hypothesized as the condensation reactions happened through covalent association of these compounds which was favored by HPP (Rivas-Gonzalo et al., 1995).

5.5.3 Carotenoids

Fruits are the major sources of carotenoids in our diet and the bright appearance of many commodities is mainly due to these compounds. Although processing leads to reduction in the carotenoid content, however, in some cases, the bioavailability of

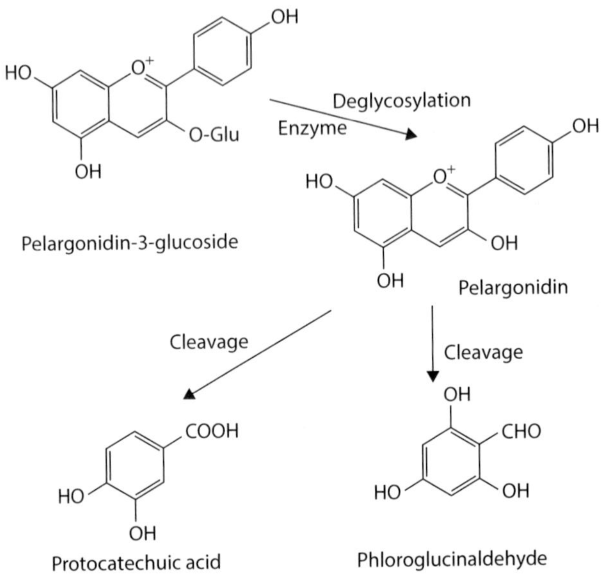

FIGURE 5.6 Possible degradation mechanism of pelargonidin-3-glucoside (anthocyanin). (From Patras, A. et al., *Trends Food Sci. Technol.*, 21, 3–11, 2010. With permission.)

TABLE 5.3
Recent Studies Describing the Changes in Carotenoids in High-Pressure-Treated Fruit Purées

Commodity	Processing Conditions	Major Findings	Reference
Tomato purée	500 and 800 MPa/ 20°C/5 min	• No changes in lycopene content • Solvent extractability of carotenoids decreased with HPP	Garcia et al. (2001)
	300–700 MPa/ 20°C–40°C/2 min	• No changes after HPP • 40% loss after thermal processing	Krebbers et al. (2003)
	50–400 MPa/25°C/ 15 min	• Increase extraction of lycopene (>14%) • Higher extractability of β-carotene (20%) and total carotenoids (10%)	Sánchez-Moreno et al. (2004)
	400–600 MPa/ 20°C/15 min	• Increased carotenoid content (up to 58%)	Patras et al. (2009b)
Gooseberry pulp	300–500 MPa/ 20°C/5 min	• 5% increase in β-carotene at 500 MPa	Torres-Ossandón et al. (2015)
Avocado pulp	600 MPa/23°C/ 3 min	• 56% increase in carotenoid content	Jacobo-Velázquez and Hernández-Brenes (2012)
Persimmon purée	50–400 MPa/25°C/ 15 min	• 27% increase in carotenoids at 400 MPa	de Ancos et al. (2000)
Nectarine purée	400–600 MPa/ 25°C/5 min	• Up to 21% increase in carotenoids at 400 MPa	García-Parra et al. (2014)

carotenoids increases due to the processing (Bowen et al., 2015). The effect of HPP on carotenoid content of different purée-based samples is presented in Table 5.3. In majority of the samples increased carotenoid content is evidenced after HPP treatment, the possible reason for this increase is attributed to enhanced extractability of the carotenoid pigments. High-pressure-induced changes in cell walls are more uniform compared to thermal processing.

In fruits, carotenoids are present into a very precise orientation like in chromoplasts. Among the carotenoids, lycopene is bound to chromoplast membranes in tomato, whereas lipophilic β-carotene is located in the lipid vacuoles (Sánchez-Moreno et al., 2004). Qiu et al. (2006) studied the pressure-induced (100–600 MPa/20°C/12 min) changes in lycopene, which is a carotenoid compound found mainly in tomato and is responsible for its red color. Though it does not show provitamin-A activity, but, it has a higher antioxidant potential compared to other carotenoids. The authors investigated structural changes in lycopene standard and lycopene in purée separately. In lycopene standard samples, they observed formation of lycopene isomers, which was higher in samples treated at 500 and 600 MPa, however, it was statistically not significant. Isomerization leads to conversion of *-trans* isomers to *-cis* isomers, which might have occurred in HPP-treated tomato purée (Shi and Maguer, 2000). Sánchez-Moreno et al. (2004) and Patras et al. (2009b) reported an increase in carotenoid content in tomato purée after 400 MPa/25°C/15 min. A maximum of 58% extra carotenoid content was obtained at 600 MPa (Patras et al., 2009b). The application of high pressure modified the cellular matrix in tomato and modulated reversible denaturalization of protein or lipid which led to different release of lycopene and β-carotene. Krebbers et al. (2003) claimed that HPP (300–700 MPa/2 min) at ambient (20°C) or elevated temperature (40°C) had no effect on the total lycopene content in the tomato purée. On the other hand, Garcia et al. (2001) recommended that extractability of lipophilic carotenoids was reduced at 800 MPa/20°C/5 min possibly due to higher exposure of hydrophilic surroundings resulting from pressure-induced structural changes in the tomato matrix.

Apart from tomato, the carotenoid content in high-pressure-treated fruit samples was investigated in gooseberry and avocado pulp, persimmon, and nectarine purée. Torres-Ossandón et al. (2015) found a significant decrease in β-carotene content in pressure-treated gooseberry pulp at 300 and 400 MPa/20°C/5 min. It was hypothesized that pressure-induced degradation of unsaturated carotenoids resulted in lower amount of β-carotene in the sample. However, pressure-induced extraction showed 5% extra β-carotene content at 500 MPa treatment. Jacobo-Velázquez and Hernández-Brenes (2012) also reported that HPP treatment of 600 MPa/20°C/3 min resulted in 56% increased carotenoid content in avocado paste. Similarly, increased extractability of carotenoids under high pressure treatment has been reported in case of persimmon purée at 50–400 MPa/25°C/15 min (de Ancos et al., 2000). García-Parra et al. (2014) studied the effect of two different HPP conditions (400 MPa/20°C/5 min and 600 MPa/25°C/5 min) on carotenoids in nectarine purée. The treatment at 400 MPa resulted in only 19% loss in lutein content in the purée, whereas, 70% β-carotene was retained after the same condition. The authors concluded that residual polyphenoloxidase activity was responsible for this degradation.

5.5.4 Total Phenolics

The bioactive potential of phenolic compounds is generally implicated in reduction of degenerative diseases in human beings. It has been reported that extraction yield of polyphenols in fruit purée was influenced by high-pressure–temperature treatments. Phenolic compounds seem to be quite resistant to HPP and in some cases the pressure treatment applied slightly increased or decreased the initial levels of the purées. Differences in total phenolic content between untreated and pressure-treated purées have been reported in other studies and it has been summarized in Table 5.4.

Patras et al. (2009b) studied the effect of HPP (400–600 MPa/20°C/15 min) on the total phenolic content in tomato purée. A maximum of 3% increase in total phenolic compounds in the purée was obtained at 500 and 600 MPa which was assumed due to high pressure extractability. García-Parra et al. (2014) obtained 30% increase in phenolic content at 600 MPa/25°C/5 min. Increase in phenolics by the application of high pressure at <40°C has also been reported in mango pulp (Kaushik et al., 2014). In case of strawberry purée, Marszałek et al. (2015) also found an increase in free ellagic acid, quercetin, and total phenolic content. HPP at 50°C increased the phenolic content in the sample irrespective of the pressure–time combinations. However, pressure and dwell time did not show a significant effect on phenolics. The increase in phenolic compounds might be attributed to the higher extractability of some phenolics by the application of instantaneous pressure. It can be hypothesized that in the view of isostatic principle, during pressurization the solvent becomes more susceptible to come in contact with bioactive phenolics present in the cell as the membrane permeability increases by HPP.

On the other hand, Landl et al. (2010) studied the effect of two commercially feasible HPP treatments (400 and 600 MPa/20°C/5 min) on the phenolics in apple purée. They obtained a maximum of 25% loss in phenolic content for samples processed at 600 MPa. The authors correlated the levels of phenolic compounds with the residual polyphenoloxidase activity after treatment. Cao et al. (2011) reported that phenolic content in strawberry pulp decreased at 400 MPa/25°C/5–25 min; whereas, 17% and 20% higher phenolic content was obtained at 25 min for 500 and 600 MPa, respectively. The presence of residual amount of oxidative enzymes at 400 MPa might be responsible for the reduced content of the phenolics. Increase in temperature up to 70°C significantly reduced the phenolic content in the purée sample. Chakraborty et al. (2015) studied the effect of combined high-pressure–temperature treatments (200–600 MPa/30°C–70°C/5–20 min) on phenolic content of pineapple purée and reported that these compounds were relatively stable up to 60°C. A maximum of 11% increase in total phenolics was obtained at 500 MPa/60°C/15 min (Figure 5.7). At 70°C, the phenolic content was significantly ($p < 0.05$) reduced at all pressure levels and a maximum of 10% loss was obtained at 500 MPa/70°C/10 min.

TABLE 5.4
Summary of Some Studies Describing the Changes in Total Phenolics and Antioxidant Capacity of High-Pressure-Treated Fruit Purées

Purée	HPP Conditions	Response	Major Findings	Reference
Strawberry purée	300–500 MPa/ 0°C–50°C/ 1–15 min	TPC	• Increase in free ellagic acid, quercetin and total content of phenolics • HPP at 50°C increased the phenolics content	Marszałek et al. (2015)
	400–600 MPa/ 10°C–30°C/ 15 min	TPC, TAC	• HPP did not affect TAC of the sample • 9% increase in TPC at 600 MPa	Patras et al. (2009a)
Strawberry (pulp)	400–600 MPa, 25°C, 5–25 min	TPC	• Total phenols decreased at 400 MPa • TPC increased at 500 or 600 MPa	Cao et al. (2011)
Tomato purée	400–600 MPa/ 20°C/15 min	TPC, TAC	• Minimal changes in phenolic content • Up to 27% increase in antiradical capacity after HPP	Patras et al. (2009b)
	500, 800 MPa/ 20°C/5 min	TAC	• No changes immediately after processing except for samples treated at 800 MPa	Garcia et al. (2001)
Pineapple purée	200–600 MPa/ 50°C–70°C/ 10–20 min	TPC, TAC	• TPC increased at 50°C (up to 11%), whereas decreased at 60°C and 70°C • HPP led to 35% increase in TAC	Chakraborty et al. (2015)
Apple purée	400 and 600 MPa/ 20°C/5 min	TPC	• 25% loss of phenolic content for samples processed at 600 MPa	Landl et al. (2010)
Gooseberry pulp	300–500 MPa/ 20°C/5 min	TAC	• ORAC assay showed loss of antioxidant capacity	Torres-Ossandón et al. (2015)
Nectarine purée	400–600 MPa/ 25°C/5 min	TAC	• Up to 50% increase in antioxidant activity	García-Parra et al. (2014)
Plum	400–600 MPa, 25°C, 0–5 min	TAC	• TAC in samples were less at 600 MPa than 400 MPa	González-Cebrino et al. (2013)
Mango pulp	100–600 MPa/ 30°C/ 1 s–20 min	TPC, TAC	• Up to 8% loss of phenolics content • Up to 10% loss of antioxidant capacity	Kaushik et al. (2014)

TPC, total phenolic content; TAC, total antioxidant capacity.

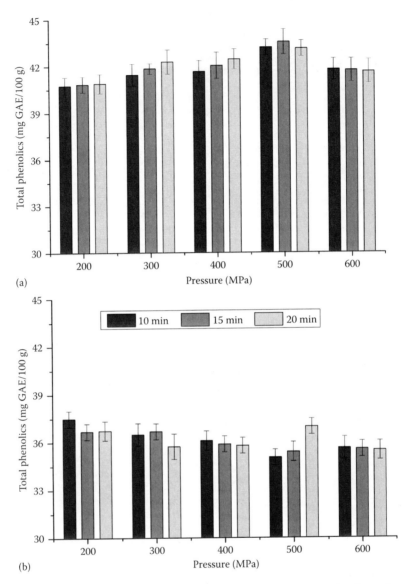

FIGURE 5.7 Changes in total phenolic content (mg GAE per 100 g) in pineapple purée obtained at different high pressure–temperature treatments: (a) 60°C and (b) 70°C. (From Chakraborty, S. et al., *Innov. Food Sci. Emerg. Technol.*, 28, 10–21, 2015. With permission.)

Overall, the stability of phenolics is governed by a cumulative effect consisting of individual contributions like higher extractability of phenolics at higher pressure, increased solubility and degradability of those phenolics both usually occur at elevated temperature. In other words, at higher pressures more phenols are extracted from the tissues and hence, they become readily susceptible to the degradation effects of high temperatures with increase in dwell time.

5.5.5 Antioxidant Capacity

Antioxidant compounds in fruit purée samples are generally radical in nature and are highly unstable and susceptible to oxidation at higher temperatures. The phenolics, flavonoids, and even the AA, which can generate radicals to scavenge the reactive species contribute to the overall antioxidant capacity in the purée sample. It has been reported that the application of high pressure does not induce any degradation of antioxidant compounds in fruit purée sample rather it can increase the extractability at certain pressure levels. On the other hand, the thermal sensitivity of those compounds counteracts its final activity in the sample.

The stability of antioxidant compounds during HPP depends not only on the pressurization conditions but also on the type of purée matrix studied, as presented in Table 5.4. In this sense, Patras et al. (2009a) observed a 9% decrease in the antioxidant activity of strawberry purée treated at 600 MPa/20°C/15 min, while no variations were reported at 400 MPa. However, in the same study it was reported that blackberry purée processed at 600 MPa obtained an increase in antioxidant capacity. In nectarine purée, García-Parra et al. (2014) obtained 50% increase in antioxidant capacity after the treatment of 600 MPa/25°C/5 min. Increased antioxidant capacity was correlated with the pressure-induced extraction of phenolics in the purée matrix. Stability of antioxidant compounds during HPP at lower temperatures (<40°C) has also been reported in case of tomato purée (Sánchez-Moreno et al., 2006). Garcia et al. (2001) reported that the water-soluble antioxidant compounds in tomato purée were stable within the domain of 500 and 800 MPa/20°C/5 min. In the same purée matrix, Patras et al. (2009b) obtained 27% increased antioxidant capacity after the treatment of 600 MPa/20°C/15 min. However, at ambient temperature, Kaushik et al. (2014) reported a 10% loss in antioxidant capacity of mango pulp after the HPP treatment of 500 MPa/10 min.

González-Cebrino et al. (2013) conveyed that the loss in antioxidant capacity of HPP-treated plum purée varied between 6% and 15% with a maximum loss obtained at 500 MPa/20°C/2.5 min. Torres-Ossandón et al. (2015) reported that HPP at 300 and 400 MPa/20°C/5 min resulted in decreased DPPH anti-radical scavenging capacity of gooseberry pulp; whereas, 35% increased activity was observed for 500 MPa sample. HPP treatments were thought to influence the extraction yield of bioactive compounds with antioxidant activity. Presumably, HPP led to an increased permeability of the pressurized cells and a subsequent improvement of the extraction effect.

Similar to phenolics, an increase in temperature resulted in higher loss in antioxidant capacity of the fruit purée. As for example, Chakraborty et al. (2015) reported a maximum of 30% loss in antioxidant capacity in pineapple purée obtained at 500 MPa/70°C/10 min. At higher temperatures like 60°C and 70°C, the thermal degradation of phenolic compounds and AA contributed to a much lower antioxidant capacity in the sample at all the pressure levels.

5.6 STORAGE STABILITY OF BIOACTIVES IN HPP-TREATED PURÉE

Retention of bioactive compounds during storage of pressurized fruit purée is very crucial for claiming a good nutritional profile of the product. However, the reported literature stating the effects of storage on the stability of bioactive compounds in pressurized fruit purée have been very scarce in the recent past. Some of these storage studies have been summarized in Table 5.5. Most of the storage studies on the bioactives in fruit purée have been conducted at refrigerated storage (4°C–5°C) conditions and it seemed to be successful for the retention of bioactives in the product as of day one. Many of the bioactives analyzed in the purée samples affected during the storage are interrelated to each other.

García-Parra et al. (2014) studied the storage stability of carotenoids, phenolic content, and antioxidant capacity in two different HPP-treated (400 MPa/20°C/5 min and 600 MPa/25°C/5 min) nectarine purée samples at 4°C. They reported that antioxidant capacity was well retained for 600 MPa samples after 30 days of refrigerated storage; whereas, 20% loss in phenolic content was obtained after the same duration. On the other hand, β-carotene content was reduced to 51% at the end of storage period. The combination of initial high-pressure extractability, oxygen concentrations, and metal ion interactions was the contributor for these observations.

In another work, Jacobo-Velázquez and Hernández-Brenes (2012) studied the storage stability of carotenoids in HPP-processed (600 MPa/23°C/3 min) avocado pulp at 4°C for 40 days. They found that up to 10 days there was a slight increase in carotenoids. However, after 15 days of storage, they observed significant decrease in carotenoids, which was mainly attributed to residual lipoxygenase activity catalyzing the degradation of carotenoids (Robinson et al., 1995).

During the storage of high-pressure-treated (400 and 600 MPa/20°C/5 min) Granny Smith apple purée, the AA degradation followed first-order decay retaining only 21% vitamin C after day 21 at 5°C (Landl et al., 2010). The total polyphenolic content in the sample at the end of the storage period were 39% for 400 MPa/20°C/5 min and 75% for 600 MPa/20°C/5 min. The higher stability of polyphenols in the sample treated at 600 MPa was correlated with the residual polyphenoloxidase enzyme activity. Fernandez-Sestelo et al. (2013) evaluated the storage stability of AA in HPP-treated (500 MPa/25°C/3 min) kiwifruit purée at 4°C up to 70 days. They obtained 94% retention in AA content after 40 days. Initially there was 25% drop in its activity after second day of storage which was correlated with the nonenzymatic browning in the sample.

An increase in storage temperature facilitates the degradation of bioactives in the purée samples. Suthanthangjai et al. (2005) reported that the anthocyanin in high-pressure-treated (200–800 MPa/20°C/15 min) raspberry purée was stable up to 9 days at 4°C storage. With an increase in storage temperature up to 30°C led to 71% reduction in cyanidin-3-glucoside content in the high-pressure treated-sample (800 MPa) after day 9. After the same storage duration, the losses in cyanidin-3-sophoroside were 60%, 73%, and 90% in the samples treated at 400 MPa and stored at 4°C, 20°C, and 30°C, respectively. On a similar note, Chakraborty et al. (2016) reported that more than 55% of AA in high-pressure-processed (600 MPa/50°C/13 min) pineapple purée was retained after 120 days of storage at 5°C; whereas for the same sample, 63% and 60% of initial AA was degraded after 55 and 30 days of storage at 15°C and 25°C, respectively. This trend has been presented in Figure 5.8 where both HPP-treated samples followed a steady decline in AA content in the pineapple purée samples. In addition, for the same sample, a 17% loss in phenolic content was

TABLE 5.5
Summary of Some Studies on Changes in Bioactive Components in High-Pressure-Treated Fruit Purée during Storage

Purée	HPP	Response	ST (°C)/day	Major Inference	Reference
Apple	400, 600 MPa, 20°C, 5 min	AA, TPC	5/21	• 72% retention in TPC for 600 MPa sample after 21 days • 50% retention in AA for 400 MPa sample after day 10 • AA decay during storage followed first-order reaction	Landl et al. (2010)
Kiwi fruit	500 MPa, 25°C, 3 min	AA content	4/40	• 15% drop in AA content after day 1 • 94% retention in AA content after 40 days	Fernandez-Sestelo et al. (2013)
Nectarine	400 and 600 MPa, 20–25°C, 5 min	Carotenoids, TPC, TAC	4/30	• 10% loss in TPC for 400 MPa sample after 30 days • 10% loss in TAC for 600 MPa sample after 30 days • 49% decrease in β-carotene content of 600 MPa sample after 30 days	García-Parra et al. (2014)
Raspberry	200–800 MPa, 20°C, 15 min	Anthocyanin content	4, 20 and 30/9	• Anthocyanin were stable when stored at 4°C • For 800 MPa samples after 9 days, 55% and 71% losses in cyanidin-3-glucoside when stored at 20°C and 30°C, respectively	Suthanthangjai et al. (2005)
Pineapple	600 MPa/50°C/13 min	AA, TPC, TAC	5, 15, 25/120	• 55% of AA was retained after 120 days at 5°C • 80% TPC was retained after 30 days at 25°C • 56% loss in TAC value after 55 days at 15°C	Chakraborty et al. (2016)
Avocado (paste)	600 MPa, 23°C, 3 min	Carotenoid content	4/45	• A slight increase after 10 days • Significant decrease after 15 days	Jacobo-Velazquez and Hernandez-Brenes (2010)

HPP, high-pressure processing; ST, storage temperature; AA, ascorbic acid; TPC, total phenolic content; TAC, total antioxidant capacity.

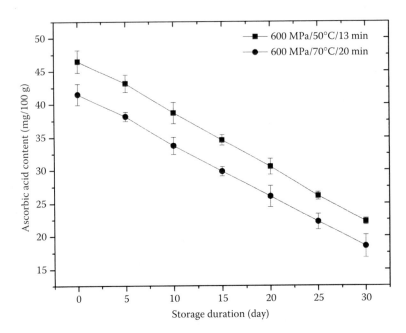

FIGURE 5.8 Changes in ascorbic acid content (mg AA per 100 g) in high pressure–temperature treated pineapple purée samples during storage at 25°C. (From Chakraborty, S. et al., *Food Bioproc. Technol.*, 9(5), 768–791, 2016. With permission.)

obtained after 120 days of storage at 5°C, whereas, 76% and 80% were retained after 55 and 30 days at 15°C and 25°C, respectively. The total antioxidant capacity loss in the sample followed the same trend as of the phenolics loss during storage.

5.7 KINETIC ANALYSES FOR DEGRADATION OF BIOACTIVES AFTER HPP

Several attempts have been made to individually quantify the effect of each of the three steps of a HPP cycle on the quality of the bioactive compounds. It is recommended that kinetics is applicable for the pressure-hold period only, and separating the effects of pressurization and depressurization is complicated. This is due to the fact that, with present technology, characterization of the sample during the high-pressure process is relatively more difficult than measuring it at the end of the treatment. Integration of these kinetic data may prove to be useful for optimization and designing of the high-pressure processes in the case of many fruit purées.

Some of the literatures are available on the degradation of nutrient or bioactive compounds in fruit products during combined high-pressure and -temperature treatments. The majority of the isobaric or isothermal degradation of any bioactive compound is characterized as pseudozero or pseudo-first order. The general forms of the quality loss reactions and their half-lives have been presented in Table 5.6.

The environmental factors strongly affect the deteriorating reactions in the food system during storage are temperature, relative humidity, total and partial pressure

TABLE 5.6
Function Forms of Several Types of Biochemical Reactions and Their Half-Life Times

Apparent Order of Reaction	Function	Half-Life
0	$C_t = C_0 - k_0 t$	$C_0/2k_0$
1	$C_t = C_0 e^{-k_1 t}$	$\ln 2 / k_1$
2	$\dfrac{1}{C_t} - \dfrac{1}{C_0} = k_2 t$	$1/(k_2 C_0)$
n ($n \neq 1$)	$\dfrac{1}{n-1}\left[C_t^{1-n} - C_0^{1-n}\right] = k_n t$	$C_0^{1-n}\left[\dfrac{2^{n-1}-1}{k_n(n-1)}\right]$

Note: C_t and C_0 are the concentration of the compounds at treatment time of "t" min and "0" min (untreated), respectively; k_j is the rate constant ($j = 0, 1, 2, \ldots, n$); and t is the treatment time in min.

of different gases, light, and mechanical stresses. The storage temperature is the important factor as it not only affects the reaction rates but also modify the quality directly unlike other factors which are generally controlled by the food packaging. The most prevalent and widely used model to describe the temperature dependency of the reaction rate is the Arrhenius relation (Equation 5.1), derived from thermodynamic laws as well as statistical mechanics principles.

$$\ln k_i = \ln k_{\text{ref}} + \frac{E_a}{R}\left(\frac{1}{T_{\text{ref}}} - \frac{1}{T}\right) \quad (5.1)$$

where:
k_{ref} is the rate constant at reference temperature
E_a is the activation energy (kJ mol^{-1})
T_{ref} and R represent reference temperature (323 K) and universal gas constant (8.314 J mol^{-1} K^{-1}), respectively

The experimental error associated with the calculated values of k can lead an amplification of a substantial error in E_a from only two points. Therefore, it is recommended to have at least three points in $\ln k$ versus ($1/T$) plot, so that from the slope, activation energy value can be calculated. In most of the quality reactions, the activation energy ranges between 50 and 150 kJ mol^{-1}.

On the other hand, the change in the molar volume (ΔV) during the biochemical reactions is influenced under a high-pressure environment finally affecting the reaction mechanism. The ΔV at constant temperature is quantified as the volume difference from the product to reactant; whereas, the activation volume (V_a) is the volume difference between the final product and the transition state molecule. According to Le Chatelier's principle, the reaction rate under a high-pressure environment increases when activation volume becomes negative ($V_a < 0$). The barosensitivity of the degradation reaction rate (k) at a constant temperature (T, K) is characterized by activation volume (V_a, mL·mol⁻¹). The relation between k and V_a is quantified by Eyring equation (Equation 5.2), where the activation volume for a reaction is calculated from the slope of the linear plot between $\ln k$ and pressure (P, MPa).

$$\ln k = \ln k_{ref} + \frac{V_a}{RT}(P_{ref} - P) \tag{5.2}$$

Where, k_{ref} is the rate constant at reference pressure (P_{ref}). The magnitude of activation volume for a degradation reaction varies from −5 to −30 mL·mol⁻¹. It is dependent on the reaction type and environment surrounding the molecule.

The degradation of vitamin C or AA content in the fruit sample was mostly modeled during HPP. Landl et al. (2010) conveyed that the AA degradation in high-pressure-treated (400 and 600 MPa/20°C/5 min) Granny Smith apple purée followed exponential decay during refrigerated storage (5°C) up to 21 days. Rodrigo et al. (2007) attempted to develop color degradation kinetics in tomato purée during combined high-pressure and -thermal treatments within the domain of 300–700 MPa, 65°C and 60 min but failed. Chakraborty et al. (2016) studied the storage stability of AA, total phenolics, and antioxidant activity in pineapple purée after HPP (600 MPa/50°C/13 min and 600 MPa/70°C/20 min). They reported a pseudo-first-order decay in all of these bioactive compounds during storage for both the HPP samples. The range of k values for total antioxidants at 5°C was 3.1–4.3 × 10⁻³ day⁻¹ and the corresponding scales at 15°C and 25°C were 1.2–1.7 × 10⁻² day⁻¹ and 2.1–2.8 × 10⁻² day⁻¹, respectively. For both the samples, the range of E_a for antioxidant degradation (65– 69 kJ·mol⁻¹) was higher compared to that for total phenolics (55–59 kJ·mol⁻¹), suggesting a greater thermal sensitivity of antioxidant compounds than the phenolics.

5.8 CONCLUDING REMARKS AND FUTURE RESEARCH NEEDS

Increasing market demand for fruit purée induced the industry people to ensure their stability, and also to avoid unpleasant aspects while extending the shelf-life of the final produce. This triggers the necessity of searching for new ways of processing like HPP of these products. Different studies on the fruit purées presented guarantee high pressure as a promising technique being nondetrimental and even advantageous for the bioactive compounds in the purée. These compounds may be affected at higher pressures like more than 800 MPa which results in degradation in vitamins and fragmentation of larger flavoring compounds. The temperature associated with high pressure is the major contributing factor behind the thermal degradation of these bioactive components.

Undoubtedly, high-pressure-treated fruit purée will attract consumer attention for their *natural* or *fresh-like* characteristics. However, it should also be noted that initial installation of the high-pressure equipment may add an extra cost to the product. In addition, the process optimization for HPP of fruit purée should also consider the impact of processing on overall quality parameters, particularly on bioactive compounds. Though the potential of combined high-pressure and -temperature processing has been shown to be substantial in the context of retaining the bioactive compounds in fruit purée, predicting the extent of degradation after the treatment is very difficult. As a result, integration of the kinetic data into the process optimization for the processing of fruit purée is complex. Cost-effective and productive high-pressure processes have to be optimized keeping in mind consumer awareness and their willingness to pay for high-pressure-processed fruit products. In addition, more data regarding stability of the bioactives in pressurized fruit purée should be generalized.

REFERENCES

Awuah, G. B., Ramaswamy, H. S., and Economides, A. Thermal processing and quality: Principles and overview. *Chemical Engineering and Processing: Process Intensification*, 46(6), (2007) 584–602.

Barba, F. J., Cortés, C., Esteve, M. J., and Frígola, A. Study of antioxidant capacity and quality parameters in an orange juice–milk beverage after high-pressure processing treatment. *Food and Bioprocess Technology*, 5(6), (2012) 2222–2232.

Bernhoft, A. (ed.). A brief review on bioactive compounds in plants. In *Bioactive Compounds in Plants: Benefits and Risks for Man and Animals* (pp. 11–18). The Norwegian Academy of Science and Letters, Oslo, Norway, (2010).

Black, E. P., Setlow, P., Hocking, A. D., Stewart, C. M., Kelly, A. L., and Hoover, D. G. Response of spores to high-pressure processing. *Comprehensive Reviews in Food Science and Food Safety*, 6(4), (2007) 103–119.

Bodelón, O. G., Avizcuri, J. M., Fernández-Zurbano, P., Dizy, M., and Préstamo, G. Pressurization and cold storage of strawberry purée: Colour, anthocyanins, ascorbic acid and pectin methylesterase. *LWT—Food Science and Technology*, 52(2), (2013) 123–130.

Bowen, P. E., Stacewicz-Sapuntzakis, M., and Diwadkar-Navsariwala, V. Carotenoids in human nutrition. In C. Chen (ed.), *Pigments in Fruits and Vegetables* (pp. 31–67). Springer, New York, (2015).

Buzrul, S., Alpas, H., Largeteau, A., and Demazeau, G. Inactivation of *Escherichia coli* and *Listeria innocua* in kiwifruit and pineapple juices by high hydrostatic pressure. *International Journal of Food Microbiology*, 124(3), (2008) 275–278.

Cao, X., Zhang, Y., Zhang, F., Wang, Y., Yi, J., and Liao, X. Effects of high hydrostatic pressure on enzymes, phenolic compounds, anthocyanins, polymeric color and color of strawberry pulps. *Journal of the Science of Food and Agriculture*, 91(5), (2011) 877–885.

Chakraborty, S., Kaushik, N., Rao, P. S., and Mishra, H. N. High-pressure inactivation of enzymes: A review on its recent applications on fruit purees and juices. *Comprehensive Reviews in Food Science and Food Safety*, 13(4), (2014) 578–596.

Chakraborty, S., Rao, P. S., and Mishra, H. N. Effect of combined high pressure–temperature treatments on color and nutritional quality attributes of pineapple (*Ananas comosus* L.) puree. *Innovative Food Science and Emerging Technologies*, 28, (2015) 10–21.

Chakraborty, S., Rao, P.S., and Mishra, H.N. Changes in quality attributes of high pressure and thermally processed pineapple (*Ananas comosus* L.) puree during storage. *Food and Bioprocess Technology*, 9(5), (2016) 768–791.

Cheah, P. B. and Ledward, D. A. Catalytic mechanism of lipid oxidation following high pressure treatment in pork fat and meat. *Journal of Food Science*, 62(6), (1997) 1135–1139.

Davey, M. W., Montagu, M. V., Inzé, D., Sanmartin, M., Kanellis, A., Smirnoff, N., Benzie, I. J. J., Strain, J. J., Favell, D., and Fletcher, J. Plant L-ascorbic acid: Chemistry, function, metabolism, bioavailability and effects of processing. *Journal of the Science of Food and Agriculture*, 80(7), (2000) 825–860.

de Ancos, B., Gonzalez, E., and Cano, M. P. Effect of high-pressure treatment on the carotenoid composition and the radical scavenging activity of persimmon fruit purees. *Journal of Agricultural and Food Chemistry*, 48(8), (2000) 3542–3548.

Deliza, R., Rosenthal, A., Abadio, F. B. D., Silva, C. H., and Castillo, C. Application of high pressure technology in the fruit juice processing: Benefits perceived by consumers. *Journal of Food Engineering*, 67(1), (2005) 241–246.

Fernández García, A., Butz, P., and Tauscher, B. Does the antioxidant potential of high pressure treated apple juice change during storage? *International Journal of High Pressure Research*, 19(1–6), (2000) 153–160.

Fernández-Sestelo, A., de Saá, R. S., Pérez-Lamela, C., Torrado-Agrasar, A., Rúa, M. L., and Pastrana-Castro, L. Overall quality properties in pressurized kiwi purée: Microbial, physicochemical, nutritive and sensory tests during refrigerated storage. *Innovative Food Science and Emerging Technologies*, 20, (2013) 64–72.

Garcia, A. F., Butz, P., and Tauscher, B. Effects of high-pressure processing on carotenoid extractability, antioxidant activity, glucose diffusion, and water binding of tomato puree (*Lycopersicon esculentum* Mill.). *Journal of Food Science*, 66(7), (2001) 1033–1038.

García-Parra, J., Contador, R., Delgado-Adámez, J., González-Cebrino, F., and Ramírez, R. The applied pretreatment (blanching, ascorbic acid) at the manufacture process affects the quality of nectarine purée processed by hydrostatic high pressure. *International Journal of Food Science and Technology*, 49(4), (2014) 1203–1214.

González-Cebrino, F., Durán, R., Delgado-Adámez, J., Contador, R., and Ramírez, R. Changes after high-pressure processing on physicochemical parameters, bioactive compounds, and polyphenol oxidase activity of red flesh and peel plum purée. *Innovative Food Science and Emerging Technologies*, 20, (2013) 34–41.

Guerrero-Beltrán, J. A., Barbosa-Cánovas, G. V., and Swanson, B. G. High hydrostatic pressure processing of fruit and vegetable products. *Food Reviews International*, 21(4), (2005) 411–425.

Haminiuk, C. W., Maciel, G. M., Plata-Oviedo, M. S., and Peralta, R. M. Phenolic compounds in fruits - an overview. *International Journal of Food Science and Technology*, 47(10), (2012) 2023–2044.

Heinz, V. and Buckow, R. Food preservation by high pressure. *Journal für Verbraucherschutz und Lebensmittelsicherheit*, 5(1), (2009) 73–81.

Hendrickx, M., Ludikhuyze, L., Van den Broeck, I., and Weemaes, C. Effects of high pressure on enzymes related to food quality. *Trends in Food Science and Technology*, 9(5), (1998) 197–203.

Holdsworth, S. D. and Simpson, R. Sterilization, pasteurization and cooking criteria. In S. D. Holdsworth, R. Simpson, and Gustavo V. Barbosa-Canovas (eds.), *Thermal Processing of Packaged Foods* (pp. 125–148). New York: Blackie Academic and Professional (2016).

Indrawati, Arroqui, C., Messagie, I., Nguyen, M. T., Van Loey, A., and Hendrickx, M. Comparative study on pressure and temperature stability of 5-methyltetrahydrofolic acid in model systems and in food products. *Journal of Agricultural and Food Chemistry*, 52(3), 485–492. (2004).

Jacobo-Velázquez, D. A. and Hernández-Brenes, C. Stability of avocado paste carotenoids as affected by high hydrostatic pressure processing and storage. *Innovative Food Science and Emerging Technologies*, 16, (2012) 121–128.

Kaushik, N., Kaur, B. P., Rao, P. S., and Mishra, H. Effect of high pressure processing on color, biochemical and microbiological characteristics of mango pulp (*Mangifera indica* cv. Amrapali). *Innovative Food Science and Emerging Technologies*, 22, (2014) 40–50.

Knorr, D., Heinz, V., and Buckow, R. High pressure application for food biopolymers. *Biochimica et Biophysica Acta (BBA)—Proteins and Proteomics*, 1764(3), (2006) 619–631.

Krebbers, B., Matser, A. M., Hoogerwerf, S. W., Moezelaar, R., Tomassen, M. M., and van den Berg, R. W. Combined high-pressure and thermal treatments for processing of tomato puree: Evaluation of microbial inactivation and quality parameters. *Innovative Food Science and Emerging Technologies, 4*(4), (2003) 377–385.

Lambert, Y., Demazeau, G., Largeteau, A., and Bouvier, J. M. Changes in aromatic volatile composition of strawberry after high pressure treatment. *Food Chemistry, 67*(1), (1999) 7–16.

Landl, A., Abadias, M., Sárraga, C., Viñas, I., and Picouet, P. Effect of high pressure processing on the quality of acidified Granny Smith apple purée product. *Innovative Food Science and Emerging Technologies, 11*(4), (2010) 557–564.

Marszałek, K., Mitek, M., and Skąpska, S. The effect of thermal pasteurization and high pressure processing at cold and mild temperatures on the chemical composition, microbial and enzyme activity in strawberry purée. *Innovative Food Science and Emerging Technologies, 27,* (2015) 48–56.

Oey, I., Loey, V. A., and Hendrick, M. Pressure and temperature stability of water soluble antioxidants in orange and carrot juice: A kinetic study. *European Journal of Food Research Technology, 219,* (2004) 161–166.

Oey, I., Van der Plancken, I., Van Loey, A., and Hendrickx, M. Does high pressure processing influence nutritional aspects of plant based food systems? *Trends in Food Science and Technology, 19*(6), (2008) 300–308.

Oksman-Caldentey, K. M. and Inzé, D. Plant cell factories in the post-genomic era: New ways to produce designer secondary metabolites. *Trends in Plant Science, 9*(9), (2004) 433–440.

Patras, A., Brunton, N. P., Da Pieve, S., and Butler, F. Impact of high pressure processing on total antioxidant activity, phenolic, ascorbic acid, anthocyanin content and colour of strawberry and blackberry purées. *Innovative Food Science and Emerging Technologies, 10*(3), (2009a) 308–313.

Patras, A., Brunton, N.P., Da Pieve, S., Butler, F., and Downey, G. Effect of thermal and high pressure processing on antioxidant activity and instrumental colour of tomato and carrot purées. *Innovative Food Science and Emerging Technologies, 10*(1), (2009b) 16–22.

Patras, A., Brunton, N. P., O'Donnell, C., and Tiwari, B. K. Effect of thermal processing on anthocyanin stability in foods; mechanisms and kinetics of degradation. *Trends in Food Science and Technology, 21*(1), (2010) 3–11.

Pellegrini, N., Serafini, M., Colombi, B., Del Rio, D., Salvatore, S., Bianchi, M., and Brighenti, F. Total antioxidant capacity of plant foods, beverages and oils consumed in Italy assessed by three different in vitro assays. *The Journal of Nutrition, 133*(9), (2003) 2812–2819.

Polydera, A. C., Stoforos, N. G., and Taoukis, P. S. Quality degradation kinetics of pasteurised and high pressure processed fresh Navel orange juice: Nutritional parameters and shelf life. *Innovative Food Science and Emerging Technologies, 6*(1), (2005) 1–9.

Qiu, W., Jiang, H., Wang, H., and Gao, Y. Effect of high hydrostatic pressure on lycopene stability. *Food Chemistry, 97*(3), (2006) 516–523.

Rastogi, N. K., Raghavarao, K. S. M. S., Balasubramaniam, V. M., Niranjan, K., and Knorr, D. Opportunities and challenges in high pressure processing of foods. *Critical Reviews in Food Science and Nutrition, 47*(1), (2007) 69–112.

Rivas-Gonzalo, J. C., Bravo-Haro, S., and Santos-Buelga, C. Detection of compounds formed through the reaction of malvidin 3-monoglucoside and catechin in the presence of acetaldehyde. *Journal of Agricultural and Food Chemistry, 43*(6), (1995) 1444–1449.

Robinson, D. S., Wu, Z., Domoney, C., and Casey, R. Lipoxygenases and the quality of foods. *Food Chemistry, 54*(1), (1995) 33–43.

Rodrigo, D., Jolie, R., Van Loey, A., and Hendrickx, M. Thermal and high pressure stability of tomato lipoxygenase and hydroperoxide lyase. *Journal of Food Engineering, 79*(2), (2007) 423–429.

Sanchez-Moreno, C., De Ancos, B., Plaza, L., Elez-Martínez, P., and Cano, M. P. Nutritional approaches and health-related properties of plant foods processed by high pressure and pulsed electric fields. *Critical Reviews in Food Science and Nutrition*, 49(6), (2009) 552–576.

Sánchez-Moreno, C., Plaza, L., de Ancos, B., and Cano, M. P. Effect of combined treatments of high-pressure and natural additives on carotenoid extractability and antioxidant activity of tomato puree (*Lycopersicum esculentum* Mill.). *European Food Research and Technology*, 219(2), (2004) 151–160.

Sánchez-Moreno, C., Plaza, L., De Ancos, B., and Cano, M. P. Impact of high-pressure and traditional thermal processing of tomato purée on carotenoids, vitamin C and antioxidant activity. *Journal of the Science of Food and Agriculture*, 86(2), (2006) 171–179.

Sánchez-Moreno, C., Plaza, L., Elez-Martínez, P., De Ancos, B., Martín-Belloso, O., and Cano, M. P. Impact of high pressure and pulsed electric fields on bioactive compounds and antioxidant activity of orange juice in comparison with traditional thermal processing. *Journal of Agricultural and Food Chemistry*, 53(11), (2005) 4403–4409.

San Martin, M. F., Barbosa-Cánovas, G. V., and Swanson, B. G. Food processing by high hydrostatic pressure. *Critical Reviews in Food Science and Nutrition*, 42(6), (2002) 627–645.

Shi, J. and Maguer, M. L. Lycopene in tomatoes: Chemical and physical properties affected by food processing. *Critical Reviews in Food Science and Nutrition*, 40(1), (2000) 1–42.

Sumitani, H., Suekane, S., Nakatani, A., and Tatsuka, K. Changes in composition of volatile compounds in high pressure treated peach. *Journal of Agricultural and Food Chemistry*, 42(3), (1994) 785–790.

Sun, J., Chu, Y. F., Wu, X., and Liu, R. H. Antioxidant and antiproliferative activities of common fruits. *Journal of Agricultural and Food Chemistry*, 50(25), (2002) 7449–7454.

Suthanthangjai, W., Kajda, P., and Zabetakis, I. The effect of high hydrostatic pressure on the anthocyanins of raspberry (*Rubus idaeus*). *Food Chemistry*, 90(1), (2005) 193–197.

Szajdek, A. and Borowska, E. J. Bioactive compounds and health-promoting properties of berry fruits: A review. *Plant Foods for Human Nutrition*, 63(4), (2008) 147–156.

Torres-Ossandón, M. J., López, J., Vega-Gálvez, A., Galotto, M. J., Perez-Won, M., and Di Scala, K. Impact of high hydrostatic pressure on physicochemical characteristics, nutritional content and functional properties of Cape Gooseberry pulp (*Physalis peruviana* L.). *Journal of Food Processing and Preservation*, 39(6), (2015) 2844–2855.

Verbeyst, L., Hendrickx, M., and Van Loey, A. Characterisation and screening of the process stability of bioactive compounds in red fruit paste and red fruit juice. *European Food Research and Technology*, 234(4), (2012) 593–605.

Wu, S. and Chappell, J. Metabolic engineering of natural products in plants; tools of the trade and challenges for the future. *Current Opinion in Biotechnology*, 19(2), (2008) 145–152.

Yen, G. C. and Lin, H. T. Comparison of high pressure treatment and thermal pasteurization effects on the quality and shelf life of guava puree. *International Journal of Food Science and Technology*, 31(2), (1996) 205–213.

6 Sonication Processing on Bioactive Compound of Fluid Foods

Francisco J. Barba

CONTENTS

6.1 Introduction 131
6.2 Materials and Methods 134
    6.2.1 Sample 134
    6.2.2 Chemicals and Reagents 135
    6.2.3 Ultrasonics Treatment System 135
    6.2.4 Methods 136
        6.2.4.1 Vitamin C Determination 136
        6.2.4.2 Total Phenolics 136
        6.2.4.3 Total Antioxidative Capacity 136
    6.2.5 Statistical Analysis 137
6.3 Results and Discussion 137
    6.3.1 Impact of USNs on Ascorbic Acid Content 137
    6.3.2 Impact of USNs on Total Phenolics 139
    6.3.3 Impact of USNs on Color 139
    6.3.4 Impact of USNs on Antioxidant Capacity (TEAC) 142
    6.3.5 Relationship between Bioactive Compounds and Antioxidant Capacity (TEAC) after USN Processing 143
6.4 Conclusions 144
References 144

6.1 INTRODUCTION

During the last decade, minimal processing has been well established as an essential strategy for modern food preservation in order to meet growing consumer demands for safe and durable food products offering high nutritional and sensory value (Toepfl et al. 2006; Barba et al. 2012b; Deng et al. 2014; Zinoviadou et al. 2015).

In this line, ultrasound (USN) technology combined with mild temperatures (<60°C) and/or with other techniques can be a promising tool to meet the required microbial reduction of 5-log reduction in the numbers of the most resistant pathogens in finished products, required by the U.S. FDA recent regulations for using USN, thus avoiding the detrimental effects of thermal treatment at high temperatures on nutritional and quality attributes of fluid foods (Zenker et al. 2003; Knorr et al. 2004; Mason et al. 2005; Kentish and Feng 2014; Zinoviadou et al. 2015).

USN can be defined as inaudible sound waves at a frequency above 20 kHz. Being USN waves of low frequency (18–100 kHz; λ = 145 mm) and high intensity (10–1000 W/cm^2) the most effective for food preservation (Dujmic et al. 2013; Rosello-Soto et al. 2015; Zinoviadou et al. 2015).

The ability of USN processing to preserve fluid foods is based on cavitation phenomena, which explains the generation and evolution of microbubbles in a liquid medium (Galanakis 2012), causing temperature and pressure peaks as well as formation of free radicals (Piyasena et al. 2003; Mason et al. 2005).

Cavitation occurs in the vicinity of a liquid when alternating high-amplitude pressures take place rapidly. During the negative half of the pressure cycle, the liquid is subjected to a tensile force, and during the positive half it undergoes compression, thus resulting in the formation of microbubbles. The size of the microbubbles augments thousands of times during the alternation of the pressure cycles (Chemat et al. 2004a, b; Rosello-Soto et al. 2015; Zinoviadou et al. 2015).

Microbubbles that reach a critical size implode violently and return to their original size, which causes the release of all the accumulated energy, producing instantaneous local temperature increases that are dissipated without substantially raising the temperature of the liquid treated. However, the energy released and the mechanical shock associated with this phenomenon affect the structure of the cells located in the microenvironment (Zinoviadou et al. 2015).

The conventional equipment used for discontinuous (more common) or continuous process is formed by a treatment chamber containing an USN source capable of generating vibration amplitudes of the order of various tens of microns, together with an impedance adaptation box and an electronic generation unit.

This unit consists of two parts: a power amplifier and a system for controlling and monitoring the resonance frequency which keeps the power applied to the transducer steady during the process. Figure 6.1 shows some common types of ultrasonic systems employed for fluid food processing.

The effectiveness of USN processing depends on critical factors that include certain parameters that are related with the treatment and others that depend on the product to be treated. The processing factors include resonance frequency, amplitude, power, treatment time, energy, and temperature applied.

In order to capture and store the various parameters of the transducer excitation signal (resonance frequency, voltage, current, phase, and power), a computer is used (Adekunte et al. 2010; Soria and Villamiel 2010). Current USN technology comes from the exploitation of two properties possessed by certain materials: piezoelectricity and magnetostriction.

Piezoelectric USN generator is based on the generation of electric oscillations of a particular frequency, which a material with piezoelectric properties transforms into mechanical oscillations (transducer).

Another method for producing USN vibrations is by the use of magnetostrictive transducers. The functioning of these devices is based on the mechanical distortions experienced by certain materials when they are subjected to an intense magnetic field (Adekunte et al. 2010; Soria and Villamiel 2010).

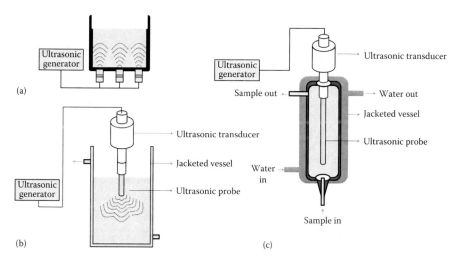

FIGURE 6.1 (a–c) Examples of ultrasound processing equipment. (From Zinoviadou, K.G. et al., *Food Res. Inter.*, 77 (2015) 743–752. With permission.)

The use of USN in food processing is of great interest because not only the potential advantages that can have be compared with conventional thermal processing at high temperatures (less thermal damage and higher overall characteristics), but also the potential generation of new ingredients and products with specific features due to action in the food matrix.

In this line, USN technology could be used for the preservation of nutritive fluid foods such as fruit juice mixed with milk beverages. This kind of food products are commercialized in several countries, particularly in Spain and they are a good source of proteins, minerals, and bioactive compounds (Barba et al. 2011, 2012a, 2013a, b). Their consumption has increased in both Europe and the United States around 30% in the past 20 years, with only 10% belonging to the group of short shelf-life beverages which need refrigerated conditions for their distribution and sale (Andlauer and Furst 2002; Cook 2003; Heckman et al. 2010).

Bioactive compounds are defined as non-nutritional substances that are found in very low concentrations in foods and that intervene in the secondary metabolism of vegetables and may have a significant effect on human health (Kris-Etherton et al. 2002), although some nutrients have also been included as bioactive compounds (Jeffery et al. 2003).

These compounds can be grouped according to their chemical structure into antioxidant vitamins, phenolic compounds, terpene derivatives, sulfur compounds, phytoestrogens, peptides and amino acids, minerals, polyunsaturated fatty acids, dietary fiber, lactic bacteria, and phytic acid. These compounds may exert their effects activating liver detoxification enzymes, blocking the activity of bacterial or viral toxins, inhibiting cholesterol adsorption, decreasing platelet aggregation, or destroying harmful gastrointestinal bacteria and acting like antioxidants among others (Pennington 2002; Barba et al. 2014; Carbonell-Capella et al. 2014).

Fruit juices are a good source of vitamin C, phenolic compounds, and carotenoids which have been shown to be good contributors to the total antioxidant capacity

(TEAC) of the foods (Chaovanalikit and Wrolstad 2004; Sanchez-Moreno et al. 2006; Barba et al. 2013a, b).

The measurement of the antioxidant capacity of food products is a matter of growing interest because it may provide a variety of information, such as resistance to oxidation, quantitative contribution of antioxidant substances, or the antioxidant activity that they may present inside the organism when ingested (Prior et al. 2003; Huang et al. 2006; Serrano et al. 2007).

Vitamin retention studies to assess the effects of food processing on the nutritional value of foods are of great importance to food technologists and consumers (Torregrosa et al. 2005; Barba et al. 2010).

Researchers have used ascorbic acid as a quality indicator in fruits and vegetables (Ayhan et al. 2001; Giannakourou and Taoukis 2003; Barba et al. 2010) because it is a sensitive bioactive compound providing an indication of the loss of other vitamins and therefore acting as a valid criterion for other organoleptic or nutritional components.

Even considering the enormous interest of vitamins as micronutrients in food samples, there exist other compounds with known biological activity that can, at the end, have a similar or even greater contribution in the functional activity of a food and/or beverage (Torregrosa et al. 2005; Zulueta et al. 2010; Barba et al. 2013a).

Phenolic compounds can play an important role in the associated bioactivity of a food product (Mendiola et al. 2008; Carbonell-Capella et al. 2014). The basic feature of all phenolic compounds is the presence of one or more hydroxylated aromatic rings, which seemed to be, in fact, responsible for their properties as radical scavengers (Lule and Xia 2005).

These compounds have been determined in high amounts in fruits and vegetables (Ismail et al. 2004; Turkmen et al. 2005; Wen et al. 2010; Barba et al. 2013a; Chipurura et al. 2013) but little information is available about the effect of nonthermal technologies on total phenolics in foods (Soliva-Fortuny et al. 2009; Patras et al. 2010; Tiwari and Mason 2012; Barba et al. 2015).

Traditionally, fruit juice mixed with milk beverages have been preserved and they are traditionally thermally pasteurized and stored under refrigeration in response to a big demand by consumer for very nutritive foods. However, thermal treatment has some detrimental effects and can reduce the nutritional and quality properties of food products.

Some previous studies have evaluated the effects of nonconventional technologies (HPP and PEF) on nutritional and quality attributes of fruit juice mixed with milk beverages. However, the literature evaluating the impact of USN processing on the parameters is scarce. Therefore, the purpose of this chapter is to evaluate the effects of USN processing on antioxidant compounds (vitamin C and total phenolics compounds).

6.2 MATERIALS AND METHODS

6.2.1 Sample

Orange juice skimmed milk beverage prepared in laboratory: fresh orange juice (500 mL/L) based on fresh oranges from Germany local supermarket Navel variety (origin South Africa), UHT skimmed milk (200 mL/L), pectine high metoxile (3 g/L), sugar (75 g/L), and distilled water (300 mL/L). The physicochemical characteristics of the beverage were analyzed according to AOAC (2003) and it had electrical conductivity = 2.99 mS/cm, pH = 3.91, and 13.8°Brix.

6.2.2 CHEMICALS AND REAGENTS

Trolox (6-hydroxy-2,5,7,8-tetramethylchroman-2-carboxylic acid), as a standard substance (2 mM) to measure TEAC, ABTS (2,2-azinobis(3-ethylbenzothiazoline 6-sulfonate)), and potassium hydroxide (85%). Sodium and disodium phosphate, acetonitrile (special grade), and magnesium hydroxide carbonate (40%–45%) were purchased from Panreac (Barcelona, Spain); L(+)-ascorbic acid, ethanol, methanol metaphosphoric acid, and sodium chloride (special grade) from Merck (Darmstadt, Germany); and tetrabutyl ammonium hydrogen sulfate from Fluka AG, Switzerland.

6.2.3 ULTRASONICS TREATMENT SYSTEM

An ultrasonic processor Hielscher UP200S (Hielscher Ultrasonics, Germany) was used, which can operate up to 400 W and 24 kHz frequency. For the experiments of the present study, the power was varied from 160–280 W and the amplitude in the range of 101.4–163.4 μm (Table 6.1).

TABLE 6.1
Ultrasound Treatments

Amplitude (%)	Distance (μm)	P (W)	Time (s)	Energy (W/mL)	UI[a] (W/cm²) $(4P/p_i{*}D^2)$	Energy (kJ/kg)
	NP	0				0
25	101.4	160	10	10.67	27.94	100
			19			200
			28			300
			56			600
50	116.2	190	8	12.67	33.18	100
			16			200
			24			300
			48			600
75	135.0	240	6	16.00	41.92	100
			12			200
			19			300
			38			600
100	163.4	280	5	18.67	48.90	100
			11			200
			16			300
			32			600

[a] UI, ultrasonics intensity.

Treatment times were varied from 5 to 56 s. The titanium sonotrode H14 with a diameter of 14 mm and a length of 100 mm is used to transmit USN inside the sample. The energy input of USN treatment was calculated as follows:

$$\text{Energy input (kJ/kg)} = \frac{P(W) * t(s)}{m(kg)} \qquad (6.1)$$

where:
  t is the total treatment duration (s)
  m is the product mass (kg) and the generator power (400 J/s)

The energy of USN treatment varied from 0 (untreated sample) to 600 kJ/kg.

6.2.4 Methods

6.2.4.1 Vitamin C Determination

Vitamin C was determined using the procedure described by Ruckemann (1980). All reagents were of analytical grade. The eluent contained 2.5 g of tetrabutyl ammonium hydrogen sulfate in 945 mL distilled water and 55 mL methanol. A standard ascorbic acid solution of 50 mg ascorbic acid in 6% metaphosphoric acid was prepared with appropriate dilutions; 400 µL of orange juice milk were diluted with 1600 µL of 6% metaphosphoric acid.

The samples prepared in this way were filtered through filters with a pore size of 0.22 µm and injected in the chromatograph (HPLC column, pump, variable wavelength monitor, Knauer GmbH, Berlin, Germany). Elution is obtained at a flow rate of 0.9 mL/min and eluate absorbance is measured at 251 nm. Ascorbic acid (Sigma, Steinheim, Germany) was used as standard for the calibration curve and results were expressed as mg/100 mL.

6.2.4.2 Total Phenolics

Total phenolics were determined using the Folin–Ciocalteau method described by Singleton and Rossi (1965). A 10-fold diluted orange juice milk of 200 µL was mixed with 1000 µL of Folin–Ciocalteau (Sigma, Steinheim, Germany) reagent (previously diluted 10-fold with distilled water) and 800 µL of sodium carbonate (105.99 g mol^{-1}).

The mixture was incubated at 40°C during 30 min after which the absorbance was measured at 760 nm in a UV–vis spectrophotometer (Uvikon 922). Catechin (Roth, Karlsruhe, Germany) was used as standard for the calibration curve and results were expressed as mg catechin/L.

6.2.4.3 Total Antioxidative Capacity

The method used was as described by Re et al. (1999), based on the capacity of a sample to inhibit the ABTS radical (ABTS$^{•+}$) compared with a reference antioxidant standard (Trolox®).

The ABTS·+ radical was generated by chemical reaction with potassium persulfate ($K_2S_2O_8$). For this purpose, 38.43 mg ABTS (7 mM) was spiked with 6.90 mg of $K_2S_2O_8$ (2.45 mM) and 10 mL of distilled water.

The mixture was allowed to stand in darkness at room temperature for 12–16 h (the time required for formation of the radical). The working solution was prepared by taking a volume of the previous solution and diluting it in ethanol until its absorbance at $\lambda = 734$ nm was 0.70 ± 0.02. The measuring was done in a UV–vis spectrophotometer (Uvikon 922).

The reaction took place directly in the measuring cuvette. For this purpose, 1 mL of the ABTS·+ radical was added; the absorbance (A_0) was measured at 734 nm, and 20 μL of sample (fivefold orange juice milk) or standard were added immediately, at which point the antioxidants present in the sample began to inhibit the radical, producing a reduction in absorbance, with a quantitative relationship between the reduction and the concentration of antioxidants present in the sample.

The reaction time was established in 6 min, after these the absorbance was measured at 734 nm (A_f). At the same time, a trolox calibration curve was prepared for a concentration range of 0.5–2 mM/L Trolox® and was used to calculate the antioxidant activity of the samples, expressed as mmol trolox eq/L. Each sample was analyzed in triplicate (Barba et al. 2012a).

6.2.5 Statistical Analysis

Significant differences between the results were calculated by analyses of the variance (ANOVA). One-way ANOVA was calculated on the three triplicate measurements. Differences at $p < .05$ were considered to be significant.

An LSD test was applied to indicate the samples between which there were differences. All statistical analyses were performed using Statgraphics Plus 5.0 (Statistical Graphics Corporation, Inc., Rockville, MD).

6.3 RESULTS AND DISCUSSION

In order to determine the impact of USN on bioactive compounds and antioxidant capacity of orange juice mixed with milk beverages, USN treatments at different amplitude (101.4–163.4 μm) and energy levels (0–600 kJ/kg) were applied.

6.3.1 Impact of USNs on Ascorbic Acid Content

The ascorbic acid values for the untreated orange juice mixed with milk beverage were 25.1 ± 0.2 mg/100 mL. These values are in the range of those previously reported by other authors in the available literature (Barba et al. 2012a, 2013b; Zulueta et al. 2013).

One-way ANOVA analysis showed that energy level had a significant impact on the ascorbic acid content (in milligrams per 100 mL), independent of the amplitude and power level used in the treatments, observing the largest AA reduction compared to untreated samples (−27%) when USN at the highest amplitude (163.4 μm) and energy input (600 kJ/kg) was used (Table 6.2).

Similar results were obtained by Tiwari et al. (2008b) in strawberry juice treated at different levels of amplitude 40%–100% during 0–10 min. Tiwari et al. (2008a) also obtained similar results for orange juice treated by sonication.

These authors found a significant ($p < .05$) decrease in orange juice and ascorbic acid content (mg/100 mL) was observed as a function of acoustic energy densities (0.20–0.81 W/mL) and treatment time (2–10 min). This fact can be explained by the ability of sonication to reduce the dissolved oxygen, which is a critical parameter influencing the stability of ascorbic acid (Solomon and Svanberg 1995).

Moreover, ascorbic acid degradation during sonication may be due to free radical formation (Portenlanger and Heusinger 1992) and hydroxyl radical formation was

TABLE 6.2
Values of Ascorbic Acid (Mean ± Standard Deviation of Three Replicates of Each Sample) in the Orange Juice Mixed with Milk Beverage Treated by Ultrasounds at Different Amplitudes (101.4–163.4 μm) and Energy Input (0–600 kJ/kg) Levels

Distance (μm)	P (W)	Time (s)	Energy (kJ/kg)	Vitamin C (mg/100 mL)
NP	0		0	25.1 ± 0.2[a]
		10	100	24.7 ± 0.3[a]
101.4	160	19	200	22.3 ± 1.0[bc]
		28	300	22.0 ± 0.5[bcde]
		56	600	21.5 ± 0.1[def]
		8	100	22.4 ± 0.2[b]
116.2	190	16	200	21.8 ± 0.1[cdef]
		24	300	20.7 ± 0.2[gh]
		48	600	19.7 ± 0.2[i]
		6	100	22.2 ± 0.1[bcd]
135.0	240	12	200	21.4 ± 0.2[cf]
		19	300	21.2 ± 0.2[fg]
		38	600	21.2 ± 0.2[fg]
		5	100	21.5 ± 0.0[ef]
163.4	280	11	200	21.4 ± 0.0[ef]
		16	300	20.5 ± 0.2[h]
		32	600	18.2 ± 0.3[j]

[a–j] Different letters indicate significant statistical differences in function of the applied treatment.

found to be at elevated levels with degasing. Therefore, sonication cavities can be filled with water vapor and gases dissolved in the juice, such as O_2 and N_2 (Korn et al. 2002).

The interactions between free radicals and ascorbic acid may occur at the gas–liquid interfaces. In summary, ascorbic acid degradation may follow one or both of the following pathways:

Ascorbic acid→thermolysis (inside bubbles) and triggering of Maillard reaction
Ascorbic acid→ reaction with OH^- → $HC-OH$ and production of oxidative products on the surface of bubbles.

Thereby, sonication can be related to advanced oxidative processes since both pathways are associated with the production and use of hydroxyl radicals (Hart and Henglein 1985; Petrier et al. 2007).

6.3.2 Impact of USNs on Total Phenolics

Total phenolics content in unprocessed orange juice skimmed milk was 671.4 ± 17.4 mg catechin/L. Values were in the range of those previously reported in the available literature by other authors (Barba et al. 2012a; Zulueta et al. 2013).

The ANOVA analysis showed that both amplitude and energy level had a significant effect on TPC. However, the trend was not clear. As can be seen in Table 6.3, USN treatments (135.0 µm, 600 kJ/kg and 163.4 µm, at 100–600 kJ/kg) increased significantly ($p < .05$) the levels of total phenolics compared to unprocessed orange juice milk.

A positive effect on the yields of phenolic compounds was observed by other authors in Satsuma mandarin peels and citrus peel extracts when the ultrasonic power of the treatment was increased (Ma et al. 2008, 2009). Similarly, Aadil et al. (2013), Zafra-Rojas et al. (2013), and Santhirasegaram et al. (2013) observed a significant increase in TPC of grapefruit, cactus pear, and mango juices after applying USN treatments (28 kHz/20°C/30–90 min, 20 kHz/20°C/3–25 min, and 40kHz/25°C/15–30 min, respectively) compared to control untreated samples. More recently, Abid et al. (2014) also found a significant increase in chlorogenic acid (40%), caffeic acid (19%), catechin (20%), epicatechin (145%), and phloridzin (76%) of USN-treated apple juice compared to untreated samples. A possible explanation for this increase could be the disruption of cell walls enhanced by USN, thus facilitating the release of the cell content.

6.3.3 Impact of USNs on Color

The color of fluid foods is an important quality attribute in consumer preferences which is also related with the nutritional quality of food products, in particular pigments such as carotenoids, polyphenols, and chlorophylls, during preservation/processing treatment. For instance, previous studies in the available literature have shown that the degradation of bioactive compounds responsible for color in

TABLE 6.3
Values of Total Antioxidant Capacity (TEAC) (Mean ± Standard Deviation of Three Replicates of Each Sample) in the Orange Juice Skimmed Milk Beverage Treated by Ultrasounds at Different Amplitudes (101.4–163.4 µm) Energy Input (0–600 KJ/kg) Levels

P (W)	Time (s)	Energy (kJ/kg)	TEAC (mM Trolox)
0		0	2.36 ± 0.07[ab]
	10	100	2.22 ± 0.03[ac]
160	19	200	2.08 ± 0.11[cdef]
	28	300	2.45 ± 0.07[b]
	56	600	2.24 ± 0.05[ac]
	8	100	2.10 ± 0.06[cd]
190	16	200	2.10 ± 0.14[cde]
	24	300	2.37 ± 0.06[ab]
	48	600	1.98 ± 0.04[defg]
	6	100	2.34 ± 0.12[ab]
240	12	200	1.93 ± 0.04[fg]
	19	300	1.75 ± 0.08[h]
	38	600	1.98 ± 0.04[defg]
	5	100	1.82 ± 0.13[gh]
280	11	200	1.93 ± 0.04[efg]
	16	300	1.95 ± 0.10[defg]
	32	600	1.76 ± 0.07[h]

[a–h] Different letters indicate significant statistical differences in the function of the applied treatment.

liquid foods can occur after USN processing. For example, degradation of (all-E)-astaxanthin into unidentified colorless compounds during USN-assisted extraction was reported when power level and processing time were increased (Zhao et al. 2006).

Therefore, the effects of USN treatments on instrumental color parameters b^* (blueness and yellowness), a^* (greenness and redness), and L^* (lightness or darkness) of the orange juice mixed with milk beverage were studied in order to establish a relationship with the bioactive compounds studied in this chapter.

As can be seen from Figure 6.2, amplitude level and energy input had a significant impact in a^* values. The a^* values changed toward a more positive direction when energy input was increased. This fact can be attributed to a better extractability of orange–red carotenoids found in orange.

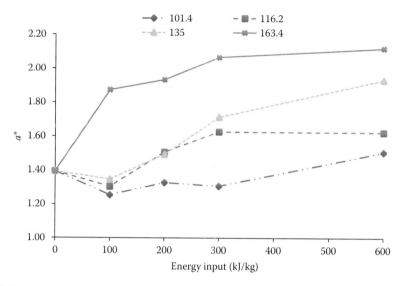

FIGURE 6.2 Impact of USN at different amplitude (101.4–163.4 μm) and energy levels (0–600 kJ/kg) on a^* values of an orange juice mixed with milk beverage.

In this line, Rawson et al. (2011) and Abid et al. (2014) found a significant increase in lycopene and total carotenoid content, respectively, when they applied USN treatments.

On the other hand, the ANOVA analysis also showed a significant impact of amplitude and energy levels on b^* values. As it is shown from Figure 6.3, nonsignificant differences or a slight increase was found in the b^* values when the lower amplitude levels were applied. However, b^* significantly decreased when energy levels were augmented ($p < .05$) to the highest amplitude level.

This fact can be attributed to nonenzymatic browning reactions that can occur between amino acids, sugars, and organic acids (Zulueta et al. 2013). Similarly, other authors also found a significant decrease in b^* when higher amplitude levels and longer treatment times were used in orange (Tiwari et al. 2008a, 2009b) and strawberry juices (Tiwari et al. 2008b).

On the other hand, the ANOVA analysis did not show any significant modifications in lightness (L^*) values at lower amplitude levels (Figure 6.4). However, L^* were increased significantly ($p < .05$) when amplitude level was augmented.

The increase in L^* values can be attributed to nonenzymatic browning. These results were in close agreement to those found by other authors when they evaluated the effects of USN processing (20 kHz, 40%–100% (24.4–61 μm), 25°–39.9°, 0–10 min, 0.30–0.81 W/mL) on color from orange (Tiwari et al. 2008a, 2009b), strawberry (Tiwari et al. 2008b), grape (Tiwari et al. 2010), and blackberry (Tiwari et al. 2009a) juices.

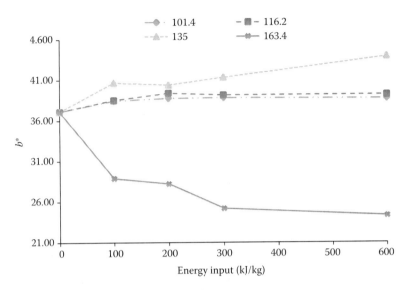

FIGURE 6.3 Impact of USN at different amplitude (101.4–163.4 μm) and energy levels (0–600 kJ/kg) on b^* values of an orange juice mixed with milk beverage.

6.3.4 Impact of USNs on Antioxidant Capacity (TEAC)

The average TEAC values for unprocessed orange juice milk is in the range 2.36 ± 0.07 mM TE. As shown in Table 6.4, when USN treatments were applied, we observed significant differences in TEAC ($p < .05$).

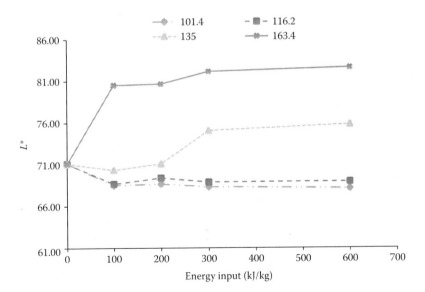

FIGURE 6.4 Impact of USN at different amplitude (101.4–163.4 μm) and energy levels (0–600 kJ/kg) on L^* values of an orange juice mixed with milk beverage.

TABLE 6.4
Values of Total Phenolic Compounds (TPC) (Mean ± Standard Deviation of Three Replicates of Each Sample) in the Orange Juice Skimmed Milk Beverage Treated by Ultrasounds at Different Amplitudes (101.4–163.4 μm) Energy Input (0–600 kJ/kg) Levels

Distance (μm)	P (W)	Time (s)	Energy (kJ/kg)	Total Phenolics (mg Catechin/L)
NP	0		0	671.4 ± 17.4[abc]
		10	100	705.1 ± 28.5[bcd]
101.4	160	19	200	719.4 ± 16.9[cde]
		28	300	729.1 ± 34.2[cde]
		56	600	722.7 ± 46.1[cde]
		8	100	752.4 ± 25.0[de]
116.2	190	16	200	750.9 ± 39.3[de]
		24	300	669.5 ± 33.1[abc]
		48	600	637.6 ± 28.6[a]
		6	100	651.6 ± 35.5[ab]
135.0	240	12	200	675.4 ± 19.5[abc]
		19	300	719.9 ± 42.2[cde]
		38	600	776.5 ± 33.8[e]
		5	100	895.0 ± 9.0[f]
163.4	280	11	200	939.6 ± 30.6[fg]
		16	300	979.5 ± 36.4[gh]
		32	600	1011.8 ± 18.6[h]

[a–h] Different letters indicate significant statistical differences in the function of the applied treatment.

These differences were correlated with ascorbic acid and total phenolics content ($p < .05$). Ma et al. 2008 found a positive correlation between TEAC and total phenolics after USN treatments of Satsuma mandarin peels.

The results obtained in this chapter were in close agreement to those found by Keenan et al. (2012) when they evaluated the impact of USN (20 kHz, 24.4–61.0 μm, 3–10 min) on total antioxidant capacity of fruit smoothies. These authors also observed higher antioxidant values when the maximum USN amplitude level was used (Table 6.4).

6.3.5 Relationship between Bioactive Compounds and Antioxidant Capacity (TEAC) after USN Processing

When the possible correlation (Pearson's test) between the various bioactive compounds that contribute to antioxidant capacity (ascorbic acid and total phenolics) was studied, it was found that there was a positive correlation between antioxidant capacity (TEAC values) and ascorbic acid ($p < .05$) (Figure 6.5).

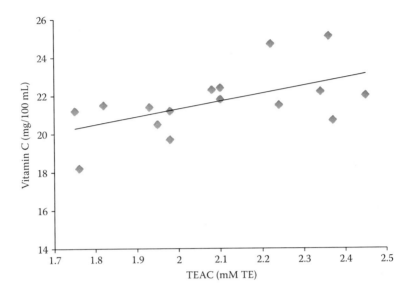

FIGURE 6.5 Correlation between total antioxidant capacity (TEAC) and vitamin C of orange juice mixed with milk beverages processed by ultrasounds.

Therefore, the antioxidant capacity decrease can be explained mainly due to a decrease in ascorbic acid content after applying USN treatments. In this line, other previous studies have demonstrated that ascorbic acid is the main contributor to antioxidant capacity (TEAC) when orange juice is studied (Barba et al. 2013b).

6.4 CONCLUSIONS

USN processing can be a useful tool to preserve fluid food when it is combined with mild temperatures. Although USN amplitude level and energy input had a significant impact on vitamin C, phenolic compounds, and antioxidant capacity of the orange juice mixed with milk beverage, the overall retention of vitamin C was around 80% compared to untreated samples.

Moreover, phenols appeared to be relatively resistant to the effect of processing or even increasing when USN treatments were applied, and the color parameters, which are related to the quality of the product were not significantly affected, except when the highest amplitude and energy input levels were used.

In addition, a strong relationship was found between vitamin C and antioxidant capacity (TEAC) when USN treatments were applied and a significant decrease in TEAC values was observed when vitamin C levels dropped.

REFERENCES

Aadil, R M, X-A Zeng, Z Han, and D-W Sun. Effects of Ultrasound Treatments on Quality of Grapefruit Juice. *Food Chemistry* 141(3): (2013) 3201–3206.

Abid, M, S Jabbar, T Wu, M M Hashim, B Hu, S Lei, and X Zeng. Sonication Enhances Polyphenolic Compounds, Sugars, Carotenoids and Mineral Elements of Apple Juice. *Ultrasonics Sonochemistry* 21(1): (2014) 93–97.

Adekunte, A O, B K Tiwari, P J Cullen, A G M Scannell, and C P O'Donnell. Effect of Sonication on Colour, Ascorbic Acid and Yeast Inactivation in Tomato Juice. *Food Chemistry* 122(3): (2010) 500–507.

Andlauer, W and P Furst. Nutraceuticals: A Piece of History, Present Status and Outlook. *Food Research International* 35(2–3): (2002) 171–176.

AOAC International. 2003. Official methods of analysis of AOAC International. 17th edition. 2nd revision. Gaithersburg, MD, USA, Association of Analytical Communities.

Ayhan, Z, H W Yeom, Q H Zhang, and D B Min. Flavor, Color, and Vitamin C Retention of Pulsed Electric Field Processed Orange Juice in Different Packaging Materials. *Journal of Agricultural and Food Chemistry* 49(2): (2001) 669–674.

Barba, F J, C Cortés, M J Esteve, and A Frígola. Study of Antioxidant Capacity and Quality Parameters in an Orange Juice-Milk Beverage after High-Pressure Processing Treatment. *Food and Bioprocess Technology* 5(6): (2012a) 2222–2232.

Barba, F J, M J Esteve, and A Frigola. Ascorbic Acid Is the Only Bioactive That Is Better Preserved by High Hydrostatic Pressure Than by Thermal Treatment of a Vegetable Beverage. *Journal of Agricultural and Food Chemistry* 58(18): (2010) 10070–10075.

Barba, F J, M J Esteve, and A Frigola. Determination of Vitamins E (Alpha, Delta and Gamma-Tocopherol) and D (cholecalciferol and Ergocalciferol) by Liquid Chromatography in Milk, Fruit Juice and Vegetable Beverage. *European Food Research and Technology* 232(5): (2011) 829–836.

Barba, F J, M J Esteve, and A Frígola. High Pressure Treatment Effect on Physicochemical and Nutritional Properties of Fluid Foods during Storage: A Review. *Comprehensive Reviews in Food Science and Food Safety* 11(3): (2012b) 307–322.

Barba, F J, M J Esteve, and A Frigola. *Gallic Acid as a Useful Indicator of the Antioxidant Capacity of Liquid Foods? Handbook on Gallic Acid: Natural Occurrences, Antioxidant Properties and Health Implications.* Edited by M A Thompson and P B Collins. Nova Science Publishers, (2013a) 247–263.

Barba, F J, M J Esteve, and A Frigola. Bioactive Components from Leaf Vegetable Products. *Studies in Natural Products Chemistry* 41: (2014) 321–346.

Barba, F J, N S Terefe, R Buckow, D Knorr, and V Orlien. New Opportunities and Perspectives of High Pressure Treatment to Improve Health and Safety Attributes of Foods: A Review. *Food Research International.* 77: (2015) 725–742.

Barba, F J A, M J A Esteve, P B Tedeschi, V B Brandolini, and A A Frígola. A Comparative Study of the Analysis of Antioxidant Activities of Liquid Foods Employing Spectrophotometric, Fluorometric, and Chemiluminescent Methods. *Food Analytical Methods* 6(1): (2013b) 317–327.

Carbonell-Capella, J M, M Buniowska, F J Barba, M J Esteve, and A Frígola. Analytical Methods for Determining Bioavailability and Bioaccessibility of Bioactive Compounds from Fruits and Vegetables: A Review. *Comprehensive Reviews in Food Science and Food Safety* 13(2): (2014) 155–171.

Chaovanalikit, A and R E Wrolstad. Total Anthocyanins and Total Phenolics of Fresh and Processed Cherries and Their Antioxidant Properties. *Journal of Food Science* 69(1): (2004) FCT67–FCT72.

Chemat, F, I Grondin, P Costes, L Moutoussamy, A S C Sing, and J Smadja. High Power Ultrasound Effects on Lipid Oxidation of Refined Sunflower Oil. *Ultrasonics Sonochemistry* 11(5): (2004a) 281–285.

Chemat, F, I Grondin, A S C Sing, and J Smadja. Deterioration of Edible Oils during Food Processing by Ultrasound. *Ultrasonics Sonochemistry* 11(1): (2004b) 13–15.

Chipurura, B, M Muchuweti, and M Bhebhe. An Assessment of the Phenolic Content, Composition and Antioxidant Capacity of Selected Indigenous Vegetables of Zimbabwe. *Acta Horticulturae* 979: (2013) 611–620.

Cook, R. *Globalization and Fresh Produce Marketing: Challenges and Opportunities*. United Fresh Fruit and Vegetable Association (UFFVA) Fellowship Program, University of California, Davis, CA, (July 24, 2003).

Deng, Q, K G Zinoviadou, C M Galanakis, V Orlien, N Grimi, E Vorobiev, N Lebovka, and F J Barba. The Effects of Conventional and Non-Conventional Processing on Glucosinolates and Its Derived Forms, Isothiocyanates: Extraction, Degradation, and Applications. *Food Engineering Reviews* 1–25. 7: (2015) 357–381.

Dujmic, F, M Brncic, S Karlovic, T Bosiljkov, D Jezek, B Tripalo, and I Mofardin. Ultrasound-Assisted Infrared Drying of Pear Slices: Textural Issues. *Journal of Food Process Engineering* 36(3): (2013) 397–406.

Galanakis, C M. Recovery of High Added-Value Components from Food Wastes: Conventional, Emerging Technologies and Commercialized Applications. *Trends in Food Science & Technology* 26(2): (2012) 68–87.

Giannakourou, M C and P S Taoukis. Kinetic Modelling of Vitamin C Loss in Frozen Green Vegetables under Variable Storage Conditions. *Food Chemistry* 83(1): (2003) 33–41.

Hart, E J and A Henglein. Free Radical and Free Atom Reactions in the Sonolysis of Aqueous Iodide and Formate Solutions. *Journal of Physical Chemistry* 89: (1985) 4342–4347.

Heckman, M A, K Sherry, and E G de Mejia. Energy Drinks: An Assessment of Their Market Size, Consumer Demographics, Ingredient Profile, Functionality, and Regulations in the United States. *Comprehensive Reviews in Food Science and Food Safety* 9(3): (2010) 303–317.

Huang, E, G S Mittal, and M W Griffiths. Inactivation of *Salmonella enteritidis* in Liquid Whole Egg Using Combination Treatments of Pulsed Electric Field, High Pressure and Ultrasound. *Biosystems Engineering* 94(3): (2006) 403–413.

Ismail, A, Z M Marjan, and C W Foong. Total Antioxidant Activity and Phenolic Content in Selected Vegetables. *Food Chemistry* 87(4): (2004) 581–586.

Jeffery, E H, A F Brown, A C Kurilich, A S Keck, N Matusheski, B P Klein, and J A Juvik. Variation in Content of Bioactive Components in Broccoli. *Journal of Food Composition and Analysis* 16(3): (2003) 323–330.

Keenan, D F, B K Tiwari, A Patras, R Gormley, F Butler, and N P Brunton. Effect of Sonication on the Bioactive, Quality and Rheological Characteristics of Fruit Smoothies. *International Journal of Food Science and Technology* 47(4): (2012) 827–836.

Kentish, S and H Feng. Applications of Power Ultrasound in Food Processing. *Annual Review of Food Science and Technology* 5(1): (2014) 263–284.

Knorr, D, M Zenker, V Heinz, and D-U Lee. Applications and Potential of Ultrasonics in Food Processing. *Trends in Food Science & Technology* 15(5): (2004) 261–266.

Korn, M, P M Prim, and C S de Sousa. Influence of Ultrasonic Waves on Phosphate Determination by the Molybdenum Blue Method. *Microchemical Journal* 73(3): (2002) 273–277.

Kris-Etherton, P M, K D Hecker, A Bonanome, S M Coval, A E Binkoski, K F Hilpert, A E Griel, and T D Etherton. Bioactive Compounds in Foods: Their Role in the Prevention of Cardiovascular Disease and Cancer. *American Journal of Medicine* 113(9 SUPPL. 2): (2002) 71S–88S.

Lule, S U and W Xia. Food Phenolics, Pros and Cons: A Review. *Food Reviews International* 21(4): (2005) 367–388.

Ma, Y-Q, J-C Chen, D-H Liu, and X-Q Ye. Simultaneous Extraction of Phenolic Compounds of Citrus Peel Extracts: Effect of Ultrasound. *Ultrasonics Sonochemistry* 16(1): (2009) 57–62.

Ma, Y-Q, X-Q Ye, Z-X Fang, J-C Chen, G-H Xu, and D-H Liu. Phenolic Compounds and Antioxidant Activity of Extracts from Ultrasonic Treatment of Satsuma Mandarin (Citrus Unshiu Marc.) Peels. *Journal of Agricultural and Food Chemistry* 56(14): (2008) 5682–5690.

Mason, T J, E Riera, A Vercet, and P Lopez-Bueza. Application of Ultrasound. In *Emerging Technologies for Food Processing*. Edited by D-W Sun. Elsevier, Cambridge, MA, (2005), pp. 323–351.

Mendiola, J A, F R Marin, F J Senorans, G Reglero, P J Martin, A Cifuentes, and E Ibanez. Profiling of Different Bioactive Compounds in Functional Drinks by High-Performance Liquid Chromatography. *Journal of Chromatography A* 1188(2): (2008) 234–241.

Patras, A, N P Brunton, C O'Donnell, and B K Tiwari. Effect of Thermal Processing on Anthocyanin Stability in Foods: Mechanisms and Kinetics of Degradation. *Trends in Food Science & Technology* 21(1): (2010) 3–11.

Pennington, J A T. Food Composition Databases for Bioactive Food Components. *Journal of Food Composition and Analysis* 15: (2002) 419–434.

Petrier, C, E Combet, and T Mason. Oxygen-Induced Concurrent Ultrasonic Degradation of Volatile and Non-Volatile Aromatic Compounds. *Ultrasonics Sonochemistry* 14(2): (2007) 117–121.

Piyasena, P, E Mohareb, and R C McKellar. Inactivation of Microbes Using Ultrasound: A Review. *International Journal of Food Microbiology* 87(3): (2003) 207–216.

Portenlanger, G and H Heusinger. Chemical Reactions Induced by Ultrasound and G-Rays in Aqueous Solutions of L-Ascorbic Acid. *Carbohydrate Research* 232(2): (1992) 291–301.

Prior, R L, H Hoang, L Gu, X Wu, M Bacchiocca, L Howard, M Hampsch-Woodill, D Huang, B Ou, and R Jacob. Assays for Hydrophilic and Lipophilic Antioxidant Capacity (oxygen Radical Absorbance Capacity [ORACFL]) of Plasma and Other Biological and Food Samples. *Journal of Agricultural and Food Chemistry* 51(11): (2003) 3273–3279.

Rawson, A, B K Tiwari, A Patras, N Brunton, C Brennan, P J Cullen, and C O'Donnell. Effect of Thermosonication on Bioactive Compounds in Watermelon Juice. *Food Research International* 44(5): (2011) 1168–1173.

Re, R, N Pellegrini, A Proteggente, A Pannala, M Yang, and C Rice-Evans. Antioxidant Activity Applying an Improved ABTS Radical Cation Decolorization Assay. *Free Radical Biology and Medicine* 26(9–10): (1999) 1231–1237.

Rosello-Soto, E, C M Galanakis, M Brncic, V Orlien, F J Trujillo, R Mawson, K Knoerzer, B K Tiwari, and F J Barba. Clean Recovery of Antioxidant Compounds from Plant Foods, By-Products and Algae Assisted by Ultrasounds Processing. Modeling Approaches to Optimize Processing Conditions. *Trends in Food Science & Technology* 42(2): (2015) 134–149.

Ruckemann, H. Methoden Zur Bestimmung Vonl-Ascorbinsaure Mittels Hochleistungs-Flussigchromatographie (HPLC). *Zeitschrift Fur Lebensmittel-Untersuchung Und Forschung* 171(5): (1980) 357–359.

Sanchez-Moreno, C, L Plaza, B de Ancos, and M P Cano. Nutritional Characterisation of Commercial Traditional Pasteurised Tomato Juices: Carotenoids, Vitamin C and Radical-Scavenging Capacity. *Food Chemistry* 98(4): (2006) 749–756.

Santhirasegaram, V, Z Razali, and C Somasundram. Effects of Thermal Treatment and Sonication on Quality Attributes of Chokanan Mango (Mangifera Indica L.) Juice. *Ultrasonics Sonochemistry* 20(5): (2013) 1276–1282.

Serrano, J, I Goni, and F Saura-Calixto. Food Antioxidant Capacity Determined by Chemical Methods May Underestimate the Physiological Antioxidant Capacity. *Food Research International* 40(1): (2007) 15–21.

Singleton, V L and J A Jr Rossi. Colorimetry of Total Phenolics with Phosphomolybdic-Phosphotungstic Acid Reagents. *American Journal of Enology and Viticulture* 16: (1965) 144–158.

Soliva-Fortuny, R, A Balasa, D Knorr, and O A Martín-Belloso. Effects of Pulsed Electric Fields on Bioactive Compounds in Foods: A Review. *Trends in Food Science & Technology* 20(11–12): (2009) 544–556.

Solomon, O and U Svanberg. Effect of Oxygen and Fluorescent Light on the Quality of Orange Juice during Storage at 8°C. *Food Chemistry* 53: (1995) 363–368.

Soria, A C and M Villamiel. Effect of Ultrasound on the Technological Properties and Bioactivity of Food: A Review. *Trends in Food Science & Technology* 21(7): (2010) 323–331.

Tiwari, B K and T J Mason. Ultrasound processing of fluid foods. In *Novel Thermal and Non-Thermal Technologies for Fluid Foods*. Edited by P J Cullen, B K Tiwari, and V P Valdramidis. Academic Press, London, UK, (2012), pp. 135–167.

Tiwari, B K, K Muthukumarappan, C P O'Donnell, and P J Cullen. Effects of Sonication on the Kinetics of Orange Juice Quality Parameters. *Journal of Agricultural and Food Chemistry* 56(7): (2008a) 2423–2428.

Tiwari, B K, C P O'Donnell, and P J Cullen. Effect of Sonication on Retention of Anthocyanins in Blackberry Juice. *Journal of Food Engineering* 93(2): (2009a) 166–171.

Tiwari, B K, C P O'Donnell, K Muthukumarappan, and P J Cullen. Ascorbic Acid Degradation Kinetics of Sonicated Orange Juice during Storage and Comparison with Thermally Pasteurised Juice. *LWT—Food Science and Technology* 42(3): (2009b) 700–704.

Tiwari, B K, C P O'Donnell, A Patras, and P J Cullen. Anthocyanin and Ascorbic Acid Degradation in Sonicated Strawberry Juice. *Journal of Agricultural and Food Chemistry* 56(21): (2008b) 10071–10077.

Tiwari, B K, A Patras, N Brunton, P J Cullen, and C P O'Donnell. Effect of Ultrasound Processing on Anthocyanins and Color of Red Grape Juice. *Ultrasonics Sonochemistry* 17(3): (2010) 598–604.

Toepfl, S, A Mathys, V Heinz, and D Knorr. Review: Potential of High Hydrostatic Pressure and Pulsed Electric Fields for Energy Efficient and Environmentally Friendly Food Processing. *Food Reviews International* 22(4): (2006) 405–423.

Torregrosa, F, C Cortes, M J Esteve, and A Frigola. Effect of High-Intensity Pulsed Electric Fields Processing and Conventional Heat Treatment on Orange-Carrot Juice Carotenoids. *Journal of Agricultural and Food Chemistry* 53(24): (2005) 9519–9525.

Turkmen, N, F Sari, and Y S Velioglu. The Effect of Cooking Methods on Total Phenolics and Antioxidant Activity of Selected Green Vegetables. *Food Chemistry* 93(4): (2005) 713–718.

Wen, T N, K N Prasad, B Yang, and A Ismail. Bioactive Substance Contents and Antioxidant Capacity of Raw and Blanched Vegetables. *Innovative Food Science and Emerging Technologies* 11(3): (2010) 464–469.

Zafra-Rojas, Q Y, N Cruz-Cansino, E Ramirez-Moreno, L Delgado-Olivares, J Villanueva-Sanchez, and E Alanis-Garcia. Effects of Ultrasound Treatment in Purple Cactus Pear (Opuntia Ficus-Indica) Juice. *Ultrasonics Sonochemistry* 20(5): (2013) 1283–1288.

Zenker, M, V Heinz, and D Knorr. Application of Ultrasound-Assisted Thermal Processing for Preservation and Quality Retention of Liquid Foods. *Journal of Food Protection* 66(9): (2003) 1642–1649.

Zhao, L, G Zhao, F Chen, Z Wang, J Wu, and X Hu. Different Effects of Microwave and Ultrasound on the Stability of (All-E)-Astaxanthin. *Journal of Agricultural and Food Chemistry* 54(21): (2006) 8346–8351.

Zinoviadou, K G, C M Galanakis, M Brncic, N Grimi, N Boussetta, M J Mota, J A Saraiva, A Patras, B Tiwari, and F J Barba. Fruit Juice Sonication: Implications on Food Safety and Physicochemical and Nutritional Properties. *Food Research International*. 77: (2015) 743–752.

Zulueta, A, F J Barba, M J Esteve, and A Frígola. Effects on the Carotenoid Pattern and Vitamin A of a Pulsed Electric Field-Treated Orange Juice-Milk Beverage and Behavior during Storage. *European Food Research and Technology* 231(4): (2010) 525–534.

Zulueta, A, F J Barba, M J Esteve, and A Frigola. Changes in Quality and Nutritional Parameters during Refrigerated Storage of an Orange Juice-Milk Beverage Treated by Equivalent Thermal and Non-Thermal Processes for Mild Pasteurization. *Food and Bioprocess Technology* 6(8): (2013) 2018–2030.

7 Multilayer Nanocapsules as a Vehicle for Release of Bioactive Compounds

Ana C. Pinheiro, Ana I. Bourbon, Hélder D. Silva, Joana T. Martins, and António A. Vicente

CONTENTS

7.1 Introduction 149
7.2 Development of Multilayer Nanocapsules 151
    7.2.1 LbL Electrostatic Deposition Technique 151
    7.2.2 Factors Affecting the Layers' Properties 152
    7.2.3 Colloidal Templates 153
        7.2.3.1 Hollow Nanocapsules 153
        7.2.3.2 Liposomes 154
        7.2.3.3 Nanoemulsions 154
    7.2.4 Advantages and Limitations of Multilayer Nanocapsules 155
7.3 Multilayer Nanocapsules as Carriers of Bioactive Compounds 156
    7.3.1 Bioactive Compounds 157
    7.3.2 Encapsulation Techniques in Multilayer Nanocapsules 157
    7.3.3 EE and Loading Capacity 159
7.4 Impact of Multilayers in the Release of Bioactive Compounds 161
7.5 Behavior of Multilayer Nanocapsules in the GI Tract 165
    7.5.1 GI Tract Main Features and Challenges 166
    7.5.2 Uptake/Absorption of Bioactive Compounds 168
    7.5.3 Changes in Multilayer Nanocapsules Delivery Properties—Potential Toxicity Effects 169
7.6 Conclusions and Future Perspectives 171
Acknowledgments 173
References 173

7.1 INTRODUCTION

In the modern lifestyle, food is not intended only to satisfy hunger and provide necessary nutrients to humans, but it is also oriented to prevent nutrition-related diseases and increase physical and mental well-being of consumers (Menrad 2003). The increase in consumers' awareness of the impact that food has on health and the consequent increase in demand for functional foods are stimulating the innovation and the development of new products in the food industry. The consumption of functional foods is frequently associated to the prevention of nutrition-related diseases, such as heart disease, obesity, diabetes, hypertension, cholesterol, and cancer. The latest trend in food industry is the inclusion of bioactive compounds with potential health benefits in food products. However, the applicability of many bioactive compounds in food systems as health promoting agents is limited by their poor solubility in aqueous solution and/or high sensitivity to heat, oxygen, or light, which dramatically decreases their bioavailability. In fact, many bioactive compounds are labile substances, that may lose their bioactivity during food processing or when in contact with the harsh conditions of the gastrointestinal (GI) system. Therefore, encapsulation of bioactive compounds (such as vitamins, minerals, polyphenols, phytosterols, and probiotics) in different bio-based delivery systems has received increasing attention from academia and food industry, searching to obtain ambitious properties such as delayed release, stability, thermal protection, and improved bioavailability (Đorđević et al. 2014).

Recent advances involving the encapsulation of food bioactive compounds deal with diverse nanostructured delivery systems, which are in general reported to be promising means of improving the bioavailability of incorporated bioactive compounds (Yu and Huang 2012) and controlling their release rate (McClements and Xiao 2012). Their subcellular size allows relatively higher intracellular uptake and their permanence in the blood circulation for longer, therefore extending their biological activity compared to micro-sized systems (Mora-Huertas et al. 2010). Also, nanoencapsulation may be beneficial regarding improved stability and protection capability of labile substances against degradation factors (Preetz et al. 2008).

In particular, multilayer nanocapsules prepared through layer-by-layer (LbL) technique, have attracted considerable attention due to the possibility to readily tailor their properties (e.g., size, composition, porosity, surface functionality, stability) (Johnston et al. 2006). Also, the step-wise formation of multilayer nanocapsules allows the introduction of different functionalities in the same nanostructure.

A variety of different food-grade biopolymers can be used to assemble nanolaminated coatings around different templates, including various proteins and polysaccharides (Weiss et al. 2006). However, despite the use of only food-grade materials and methodologies for multilayer nanocapsules production, the use of very small particle sizes may alter the biological fate of the ingested bio-based materials and bioactive compounds, which could potentially have adverse effects on human health (McClements 2013). Therefore, the knowledge of the molecular, physicochemical, and physiological processes that occur during digestion and absorption of multilayer nanocapsules is crucial to evaluate the potential health risks from their use.

Supported by the increasing interest in encapsulation of bioactive compounds in order to improve their stability and bioavailability, this chapter aims at reviewing the latest achievements in the multilayer nanoencapsulation approach for the delivery of food bioactive compounds. The preparation of multilayer nanocapsules will be focused, with particular emphasis on the encapsulation of bioactive compounds, and the recent advances in releasing these compounds from multilayer nanocapsules will be discussed. Also, a special focus will be directed to the behavior of the multilayer nanocapsules in the GI tract, as a mean to develop optimized multilayer delivery systems for food applications (i.e., edible delivery systems that maximize the bioactive compounds' bioavailability) and to assure their safety. Additionally, the future trends in the multilayer nanocapsules as bioactive compounds delivery vehicles will be briefly discussed.

7.2 DEVELOPMENT OF MULTILAYER NANOCAPSULES

In order to be suitable for incorporation into a food product, an efficient delivery system for bioactive compounds must have the following characteristics (McClements et al. 2007): (1) it must be produced entirely from food-grade ingredients, using processing operations that have regulatory approval; (2) it should be able to allow the incorporation of the bioactive compounds into food products, maintaining bioactive compounds' stability, and having minimal impact on the organoleptic properties of the product; (3) should allow high encapsulation efficiency (EE), absence of cytotoxicity, and should be able to protect the bioactive compounds from interaction with other food ingredients and from degradation (e.g., by heat, pH, or oxidation) during production and shelf-life; (4) should maximize the bioavailability of bioactive compounds; (5) the bioactive compounds release should be controlled and triggered in response to a specific environmental stimulus; and (6) should be easily scalable to industrial production and economically affordable.

Nevertheless, so far nanocapsules do not meet all the above-mentioned requirements and are mostly monofunctional, lacking versatility and multifunctionality (Shchukina and Shchukin 2011). One of the most promising techniques to incorporate additional functionalities into the nanocapsules is the LbL electrostatic deposition technique (Shchukina and Shchukin 2011, Cerqueira et al. 2013, 2014). The LbL makes it possible to obtain multilayer nanocapsules with well-defined chemical and structural properties, offering a precise control over thickness and morphology, while allowing a controlled release of encapsulated hydrophilic and/or lipophilic bioactive compounds (Mora-Huertas et al. 2010, Cui et al. 2014).

7.2.1 LbL Electrostatic Deposition Technique

LbL is frequently viewed as a surface engineering technique aiming at modifying the properties of interfaces, maintaining the properties of the bulk system (Costa et al. 2015). Decher and coworkers first described LbL in the 1990s, where a protocol for development of ultrathin multilayer films on a solid surface was implemented (Decher et al. 1992, Decher 1997). Donath et al. (1998) and Sukhorukov et al. (1998) applied the LbL technique into colloidal templates, transferring this technique from planar films to multilayer capsules (Donath et al. 1998, Sukhorukov et al. 1998, Mora-Huertas et al. 2010, del Mercato et al. 2014).

Briefly, the LbL technique consists of the deposition of polyelectrolytes on charged templates due to strong electrostatic attraction between the surface and the charged polyelectrolyte. A charged polyelectrolyte is added, which adsorbs to the oppositely charged template surface, producing the *first layer* of polyelectrolyte. The adsorption of the following layer is achieved by the simple addition of an oppositely charged polyelectrolyte, promoting the *second layer* adsorption on top of the first layer of polyelectrolyte. This procedure can be repeated to form multilayer nanocapsules coated by interfaces containing three or more layers of polyelectrolytes. The number of layers will be defined by the final properties of these systems (Mora-Huertas et al. 2010, Cerqueira et al. 2014, Martins et al. 2015).

However, between the deposition of new layers around the nanocapsules, it is frequently necessary to assure that few or no free polyelectrolyte is present in the solution. The presence of free polyelectrolyte will interact with the new oppositely charged polyelectrolytes of the subsequent solution. It is therefore necessary to remove any excess of polyelectrolyte in order to avoid interferences with the multilayer formation around the colloidal templates (Guzey and McClements 2006, Szczepanowicz et al. 2015). A number of strategies have thus been developed in order to overcome particles' aggregation (Guzey and McClements 2006, Mora-Huertas et al. 2010, Cui et al. 2014), such as (1) centrifugation method, in which the free polyelectrolytes are separated by performing multiple centrifugation and washing steps; (2) filtration method, in which the excess of free polyelectrolyte is removed from the colloidal templates by membrane filtration; (3) electrophoretic method, using an agarose hydrogel to suspend the colloidal templates and achieving the layer deposition by electrophoresesis of the polyelectrolytes; and (4) saturation method, using precise concentrations of polyelectrolyte, without the need of separation methods.

Usually, the development of multilayer nanocapsules is confirmed through the measurements of the ζ-potential, after each polyelectrolyte adsorption, resulting in a *saw-like* profile (Szczepanowicz et al. 2010a, 2015). Alternatively, fluorescently labeled polyelectrolytes spectroscopy or spectrofluorimetry, confocal microscopy, or FTIR methods can be used to confirm the successful application of the LbL electrostatic technique (Caruso et al. 1998, Berth et al. 2002, Szczepanowicz et al. 2010b, 2015).

7.2.2 Factors Affecting the Layers' Properties

Factors such as pH and ionic strength significantly influence the amount of polyelectrolyte adsorbed in each layer, due to conformation variations. Also, salts are able to rearrange the conformation of the polyelectrolytes, shielding the electrostatic repulsion within the polyelectrolyte chains (del Mercato et al. 2014).

1. *pH*: One of the most essential factors responsible for the formation and properties of multilayer nanocapsules is the solution pH, either during and after formation of the interface. For an appropriate layer deposition, the pH of the solution should be properly selected in order to have sufficiently high opposite charges between the colloidal template surface and the adsorbing polyelectrolyte and also between polyelectrolytes. After multilayer nanocapsules formation, small variations in the pH may adjust the properties (e.g., thickness, packing, and integrity of the new interface) of the polyelectrolyte multilayer interface that surrounds the colloidal templates (Guzey and McClements 2006).
2. *Salt*: The presence of salt during the construction of polyelectrolyte multilayers may affect composition, structure, and thickness of the polyelectrolyte

layers, since the ionic strength of the solution will determine the strength of intra- and intermolecular electrostatic interactions (Decher et al. 1992, Guzey and McClements 2006). An increase in the ionic strength of the solution decreases the magnitude of the electrostatic interactions between polyelectrolytes and colloidal templates, as a result of accumulation of counterions surrounding the surface, this is being referred to as electrostatic screening (McClements 2005, Guzey and McClements 2006). In the presence of salt and due to weaker intramolecular repulsion, polyelectrolytes layers are often thicker, since they have a more compact chain conformation (Guzey and McClements 2006).

3. *Polyelectrolyte characteristics*: The nature of the polyelectrolytes used has a major importance in the formation, stability, and properties of the multilayer nanocapsules. Characteristics such as chain length, rigidity, degree of branching, and electrical charge influence the ability of polyelectrolytes to adsorb to colloidal templates surfaces, reflecting in different thicknesses, structures, porosities, and environmental triggers sensitivity (Guzey and McClements 2006). Characteristics like flexibility and packing will depend on the relative position of the polyelectrolytes within the multilayer, since polyelectrolytes within inner layers are prone to be packed and their motion is restricted by those in the outside layer (Guzey and McClements 2006). Environmental triggers such as pH, ionic strength, and temperature lead to changes in the electrical charge, conformation, and hydrophobicity, which influences the amount of polyelectrolyte adsorbed to the colloidal template surface. Therefore, the proper selection of polyelectrolytes allows creating new interfaces with different environmental responses and may lead to different electrical characteristics of the trapped polyelectrolyte. Finally, the amount of polyelectrolyte should be high enough to saturate the surface of the colloidal templates, reversing the charge of the colloidal templates (or previous layers), in order to create a strong electrostatic repulsion between the multilayer nanocapsules. This amount will be influenced by the molecular characteristics of the polyelectrolyte, such as chain length, conformation, flexibility, and electrical charge (Guzey and McClements 2006).

7.2.3 Colloidal Templates

The proper choice of the colloidal template and polyelectrolytes is crucial for the development of multilayer nanocapsules with distinct properties and therefore suitable for a particular application. The colloidal template choice will determine the size and shape, while the polyelectrolytes will determine the properties of the multilayer nanocapsules (Cui et al. 2014).

7.2.3.1 Hollow Nanocapsules

Multilayer nanocapsules can be obtained using hollow nanocapsules through the adsorption of the polyelectrolyte onto the colloidal template, using the LbL electrostatic deposition technique. The core (hollow nanocapsule) can then be removed by chemical, using acid or solvents, or physical methods. After chemical methods application, usually multiple centrifugation-washing cycles are performed to reassure complete removal of the core.

Polystyrene and silica particles are some examples of the most used colloidal templates. For instance, Shu et al. (2010b) developed multilayer nanocapsules, using amino-functionalized silica particles. In another work, water-soluble chitosan and dextran sulfate were successfully deposited onto the silica core and the silica core can be removed by exposure to an aqueous hydrofluoric acid/ammonium fluoride (HF/NH$_4$F) solution (Shu et al. 2010a). This approach allowed the development of a multilayer nanocapsule with a size of 160 nm.

However, the disadvantage of this method is the low efficiency of encapsulation of the bioactive compound into the hollow nanocapsules and possible presence of traces of the destructed core trapped within the capsule (Szczepanowicz et al. 2014).

7.2.3.2 Liposomes

Liposomes with their ability to encapsulate either lipophilic or hydrophilic bioactive compounds, self-assembly, small sizes (50–400 nm), large surface area, and biocompatibility are attractive templates for multilayer nanocapsules production via LbL. The formation of liposomes is based in a hydrophilic–hydrophobic interactions between phospholipids and water molecules (Fang and Bhandari 2010). This principle results in a variety of liposome development techniques, such as thin film evaporation, sonication, reverse phase evaporation, melting, and freezing-thawing (Fang and Bhandari 2010). Liposomes *per se* are unstable under *in vivo* conditions and present a lack of control over degradability, characteristics that can be minimized using the LbL electrostatic deposition technique to develop multilayer nanocapsules (Städler et al. 2009, Cui et al. 2014). Madrigal-Carballo et al. (2010) developed a delivery system using the LbL electrostatic deposition technique onto soybean lecithin liposomes. In this work, in order to form the multilayers onto the anionic liposomes, positively charged chitosan and negatively charged dextran sulfate were used as polyelectrolytes (Madrigal-Carballo et al. 2010). The resulted multilayer nanocapsules exhibited a size of 386.5 nm and better stability against temperature and pH. Cuomo and coworkers (2012) used liposome templates composed of phosphatidylcholine (65%) and didodecyldimethylammonium bromide (35%) with 80 nm of diameter for the development of multilayer nanocapsules by LbL, using alginate and chitosan. In this work, the liposome template was removed after the development of the multilayer nanocapsule, using Triton X100, a nonionic surfactant, recognized for being a solubilizing agent for membranes (Cuomo et al. 2012). This work showed that multilayer nanocapsules may be developed to be pH responsive depending on the final layer, either alginate or chitosan.

7.2.3.3 Nanoemulsions

Nanoemulsions could also be used as a colloidal template for the development of multilayer nanocapsules through LbL deposition technique. Nanoemulsions can either be formed by high-energy methods (high-pressure homogenization or ultrasonication) or low energy methods (membrane emulsification or spontaneous emulsification) (Silva et al. 2012). Adamczak and coworkers (2014) successfully developed multilayer nanocapsules using nanoemulsions as colloidal templates. Briefly, squalene nanoemulsions were prepared using the membrane emulsification method with 80 nm of diameter. The multilayer nanocapsules formation was achieved by the LbL technique, using the saturation method, until the formation of 10 layers, with 120 nm of diameter (Adamczak et al. 2014). Bazylinska and coworkers (2011) developed multilayer nanocapsules using nanoemulsion templates prepared by the conjugation of spontaneous emulsification followed by ultrasonication, achieving a template with less than 100 nm of diameter (Bazylinska et al. 2011). Nanoemulsion templates were then surrounded by LbL adsorption of poly(sodium 4-styrenesulfonate) and poly(diallyldimethylammonium chloride) using the saturation method, in order to form multilayer nanocapsules with less than 200 nm of diameter (Bazylinska et al. 2011). These authors developed multilayer nanocapsules with tunable thickness and permeability based in different polyelectrolytes.

The formation of the multilayer polyelectrolyte shell on the diverse colloidal template is schematically shown in Figure 7.1.

7.2.4 Advantages and Limitations of Multilayer Nanocapsules

A benefit of using multilayer nanocapsules as delivery systems is the fact that they can be tailor-made from natural food-grade components (lipids, proteins, polysaccharides) using simple processing operations (homogenization, mixing).

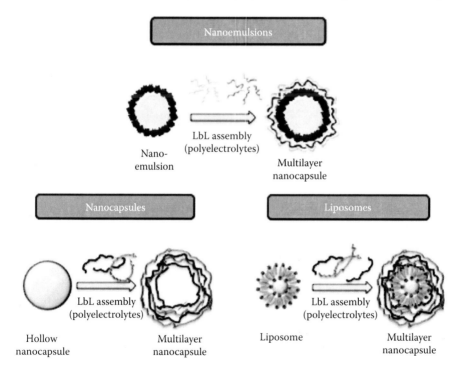

FIGURE 7.1 Representative scheme of multilayer nanocapsules after LbL deposition onto the different colloidal templates.

The physical stability to environmental stresses can be improved by controlling the composition and properties of the interfacial layers (Decher 2003, Aoki et al. 2005, Guzey and McClements 2006, Grigoriev and Miller 2009, Costa et al. 2015). The chemical stability of the bioactive compounds can also be improved by controlling the thickness and charge of the interfacial layer, which minimizes the interaction between the bioactive compounds and transition metal ions (McClements and Decker 2000, McClements et al. 2007, Klinkesorn et al. 2005). Due to the ability to control the thickness and properties of the interfacial layer by varying the number of the adsorbed layers, it is possible to build a better delivery system with a great control of the release rate (Decher 2003, Costa et al. 2015). The multilayer structure of these nanocapsules enables the combination of different properties in one system, mostly governed by the material. They can also be designed to control the digestion and absorption of bioactive compounds within the GI tract, while controlling the release of bioactive compounds at specific locations of the GI tract (mouth, stomach, small intestine, or colon), due to the ability to trigger that release in response to specific environmental conditions (pH, temperature, ionic strength) (McClements et al. 2007, Li et al. 2010, McClements 2010, McClements and Li 2010).

Therefore, the LbL technique offers a promising way to improve the stability of the actual delivery systems. Nevertheless, the optimization of this technique depends on the preparation conditions (bioactive compound concentration, pH, ionic strength, temperature), the choice of appropriate polyelectrolytes and their combination, as well as the knowledge of the physicochemical principles involved. Also, multilayer nanocapsules construction implies long times, with the assembly of a single layer taking typically a few minutes, depending on the nature of the polyelectrolytes (Costa et al. 2015) and the formation of stable multilayer nanocapsules requires meticulous control of the system composition and preparation in order to avoid droplet aggregation (Guzey and McClements 2006). Another limitation is the fact that LbL method is based on electrostatic interactions, and they can be disrupted by changes in pH and ionic strength (McClements 2010, McClements and Li 2010). Also, depending on the template used, the removal of the core template could be required, leaving traces of solvent (Guzey and McClements 2006, McClements et al. 2007, Li et al. 2010, McClements 2010, McClements and Li 2010, Szczepanowicz et al. 2015). Despite the method for producing multilayer nanocapsules being simple, additional processing steps are frequently needed over the nanocapsule formation and diluted multilayer nanocapsules are typically obtained (McClements et al. 2007).

Although LbL devices did not receive significant attention from a scale-up perspective, efforts in developing devices able to produce multilayer nanocapsules in an automated way have been reported (Priest et al. 2008, Kantak et al. 2011, Costa et al. 2015).

7.3 MULTILAYER NANOCAPSULES AS CARRIERS OF BIOACTIVE COMPOUNDS

Multilayer nanocapsules may be used to improve the stability, solubility, cellular uptake, and bioavailability of bioactive compounds, while providing their controlled release, for enhanced functionality.

7.3.1 BIOACTIVE COMPOUNDS

Poorly water-soluble bioactive compounds are the ones that can benefit more from nanoencapsulation, once it contributes to the improvement of their dispersibility in aqueous solutions, as well as to the improvement of absorption and bioavailability.

Table 7.1 shows some examples of bioactive compounds (both lipophilic and hydrophilic) of interest to be encapsulated with potential applications in food and beverage products.

The choice of the multilayer delivery system for a particular bioactive compound should be done in a case-by-case basis, considering their nature (e.g., hydrophilic or lipophilic) and the intended characteristics for the delivery system (e.g., permeability, release rate).

7.3.2 ENCAPSULATION TECHNIQUES IN MULTILAYER NANOCAPSULES

A number of techniques have been proposed to incorporate bioactive compounds within multilayer nanocapsules. Loading of the compound can be achieved during different steps of the LbL method: before the template formation, onto pre-formed multilayer nanocapsules, during the formation of the layers or using the bioactive compound as a template (in the case of bioactive compounds that form crystals) (del Mercato et al. 2014) (Figure 7.2).

TABLE 7.1
Examples of Bioactive Compounds of Interest to Be Encapsulated and Their Potential Benefits

Bioactive Compound	Type	Potential Benefit	Reference
ω-3 Fatty acids	Lipophilic	Anticarcinogenic, prevent heart diseases	Chalothorn and Warisnoicharoen (2012)
α-Tocopherol	Lipophilic	Antioxidant	Hatanaka et al. (2010)
Curcumin	Lipophilic	Antioxidant, anti-inflammatory, and anticarcinogenic	Ahmed et al. (2012)
β-Carotene	Lipophilic	Antioxidant, vitamin A precursor	Silva et al. (2011)
Phytosterols	Lipophilic	Cholesterol absorption inhibitor	Garti et al. (2005)
Quercetin	Lipophilic	Anti-inflammatory, antioxidant	Chen-yu et al. (2012)
Carvacrol	Lipophilic	Antimicrobial activity	Donsì et al. (2012)
Resveratrol	Lipophilic	Antioxidant, antimicrobial, anti-inflammatory, chemopreventive, and anti-cancer activity	Donsì et al. (2011)
Poly-L-lysine	Hydrophilic	Antimicrobial activity	Pinheiro et al. (2015)
Vitamin C	Hydrophilic	Antioxidant	Marsanasco et al. (2011)
Polyphenols	Lipophilic	Antioxidants, antimutagens, anticarcinogens, and antimicrobial agents	El Gharras (2009)
Bioactive peptides	Hydrophilic	Antihypertensive properties	Bourbon et al. (2011)

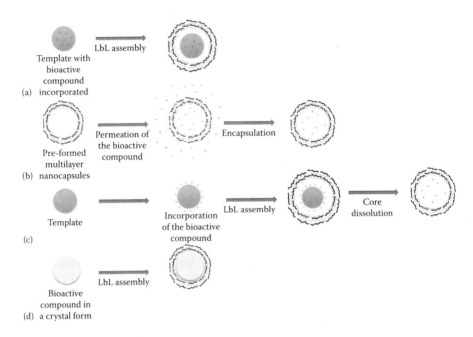

FIGURE 7.2 Different methods for encapsulation of a bioactive compound in multilayer nanocapsules: (a) before the template formation, (b) loading pre-formed multilayer nanocapsules, (c) during the LbL assembly, and (d) encapsulation of crystalline bioactive compounds by LbL technique.

In the case of using nanoemulsions as templates for multilayer formation, the bioactive compound is previously dispersed in the lipid matrix, which is then homogenized with an aqueous phase containing a water-soluble emulsifier. After nanoemulsion formation, the LbL technique is applied originating a multilayer nanoemulsion (Pinheiro et al. 2016). Similarly, for multilayer liposomes, the bioactive compound is added before the formation of the primary liposome, which is subsequently coated (Fukui and Fujimoto 2009).

One of the most used approaches is the diffusion of the bioactive molecules from the surrounding medium in which the nanocapsules are dispersed into their interior. When the polyelectrolyte nanocapsule shell is assembled from polyelectrolytes that are responsive to salt, pores within the shell can open to allow molecules to diffuse into the capsules (Sukhorukov et al. 2001). The limitations of this technique are the low loadings achieved (once the maximum concentration of bioactive compounds inside the nanocapsules is often limited to the concentration in the solution) and the molecular weight of diffusing molecules should be low enough to allow their permeation through the nanocapsule pores (Johnston et al. 2006). Another commonly used technique for encapsulating bioactive compounds involves adsorbing the bioactive compound within a porous nanoparticle that can be subsequently LbL coated and subsequently removing the porous template. This technique allows a high loading of the bioactive compound (due to the high surface area of the porous particle) and is applicable to a range of materials of different sizes (Johnston et al. 2006).

In cases where the bioactive compound forms crystals, polyelectrolyte multilayers can be directly assembled on the crystal template, encapsulating the compound (Caruso et al. 2000). The main advantage of this technique is that a very high concentration of compound is encapsulated. However, this strategy is most effective when the bioactive compound is insoluble or poorly soluble in water (once multilayers are typically deposited from aqueous solutions), and it is limited to compounds that form crystals (Johnston et al. 2006).

The selection of the loading approach should be done according to the physicochemical properties of the bioactive compound (e.g., molecular weight, solubility, and shape).

7.3.3 EE AND LOADING CAPACITY

It is intended that a delivery system encapsulate the maximum possible quantity of a bioactive compound in order to obtain optimum efficacy and cost-effectiveness. To evaluate the suitability of a multilayer nanocapsule to act as a carrier for a particular bioactive compound, the EE and the loading capacity (LC) are often determined. EE corresponds to the ratio between the bioactive compound encapsulated in a carrier system and the bioactive compound added; whereas, LC is the ratio between the bioactive compound encapsulated and the weight of total carrier system. Therefore, EE and LC can be determined as follows (Shu et al. 2010a):

$$EE\% = \frac{\text{total bioactive compound} - \text{free bioactive compound}}{\text{total bioactive compound}} \times 100 \quad (7.1)$$

$$LC\% = \frac{\text{total bioactive compound} - \text{free bioactive compound}}{\text{nanostructure weight}} \times 100 \quad (7.2)$$

Some examples of bio-based multilayer nanocapsules encapsulating bioactive compounds, as well as their size, EE and LC can be seen in Table 7.2.

The encapsulation efficiency can be improved by optimizing the multilayer nanocapsules formulation and encapsulation procedure, by evaluating the variables that affect the bioactive compounds' entrapment. Pinheiro et al. (2015) has shown that the EE and the LC of chitosan/fucoidan nanocapsules strongly depend on the initial concentration of the bioactive compound and also on the encapsulation method. Generally, the loadings achieved were higher when the bioactive compound was loaded during the nanocapsules' preparation (LbL assembly) in comparison with loading in a pre-formed multilayer nanocapsule. Other authors showed the effect of the template used: the EE of BSA on multilayer nanocapsules was higher using amino-functionalized silica as template when compared with bare silica, amino-functionalized silica adsorbed BSA by electrostatic interaction, while the bare silica did it mainly by physical adsorption through many hydrogen bonds (Shu et al. 2010a).

TABLE 7.2
Selected Examples of the Encapsulation of Bioactive Compounds in Multilayer Nanocapsules for Food Applications

Type of Nanocapsule	Template	Multilayer Compositions	Size (nm)	Bioactive Compound Encapsulated	EE (%)	LC (%)	Reference
Hollow multilayer nanocapsules	Polystyrene nanoparticles	Chitosan and fucoidan	≈50	Poly-L-lysine	45.1 ± 1.5	16.0 ± 1.0	Pinheiro et al. (2015)
	Amino-functionalized silica particles	Chitosan and alginate	≈147	Glycomacropeptide	49.4 ± 9.2	—	Rivera et al. (2015)
	Silica nanoparticles	Water-soluble chitosan and dextran sulfate	≈170	Bovine serum albumin	90–60	30–40	Shu et al. (2010a)
		Chitosan and dextran sulfate	180 ± 20	Ciprofloxacin and ceftriaxone	—	—	Gnanadhas et al. (2013b)
Multilayer liposomes	Liposomes composed of DLPA and DMPC	Chitosan and dextran sulfate	≈100	Glucose	—	—	Fukui and Fujimoto (2009)
Multilayer nanoemulsions	Nanoemulsions stabilized by OSA-modified starch	Chitosan and carboxymethylcellulose	159.85 ± 0.92	Curcumin	—	—	Abbas et al. (2015)

DLPA, dilauroyl phosphatidic acid; DMPC, dimyristoyl phosphatidylcholine; OSA, octenyl succinic anhydride.

7.4 IMPACT OF MULTILAYERS IN THE RELEASE OF BIOACTIVE COMPOUNDS

Controlled release technologies are becoming more and more relevant in food industry (Khare and Vasisht 2014). Release-tailored bioactive compounds have numerous advantageous in food products such as prolonging time of activity of bioactive compounds (e.g., vitamins, antioxidants) to protect food product and/or be released in human tract during the digestion process, protecting sensitive bioactive compounds from enzymatic or acidic degradation in the GI tract, taste masking, and so on (Soppimath et al. 2001, Pinheiro et al. 2012, 2013a, Khare and Vasisht 2014). In controlled bioactive compounds release, one of the great challenges related with carriers is the prominent initial burst and incomplete unloading of encapsulated compounds. In the former case, a large portion of the active compound is lost in a short period, which may cause acute toxicity and failure of controlled release. In the latter case, a portion of bioactive compound will remain within the carriers without release, which may affect the efficiency of the encapsulated bioactive compounds (Qiu et al. 2001).

More recently, multilayer nanocapsules have been investigated as bioactive compounds release systems (Sukhorukov et al. 1998, Shu et al. 2010a, Pinheiro et al. 2015, Rivera et al. 2015) due to the possibility to control compounds release through manipulating the layers properties and to incorporate a wide range of functional molecules without substantial loss of their biological functions. Additionally, controlled release of multilayer nanocapsules systems have been developed and studied to improve the performance of bioactive compounds and in particular to increase their effect, and be delivered at specific conditions.

A number of design options are available to control or modulate bioactive compounds release from delivery systems (Qiu et al. 2001). Physicochemical properties of coating material and bioactive compounds (molecular weight, size, and hydrophilic or hydrophobic nature), the number of layers, the position of active compound in multilayer nanocapsule, and the thickness of layer are some of the relevant parameters that are believed to have an influence on release properties (Pinheiro et al. 2015, Rivera et al. 2015). Table 7.3 shows examples from literature, revealing the numerous possibilities of using multilayer nanocapsules as controlled released systems.

The release of bioactive compounds from polymeric matrices may occur due to mechanisms of Fick's diffusion, polymer matrix swelling, polymer erosion, and degradation (Faisant et al. 2002) and a different mechanism may prevail, depending on the system and environmental conditions. Understanding the transport properties and how they are related with systems' structural rearrangement may be crucial. Moreover, understanding the transport of bioactive compounds in multilayer nanocapsules can provide valuable information for the incorporation of compounds that are able to extend the products' shelf life.

To date, there is scarce information of release mechanisms involved at the nanoscale. Few works using multilayer nanocapsules evaluate the release mechanisms involved during bioactive compounds release (Pinheiro et al. 2012, Rivera et al. 2015). The release behavior of multilayer systems depends on the permeability and on the disassembly or erosion of the multilayer structure and on other

TABLE 7.3
Selected Examples of Bioactive Compounds Release in Multilayer Nanocapsules

Main Ingredients of Multilayer Nanocapsules	Bioactive Compounds	Number of Layers	Release Mechanism	Reference
Chitosan and fucoidan	Poly-L-lysine	10	Anomalous behavior	Pinheiro et al. (2015)
Chitosan and alginate	5-aminosalycilic acid and glycomacropeptide	6	Anomalous behavior	Rivera et al. (2015)
Chitosan and sodium alginate	Acridine hydrochloride	10	Diffusion mechanism	Ye et al. (2005)
Poly(L-glutamic acid) and chitosan	5-Fluorouracil	10	Diffusion mechanism	Yan et al. (2011)
Dextran and chitosan	Polyphenols	16	Diffusion mechanism	Paini et al. (2015)
Chitosan and dextran sulfate	Bovine serum albumin	8	Diffusion mechanism	Shu et al. (2010a)

experimental variables (Jiang and Li 2009). Literature suggests that, depending on nanolayered coatings composition and on the released compound, release may follow a Fickian or an anomalous transport behavior. Also, some authors attribute the observed behavior to a Fickian transport through the nanolayers, followed by release due to polymer dissolution (Antipov et al. 2001). However, there are still a lot of unknown variables that influence the release of compounds from multilayer nanocapsules, especially regarding the effect of environmental conditions.

Mathematical models have been used to predict the release of the encapsulated molecule(s). These models add value in terms of ensuring optimal design of food formulation(s), as well as to understand release mechanism(s) through experimental verification (Peppas and Narasimhan 2014). Only combining precise experimental observations with models that capture the underlying physics will provide new insights into the release mechanism.

The mechanisms associated with mass transport in polymeric systems may be generally classified in three different types: ideal Fickian diffusion, anomalous behavior, and Case II transport. Fickian diffusion involves a substantially stochastic phenomenon (related to Brownian motion), in which the penetrant flow is exclusively driven by a concentration gradient. Case II transport, describes a behavior where compound release is due to a relaxation phenomenon driven by the distance of the local system from the equilibrium. Anomalous diffusion is used to identify behaviors departing from Fickian and Case II transport (Del Nobile et al. 1994).

Many existing models are based on diffusion equations. The diffusion of bioactive compounds is a strong function of the structure through which diffusion takes place, models need to account for polymer morphology (Peppas and Narasimhan 2014).

One of the most often used mathematical models to describe the release rate of compounds from matrix is the Higuchi equation:

$$\frac{M_t}{M_\infty} = k \times t^{0.5} \tag{7.3}$$

where:
M_t/M_∞ is the fraction of compounds released at time t
k is a constant reflecting the design variables of the system

Thus, the fraction of compound released is proportional to the square root of the time. The simplicity of this model is an advantage, however, this model also has limitations such as (1) mathematical analysis is based on one-dimensional diffusion; (2) it assumes that during the release experiment, the matrix system is not affected by swelling; (3) the diffusivity of the compound is constant; and (4) perfect sink conditions are maintained (Siepmann and Peppas 2001).

The power law model is also a semiempirical model purposed by Korsmeyer and Peppas (1981):

$$\frac{M_t}{M_\infty} = k \times t^n \tag{7.4}$$

where:
M_t and M_∞ are the cumulative amounts of compound released at time t and infinite time, respectively
k is a constant that contains the structural and geometric information about the system
n is the indicative of the bioactive compound release mechanism

Literature reports that, for spherical geometries, the following extreme values of n should be considered: $n = 0.43$ for Fickian diffusion, $0.43 < n < 0.85$ for anomalous transport, and $n = 0.85$ for Case II transport (Siepmann and Peppas 2001).

The properties of polymeric matrix are extremely relevant in release mechanism of bioactive compounds. When immersed in liquid media, hydrophilic polymers gradually start to hydrate, causing relaxation of the polymer chain with consequent volume expansion, that is, swelling, and therefore this will influence transport mechanism of compounds from the polymeric matrix (Berens and Hopfenberg 1978, Siepmann et al. 2002, Flores et al. 2007).

The linear superposition model is a mathematical model that accounts for both Fickian and Case II transport effects on the observed anomalous behavior in hydrophilic matrices (Berens and Hopfenberg 1978). The linear superposition approach assumes that the observed transport of molecules within the polymer can be described by the sum of the molecules transported due to Brownian motion and the molecules transported due to polymer relaxation (Berens and Hopfenberg 1978):

$$M_t = M_{t,F} + M_{t,R} \tag{7.5}$$

where $M_{t,F}$ and $M_{t,R}$ are the contributions of the Fickian and relaxation processes, respectively, at time t. Hence, compounds release from a hydrophilic polymer capsule can be described by

$$M_{t,F} = M_{\infty,F}\left[1 - \frac{6}{\pi^2}\sum_{n=1}^{\infty}\frac{1}{n^2}\exp(-n^2 k_F t)\right] \quad (7.6)$$

where:
$M_{\infty,F}$ is the compound released at equilibrium
k_F is the Fickian diffusion rate constant

Regarding polymer relaxation, it is dependent on the polymer swelling ability and related to the dissipation of stress induced by entry of the penetrant and can be described as a distribution of relaxation times, each following a first-order kinetic type equation (Berens and Hopfenberg 1978):

$$M_{t,R} = \sum_{i} M_{\infty,R_i}\left[1 - \exp(-k_{R_i} t)\right] \quad (7.7)$$

where, M_{∞,R_i} are the contributions of the relaxation processes for compound release and k_{R_i} are the relaxation ith rate constants. For most cases, there is only one main polymer relaxation that influences transport and thus the above Equations 7.6 or 7.7 can be simplified using the first term of the Taylor series (Jeong et al. 1999).

Therefore, the linear superposition model can be described by

$$M_t = M_{\infty,F}\left[1 - \frac{6}{\pi^2}\sum_{n=1}^{\infty}\frac{1}{n^2}\exp(-n^2 k_F t)\right] + M_{\infty,R}\left[1 - \exp(-k_R t)\right] \quad (7.8)$$

Using the appropriate modification, this *general* model can then be used to describe Fickian ($M_{\infty,F} \neq 0$ and $i = 0$); anomalous ($M_{\infty,F}$ and $i \neq 0$) or Case II transport ($M_{\infty,F} = 0$ and $i \neq 0$).

Recently, Pinheiro et al. (2015) evaluated the poly-L-lysine (PLL) release from chitosan/fucoidan multilayer nanocapsule at different pH values applying Fick's second law model (Equation 7.6) and linear superposition model (Equation 7.8). These authors observed that the physical mechanisms involved in the release of PLL were not successfully described by Brownian motion of PLL in the nanocapsules, that is, it does not strictly follow Fick's behavior but was governed by both Fickian and Case II transport (anomalous behavior), with only one main relaxation of the nanocapsules. The linear superposition model was able to adequately describe the experimental results (Figure 7.3).

Moreover, the PLL release was found to be pH-dependent: at acidic conditions (pH = 2) relaxation was the governing phenomenon and at neutral conditions (pH = 7) Fick's diffusion was the main mechanism of PLL release (Pinheiro et al. 2015).

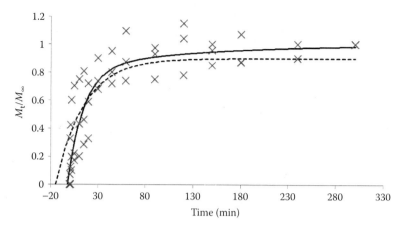

FIGURE 7.3 Profile of PLL release from chitosan/fucoidan nanocapsules at 37°C and pH 2: experimental data (×); description of Fick's model ($i = 0$) (–) and of linear superposition model ($i = 1$) (– –). (Reprinted from *Carbohydrate Polymers*, 115, Pinheiro, A.C., Bourbon, A.I., Cerqueira, M.A., Maricato, É., Nunes, C., Coimbra, M.A., and Vicente, A.A., Chitosan/fucoidan multilayer nanocapsules as a vehicle for controlled release of bioactive compounds, 1–9. Copyright 2015, with permission from Elsevier.)

Also, Rivera et al. (2015) used the linear superposition model to describe the release behavior of two bioactive compounds (5-aminosalicylic acid [5-ASA] and glycomacropeptide [GMP]) from chitosan/alginate multilayer nanocapsules. These authors concluded that the bioactive compounds release was due to both Brownian motion and the polymer relaxation of the layers. Moreover, they observed that the environmental conditions (i.e., temperature and pH) influenced the quantity of bioactive compounds (5-ASA and GMP) released. Depending on the environmental conditions, the electrostatic interactions between layers can be changed and different release behaviors can be observed. These authors also observed that the decrease of electrostatic interactions facilitates loosening of the polymer chain network, thus promoting release due to relaxation phenomena (Rivera et al. 2015).

Finally, it is extremely important to understand the mechanisms involved in the release behavior of active compounds to design new release systems triggered in specific conditions. Detailed studies will allow customization of systems' properties that provide a desired rate of release for encapsulated molecules corresponding to their optimal biological and bioactive effect, which will facilitate the application of multilayer nanocapsules in bioactive compounds formulation and delivery. In the future, this area of research should focus on simulating release in GI conditions, which are one of the applications of multilayer nanocapsules.

7.5 BEHAVIOR OF MULTILAYER NANOCAPSULES IN THE GI TRACT

Most of the multilayer nanocapsules are being developed to control the release of bioactive compounds within the GI tract after their ingestion (Liu et al. 2013, Li and McClements 2014). However, when developing nanocapsules it is necessary to consider different challenges and barriers in GI tract to prevent earlier degradation or release of bioactive compound (McClements 2010, Zhang et al. 2015):

1. Some bioactive compounds present a strong flavor so it is necessary to mask the flavor in the mouth. Multilayer nanocapsules could be an excellent option to fulfill this requirement.
2. As previously stated, many bioactive compounds need to be protected from the harsh GI tract environmental conditions (e.g., acid pH, presence of enzymes) once they are susceptible to biochemical degradation within the GI tract.
3. Ideally, the entrapped bioactive compound is ultimately released into specific body site either in the intestine, during systemic circulation, or in the cells of organ tissues. For instance, a nutraceutical should be released in the small intestine, on the other hand, a flavor should be released in the mouth.

Considering the foremost issues, it is essential to understand and to become aware of the physiological and physicochemical processes that occur when nanoencapsulated bioactive compounds pass through the different sections of the GI tract.

7.5.1 GI Tract Main Features and Challenges

GI tract delivery of bioactive compounds using multilayer nanocapsules is a great challenge owing to several physiological and morphological barriers (e.g., pH variation, enzymes present in the gut lumen, the mucus layer) that play a key role in the bioactive compounds uptake.

GI tract is essentially divided into two parts: the upper part comprises mouth, pharynx, esophagus, and stomach; and the lower part includes small intestine (duodenum, jejunum, and ileum), large intestine (cecum, colon, and rectum), and anus (Bellmann et al. 2015). More than 90% of the nutrients comprising minerals, proteins, carbohydrates, fats, and vitamins are absorbed in the small intestine (Vander et al. 2001). The main benefits associated with colon delivery are related with the colon characteristics: (1) a pH close to neutral, (2) a long transit time, (3) a reduced digestive enzyme activity, and (4) increased responsiveness to absorption enhancers (Singh et al. 2014). On the other hand, the colon is difficult to access due to its position at the distal part of GI tract. Moreover, bioactive compounds have to move through GI tract before reaching the target site, and consequently, pass through a wide range of pH values and different enzymes present in GI tract, which could further decrease the delivery efficiency. Various approaches employed for multilayer nanocapsules delivering bioactive compounds to the colon include the use of swellable polymers and polysaccharides (Benshitrit et al. 2012).

The mouth is the first entry of the food in the human body, where food (liquid or solid) undergoes a series of mechanical and physicochemical processes. Food products may be subject to (1) temperature, pH, and ionic strength changes; (2) particle size reduction (due to mastication); (3) action of several enzymes (e.g., amylase, protease); and (4) interactions with tongue and mouth receptors. In this site, an intensive mixing of saliva is essential to produce the food bolus. Saliva is a fluid that contains various proteins (e.g., digestive enzymes, glycoproteins, albumin) and minerals. Increasing the surface area of hydration and accessibility to the action of digestive enzymes will lead to an increase of overall digestion efficiency and GI tract absorption of bioactive compounds (McClements 2013).

Food fragmentation in the stomach is a process that includes mechanical action and gastric juice activity. This juice contains enzymes (pepsinogens, lipase), hydrochloric acid, and electrolytes. The denaturation of proteins and activation of pepsin is a consequence of the decrease of pH due to the presence of hydrochloric acid. Also, peristaltic movements in the stomach are responsible for size reduction of particles and food (Zhang et al. 2015). Gastric juice released from the stomach wall inactivates acid- and protease-labile bioactive compounds; protection can be achieved by entrapment in multilayer nanocapsules. The release of the bioactive compounds (e.g., omega-3 fatty acids) from multilayer nanocapsules in the stomach could maximize the potential for absorption in the small intestine (Ilyasoglu and El 2014).

After leaving the stomach, any multilayer nanocapsules remaining in the chyme enter the small intestinal lumen, where they are mixed with alkaline small intestinal fluids containing bile salts, phospholipids, digestive enzymes (e.g., proteases, amylases, and lipases), and various salts, being the higher pH responsible for the dissolution of macromolecules (e.g., peptides, proteins) (Joye et al. 2014). To enter the systemic circulation, bioactive compounds need to cross the epithelium and reach the capillary network in the lamina propria. The epithelium is a monolayer of polarized cells covered by mucus and arranged into crypts and villi with microvilli, which increase the surface area of the small intestine (Ensign et al. 2012). It is advantageous to have bioactive compounds released from nanocapsules in the small intestine in order to increase their bioavailability. For instance, lipophilic bioactive compounds (e.g., vitamins, carotenoids, and fatty acids) have limited solubility in the GI fluid, thus decreasing their bioavailability. Therefore, these compounds need to be released within the small intestine, solubilized within mixed micelles, and then transported to the surfaces of the intestinal epithelium cells where they are absorbed (Zhang et al. 2015).

Additionally, the intestinal mucosal surface of the human GI tract is colonized by a variety of bacteria. The human microbiota plays a central role in intestinal health and disease, once it is responsible for several activities (e.g., protection against exogenous pathogenic bacteria, digestion of polysaccharides, enzymatic hydrolysis, and modulation of immune system) (Bergin and Witzmann 2013, Kovatcheva-Datchary and Arora 2013). GI microbiota can also play an important role on the interaction with nanocapsules. For instance, microbiota adhesion to gut wall can result in biofilm formation due to exopolymer secretions and subsequent gut protection from contact with nanocapsules (Neal 2008).

In recent years, different *in vitro* digestion models have been used to understand the physicochemical behavior of nanostructures within the GI tract—nanoemulsions (Pinheiro et al. 2013b, Salvia-Trujillo et al. 2013). The models mostly diverge in the inclusion of different stages of digestion (i.e., mouth, gastric, small intestinal, large intestinal); digestion times; concentrations of electrolytes; bile acids' pH; and digestive enzymes involved. The models are operated in static conditions (with prefixed concentrations and volumes of digested materials, enzymes, salts, etc.), or in dynamic conditions that mimic the continuous changes of the physicochemical conditions

(Kostewicz et al. 2014). However, few studies have been conducted to assess the behavior of multilayered nanocapsules under simulated human GI conditions. The enzymatic digestion stability (i.e., particle size, surface charge, free fatty acids, and model functional component release) of polyelectrolytes delivery system (PDS) based on sodium alginate and chitosan coated on the surface of nanoliposomes using the LbL deposition technique was studied (Liu et al. 2013). These authors showed that PDS could better resist lipolytic degradation by the coverage blocking of outer layer polymers and delayed release of encapsulated content, as compared with nanoliposomes. Also, PDS facilitated a lower level of encapsulated component release in simulated GI conditions. Thus, PDS could be developed as a promising delivery system for oral administration. Other authors produced multilayered nanocapsules encapsulating insulin consisting of alginate nucleated around calcium ions and associated with dextran sulfate and poloxamer, stabilized with chitosan and coated with albumin (Woitiski et al. 2010). They found that nanocapsules retained insulin in gastric fluid for 120 min, and over 95% of the insulin was released in the intestinal fluid during the following 180 min. Also, neutral pH triggers insulin release as nanocapsules pass from acidic gastric medium to neutral intestinal fluid. Additionally, nanocapsules retained insulin bioactivity and bioavailability by protecting insulin from enzymatic degradation, and facilitating permeation through the intestinal membrane.

7.5.2 Uptake/Absorption of Bioactive Compounds

The uptake of multilayer nanocapsules in the GI tract and subsequent bioactive compounds delivery depends on accessibility to the gut epithelium, diffusion through mucus layer, and uptake/translocation processes (Martins et al. 2015).

The main route for nanocapsules uptake has been considered the follicle-associated epithelium (FAE) in small intestine due to the presence of gut-associated lymphoid tissue (Peyer's patches). Three possible transport routes for nanocapsules in FAE are (1) paracellular transport, (2) transcellular uptake across enterocytes by endocytic process, and (3) endocytosis by M cells. Particle size and surface charge are important factors related to the uptake of nanocapsules (Plapied et al. 2011). The literature described several experimental *in vitro, ex vivo,* and *in vivo* models to investigate the interaction between nanocapsules and GI tract—bioadhesion and uptake of nanocapsules (Groo and Lagarce 2014).

GI digestion is a dynamic process, in which the nutrients are constantly moving to following digestion and absorption sites. Inadequate gastric retention time could result in the deficient absorption of bioactive compounds, compound excretion, and decrease in the efficiency of compound therapeutic purpose. Moreover, mucus covering the epithelial cell surface could hinder the diffusion of bioactive compounds. Therefore, the production of nanocapsules with higher viscosity or ability to slow down the gastric movement prolongs the residence time in the GI tract and allows higher percentage of bioactive compounds to be absorbed before gastric emptying (Pawar et al. 2014).

Different approaches have been applied to enhance permeation of bioactive compounds by oral route, such as absorption enhancers, mucoadhesive polymers, and enzymatic inhibitors (Mrsny 2012). Absorption enhancers directly transport bioactive compounds through the epithelium once they act on tight junctions opening, modulating paracellular transport of bioactive compounds, or increase transcellular permeation through membrane perturbation. A range of materials such as chitosan, quercetin, and curcumin have been shown to enhance intestinal permeability once they promote changes in the physical barrier function of the intestinal wall (Martins et al. 2015). For example, chitosan, a positively charged mucoadhesive polymer affects intestinal membrane integrity and tight junction widening to allow the paracellular absorption of lipophilic compounds (Du et al. 2015). Surfactants also increase transcellular transport by disturbing the lipid bilayer and making it more permeable to bioactive compounds (Pawar et al. 2014). One of the drawbacks of absorption enhancers is the potential damage of the membrane, and consequently local inflammation, if there is a continuing usage of these compounds (Muheem et al. 2014). Mucoadhesive polymer adheres to the mucus through nonspecific (van der Waals and hydrophobic interactions) or specific interactions and increases the bioactive compound concentration gradient (Sosnik et al. 2014). The most common mucoadhesive polymers are chitosan, alginate, poly (lactic-*co*-glycolic acid) (PLGA), thiolated polymer, which have been studied extensively (Sosnik and Menaker Raskin 2015). Woitiski and his team (Woitiski et al. 2010, 2011) studied the permeation enhancing properties of the mucoadhesive multilayered nanocapsules composed of alginate, dextran sulfate, and poloxamer 188. Insulin was incorporated into the nanocapsule and subsequently coated with chitosan and albumin. It was concluded that insulin was protected against enzymatic digestion and its absorption enhanced. *In vitro* studies showed that insulin transport was increased by the multilayered nanocapsules (compared to free insulin) as much as 2.1-fold through Caco-2 cell monolayer, 3.7-fold through Caco-2/HT29 coculture (mucus secreting), and 3.9-fold through rat intestinal models. This resulted in sustained (>24 h) hypoglycemia (up to 40% of basal levels) in diabetic rats, as well as 13% bioavailability.

The insertion of enzyme inhibitors in the multilayer nanocapsules may significantly enhance the concentration level of bioactive compounds in the systemic circulation. A range of enzyme inhibitors could be used such as aprotinin and soybean trypsin inhibitors, camostat mesilate, and chromostatin. However, the use of such enzyme inhibitors for long duration may need more cautious thoughts to avoid deficiency of enzymes in humans (Truong-Le et al. 2015).

7.5.3 Changes in Multilayer Nanocapsules Delivery Properties—Potential Toxicity Effects

Different physiological and physicochemical scenarios (pH, ionic composition, digestive enzymes, etc.) within the GI tract, from consumption stage to body elimination, potentially modify original ingested multilayer nanocapsules characteristics (Figure 7.4) such as thickness, charge, and size (Yada et al. 2014).

Changes of surface properties through nanocapsules coating to prevent agglomeration or aggregation with different types and concentrations of materials (e.g., surfactants) have been shown to change their body distribution and the effects on the biological systems significantly (Mora-Huertas et al. 2010). The size of nanocapsules alone may not be the critical factor determining their toxicity; when size decreases, the surface area increases (per unit mass only) and a greater proportion of molecules

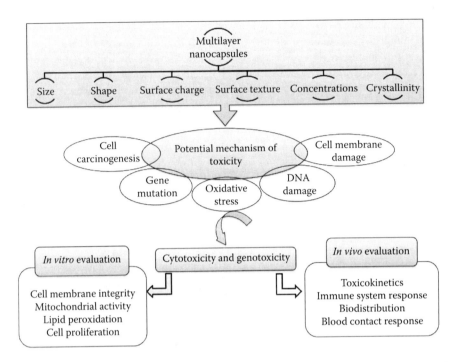

FIGURE 7.4 Potential influence of multilayer nanocapsules characteristics on toxicity and examples of evaluation methods of their behavior in biological systems.

are found at the surface. Thus, nanocapsules have a much larger specific surface area, and an increase in the surface-to-volume ratio will result in an increase of surface energy, which may lead to biologically reactive structures (Oberdörster et al. 2005). The influence of the number of polyelectrolyte layers in the nanocapsule shell and surface charge on cell viability have been studied (Łukasiewicz and Szczepanowicz 2014). The nanocapsules were based on a liquid core (docusate sodium salt and poly-L-lysine hydrobromide) encapsulated by sequential adsorption with polyelectrolyte shells, polyanion poly-L-glutamic acid sodium salt, and polycation poly-L-lysine hydrobromide. The *in vitro* cytotoxicity results suggested that (1) single layer nanocapsules are more toxic than multilayer nanocapsules and (2) the negatively charged nanocapsules exhibited less unfavorable effects on cells viability.

The multilayer nanocapsules characteristics, such as small size, high surface area, and high surface energy, may lead to effects in the GI tract that are not predictable from our knowledge on the behavior of macro- and micro-sized capsules. Nanocapsules could induce inflammation and immune responses; overproduction of 8t (ROS), which can induce oxidative stress; disturb cell normal physiological functions; inactivation of specific proteins; peroxidation of lipids; and DNA damage that result in cell death and genotoxic effects (Zolnik et al. 2010, Lai 2012) (Figure 7.4). When developing multilayer nanocapsules it is essential to produce structures that should incorporate the bioactive compounds and that do not present toxicity to the body. Thus, ideally, the biopolymers used must have no toxicity, and be efficiently removed from the body without accumulation in the tissues and organs. Verification of the safety of multilayer nanocapsules is essential before food application and ingestion (Sajid et al. 2014). Generally, the first step involves *in vitro* experiments in cell culture, examining the toxicity (e.g., cytotoxicity and genotoxicity) of the material to the cells using different techniques such as cell viability/proliferation, ROS generation, apoptosis, necrosis, DNA damaging potential, and gene expression analysis (Arora et al. 2012). Figure 7.4 presents different *in vitro* and *in vivo* evaluation methods to assess toxicity of the multilayer nanocapsules prior to their use. When considering cytotoxicity, the exposure time and concentration of the material are important parameters, as cytotoxicity generally increases with dose and time (Sharifi et al. 2012).

Currently, toxicity studies are primarily performed using a wide range of cell lines (e.g., Caco-2 cells), as well as *in vivo* models (e.g., mice). These conditions led to an extensive number of tested nanostructures, experimental designs, and model systems, which consequently result in several contradictory experimental data (Bergin and Witzmann 2013). To avoid these questions and uncertainties regarding cytotoxicity assays, species of cells used, dose selection, dose metrics, and nano-relevant controls, the European Food Safety Authority (EFSA) provides a document with key issues to be considered including characterization of nanomaterials, dose–response risk assessment, and toxicological studies used for hazard identification (EFSA 2011). The methods for *in vitro* toxicity evaluation are capable of providing adequate data for many bulk materials. On the other hand, the *in vivo* interaction of nanocapsules with the biological system is much more complex and dynamic. Animal models (e.g., rats) would be particularly useful to study *in vivo* aspects that cannot be obtained with *in vitro* systems, such as toxicokinetics in the body (i.e., absorption, distribution, metabolism, and elimination) (Fischer and Chan 2007). However, few *in vivo* works have been done to evaluate the fate of multilayer nanocapsules after ingestion intentionally designed for food uses and to correlate with their possible toxic effects. A selection of *in vitro* and *in vivo* toxicity evaluations using multilayer nanocapsules is summarized in Table 7.4.

The increase of oral bioavailability of bioactive compounds through the use of multilayer nanocapsules can be considered a positive aspect of this technology. However, it is important to determine if these compounds exhibit deleterious effects when consumed at higher levels. Usually, encapsulation of bioactive compounds could increase considerably their absorption, so it could unveil toxic effects that could not be predicted from data obtained on the material used in nanocapsules formulations at macro- and microscopic scale. This fact is mainly true if the bioactive compound is incorporated into a food product that is consumed frequently in large volumes, such as a beverage or soft drink. Therefore, multilayer nanocapsules used in food sector raised concerns about their safety, as well as policy, ethical, and regulatory issues (Garduño-Balderas et al. 2015).

7.6 CONCLUSIONS AND FUTURE PERSPECTIVES

Multilayer nanocapsules allow the introduction of different functionalities in the same delivery system and can be specifically designed to control the digestion, release, and absorption of bioactive compounds within the GI tract. These delivery systems are generally very sensitive to the environmental conditions, so changes in pH, ionic strength, or solvent type can change their permeability, and consequently

TABLE 7.4
Selected *In Vitro* and *In Vivo* Toxicity Studies of Multilayer Nanocapsules for Bioactive Compounds Delivery

Type	Nanocapsule Compositions	Bioactive Compounds	Cell or Animal Models	Assay Methods	Major Observations	Reference
In vitro	Cysteamine conjugated chitosan/dextran sulfate	Bovine serum albumin	Caco-2	MTT	No apparent cytotoxic effect of nanocapsules up to 0.15 mg/mL (cell viability up to 90%)	Shu et al. (2010b)
	Chitosan/dextran sulfate	Ciprofloxacin	RAW264.7	Genotoxicity damage test (mitotic index)	Nanocapsules did not affect the viability of cell lines up to a concentration of 40 mg/mL	Gnanadhas et al. (2013a)
	AOT/PLL/PGA	Coumarine 6	HEK 293	MTT, LDH, ATP detection assay	For the most toxic nanocapsules (with only one polycation layer) about 90% of cells could survive when the concentration of nanocapsules was below 0.2×10^6 per one cell	Łukasiewicz and Szczepanowicz (2014)
In vivo	Polyacrylic acid/polyallylamine hydrochloride	Paclitaxel	Female SD rats	Plasma levels of various toxicity markers (such as aspartate, aminotransferase)	The formulation showed significantly lesser hepatotoxicity and nephrotoxicity in comparison with Taxol® (marketed intravenous product)	Jain et al. (2012)
	Alginate/dextran/chitosan/albumin	Insulin	Wistar rats	MTT	No toxicity; intestinal mucosa biocompatibility	Woitiski et al. (2011)
	Chitosan/heparin	Insulin	ICR mice	Hematology (e.g., blood glucose levels)	No apparent toxicity observed for the animals treated with the nanocapsules	Song et al. (2014)
	o-Carboxymeymethy chitosan/chitosan	DOX	Male SD rats	SOD, CAT activities, and MDA level	Cardiac and renal toxicities reduced with nanocapsules	Feng et al. (2013)

AOT, docusate sodium salt; ATP, adenosine triphosphate; CAT, catalase; DOX, doxorubicin hydrochloride; ICR, imprinting control region; LDH, lactate dehydrogenase; MDA, malondialdehyde; MTT, 3-(4,5-dimethylthiazol-2-yl)-2,5-diphenyltetrazolium bromide; PGA, poly-L-glutamic acid sodium salt; PLL, poly-L-lysine hydrobromide; SD, Sprague–Dawley; SOD, superoxide dismutase.

the release of bioactive compounds can be triggered. Also, both mechanical and physicochemical properties of the multilayer nanocapsules can be tailored by varying their constituents or their thickness.

Work so far has shown that stable multilayer nanocapsules can be prepared only with food-grade ingredients using a simple cost-effective method, the LbL method, and they can act as carriers of different bioactive compounds. Despite the fact that the low cost, preparation simplicity, food-grade composition, and versatility of multilayer nanocapsules make them attractive candidates for application in food products, there are several challenges that must be overcome before these delivery systems can be entirely embraced by food industry. A deeper understanding of the relationship between the structure and the properties of the multilayer nanocapsule is required to develop new responsive nanocapsules. Moreover, nano-based delivery systems are raising concerns regarding their proper functionality after ingestion once at nanoscale, and also about their safety. In fact, a prerequisite for application of delivery systems in food formulations is their lack of toxicity, therefore the assessment of their safety has become a serious demand. The great challenge for characterizing multilayer nanocapsules with regard to human toxicity is due to the variety of their size and surface chemistry. Nevertheless, uptake, biodistribution, degradation, agglomeration, or mucus interaction of nanocapsules as a function of biomaterial composition and surface characteristics will shape the multilayer nanocapsules safety profile.

Various efforts are being made to unraveling the behavior of multilayer nanocapsules in the GI tract, however, further work is still needed. Also, since the use of *in vivo* models is unfeasible (high costs and ethical constrains often involved), there is a need of using standardized and more realistic *in vitro* GI models, that is, models that can accurately simulate the complex physicochemical and physiological processes that occur within the human GI tract, in order to reliably evaluate the performance/safety of these delivery systems.

ACKNOWLEDGMENTS

Authors Ana C. Pinheiro, Ana I. Bourbon, Hélder D. Silva, and Joana T. Martins (SFRH/BPD/101181/2014, SFRH/BD/73178/2010, SFRH/BD/81288/2011, SFRH/BPD/89992/2012, respectively) are recipients of a fellowship from the Fundação para a Ciência e Tecnologia (FCT, Portugal). The authors thank the FCT Strategic Project of UID/BIO/04469/2013 unit, the project RECI/BBB-EBI/0179/2012 (FCOMP-01-0124-FEDER-027462), and the project "BioInd—Biotechnology and Bioengineering for Improved Industrial and Agro-Food Processes," REF. NORTE-07-0124-FEDER-000028 Co-funded by the Programa Operacional Regional do Norte (ON.2—O Novo Norte), QREN, FEDER.

REFERENCES

Abbas, Shabbar, Eric Karangwa, Mohanad Bashari, Khizar Hayat, Xiao Hong, Hafiz Rizwan Sharif, and Xiaoming Zhang. 2015. Fabrication of polymeric nanocapsules from curcumin-loaded nanoemulsion templates by self-assembly. *Ultrasonics Sonochemistry* 23:81–92.

Adamczak, Małgorzata, Anna Kupiec, Ewelina Jarek, Krzysztof Szczepanowicz, and Piotr Warszyński. 2014. Preparation of the squalene-based capsules by membrane emulsification method and polyelectrolyte multilayer adsorption. *Colloids and Surfaces A: Physicochemical and Engineering Aspects* 462:147–152.

Ahmed, Kashif, Yan Li, David J. McClements, and Hang Xiao. 2012. Nanoemulsion- and emulsion-based delivery systems for curcumin: Encapsulation and release properties. *Food Chemistry* 132(2):799–807.

Antipov, Alexei A., Gleb B. Sukhorukov, Edwin Donath, and Helmuth Möhwald. 2001. Sustained release properties of polyelectrolyte multilayer capsules. *The Journal of Physical Chemistry B* 105(12):2281–2284.

Aoki, Tomoko, Eric A. Decker, and David J. McClements. 2005. Influence of environmental stresses on stability of O/W emulsions containing droplets stabilized by multilayered membranes produced by a layer-by-layer electrostatic deposition technique. *Food Hydrocolloids* 19(2):209–220.

Arora, Sumit, Jyutika M. Rajwade, and Kishore M. Paknikar. 2012. Nanotoxicology and in vitro studies: The need of the hour. *Toxicology and Applied Pharmacology* 258(2):151–165.

Bazylinska, Urszula, Renata Skrzela, Krzysztof Szczepanowicz, Piotr Warszynski, and Kazimiera A. Wilk. 2011. Novel approach to long sustained multilayer nanocapsules: Influence of surfactant head groups and polyelectrolyte layer number on the release of hydrophobic compounds. *Soft Matter* 7(13):6113–6124.

Bellmann, Susann, David Carlander, Alessio Fasano, Dragan Momcilovic, Joseph A. Scimeca, Waldman W. James, Lourdes Gombau et al. 2015. Mammalian gastrointestinal tract parameters modulating the integrity, surface properties, and absorption of food-relevant nanomaterials. *Wiley Interdisciplinary Reviews: Nanomedicine and Nanobiotechnology* 7(5):609–622.

Benshitrit, Revital C., Carmit S. Levi, Sharon L. Tal, Eyal Shimoni, and Uri Lesmes. 2012. Development of oral food-grade delivery systems: Current knowledge and future challenges. *Food Funct.* 3(1):10–21.

Berens, Alan R. and Harold B. Hopfenberg. 1978. Diffusion and relaxation in glassy polymer powders.2. Separation of diffusion and relaxation parameters. *Polymer* 19(5):489–496.

Bergin, Ingrid L. and Frank A. Witzmann. 2013. Nanoparticle toxicity by the gastrointestinal route: Evidence and knowledge gaps. *International Journal of Biomedical Nanoscience and Nanotechnology* 3(1/2):163.

Berth, Gisela, Andreas Voigt, Herbert Dautzenberg, Edwin Donath, and Helmuth Möhwald. 2002. Polyelectrolyte complexes and layer-by-layer capsules from chitosan/chitosan sulfate. *Biomacromolecules* 3(3):579–590.

Bourbon, Ana I., Ana C. Pinheiro, Miguel A. Cerqueira, Cristina M. R. Rocha, Maria C. Avides, Mafalda A. C. Quintas, and António A. Vicente. 2011. Physico-chemical characterization of chitosan-based edible films incorporating bioactive compounds of different molecular weight. *Journal of Food Engineering* 106(2):111–118.

Caruso, Frank, D. Neil Furlong, Katsuhiko Ariga, Izumi Ichinose, and Toyoki Kunitake. 1998. Characterization of polyelectrolyte–protein multilayer films by atomic force microscopy, scanning electron microscopy, and Fourier transform infrared reflection–absorption spectroscopy. *Langmuir* 14(16):4559–4565.

Caruso, Frank, Wenjun Yang, Dieter Trau, and Reinhard Renneberg. 2000. Microencapsulation of uncharged low molecular weight organic materials by polyelectrolyte multilayer self-assembly. *Langmuir* 16(23):8932–8936.

Cerqueira, Miguel A., Ana I. Bourbon, Ana C. Pinheiro, Hélder D. Silva, Mafalda A. C. Quintas, and António A. Vicente. 2013. Edible nano-laminate coatings for food applications. In *Ecosustainable Polymer Nanomaterials for Food Packaging*, 221–252. CRC Press.

Cerqueira, Miguel A., Ana C. Pinheiro, Hélder D. Silva, Philippe E. Ramos, Maria A. Azevedo, María L. Flores-López, Melissa C. Rivera, Ana I. Bourbon, Óscar L. Ramos, and António A. Vicente. 2014. Design of bio-nanosystems for oral delivery of functional compounds. *Food Engineering Reviews* 6(1–2):1–19.

Chalothorn, Kunyanatt and Warangkana Warisnoicharoen. 2012. Ultrasonic emulsification of whey protein isolate-stabilized nanoemulsions containing omega-3 oil from plant seed. *American Journal of Food Technology* 7(9):532–541.

Chen-yu, Guo, Yang Chun-fen, Li Qi-lu, Tan Qi, Xi Yan-wei, Liu Wei-na, and Zhai Guang-Xi. 2012. Development of a Quercetin-loaded nanostructured lipid carrier formulation for topical delivery. *International Journal of Pharmaceutics* 430(1–2):292–298.

Costa, Rui R., Manuel Alatorre-Meda, and João F. Mano. 2015. Drug nano-reservoirs synthesized using layer-by-layer technologies. *Biotechnology Advances* 33(6):1310–1326.

Cui, Jiwei, Martin P. van Koeverden, Markus Müllner, Kristian Kempe, and Frank Caruso. 2014. Emerging methods for the fabrication of polymer capsules. *Advances in Colloid and Interface Science* 207:14–31.

Cuomo, Francesca, Francesco Lopez, Andrea Ceglie, Lucia Maiuro, Maria G. Miguel, and Bjorn Lindman. 2012. pH-responsive liposome-templated polyelectrolyte nanocapsules. *Soft Matter* 8(16):4415–4420.

Decher, Gero. 1997. Fuzzy nanoassemblies: Toward layered polymeric multicomposites. *Science* 277(5330):1232–1237.

Decher, Gero. 2003. Polyelectrolyte multilayers, An overview. In *Multilayer Thin Films: Sequential Assembly of Nanocomposite Materials*, edited by Gero Decher and Joseph B. Schlenoff, pp. 1–46. Weinheim, Germany: Wiley-VCH.

Decher, Gero, Jong-Dal Hong, and Johannes Schmitt. 1992. Buildup of ultrathin multilayer films by a self-assembly process: III. Consecutively alternating adsorption of anionic and cationic polyelectrolytes on charged surfaces. *Thin Solid Films* 210–211(Part 2):831–835.

Del Mercato, Loretta L., Marzia M. Ferraro, Francesca Baldassarre, Serena Mancarella, Valentina Greco, Ross Rinaldi, and Stefano Leporatti. 2014. Biological applications of LbL multilayer capsules: From drug delivery to sensing. *Advances in Colloid and Interface Science* 207:139–154.

Del Nobile, Matteo A, Giuseppe Mensitieri, Pablo A. Netti, and Luigi Nicolais. 1994. Anomalous diffusion in poly-ether–ether-ketone (PEEK). *Chemical Engineering Science* 49(5):633–644.

Donath, Edwin, Gleb B. Sukhorukov, Frank Caruso, Sean A. Davis, and Helmuth Möhwald. 1998. Novel hollow polymer shells by colloid-templated assembly of polyelectrolytes. *Angewandte Chemie International Edition* 37(16):2201–2205.

Donsì, Francesco, Marianna Annunziata, Mariarosaria Vincensi, and Giovanna Ferrari. 2012. Design of nanoemulsion-based delivery systems of natural antimicrobials: Effect of the emulsifier. *Journal of Biotechnology* 159(4):342–350.

Donsì, Francesco, Mariarenata Sessa, Houda Mediouni, Arbi Mgaidi, and Giovanna Ferrari. 2011. Encapsulation of bioactive compounds in nanoemulsion- based delivery systems. *Procedia Food Science* 1(0):1666–1671.

Đorđević, Verica, Bojana Balanč, Ana Belščak-Cvitanović, Steva Lević, Kata Trifković, Ana Kalušević, Ivana Kostić, Draženka Komes, Branko Bugarski, and Viktor Nedović. 2014. Trends in encapsulation technologies for delivery of food bioactive compounds. *Food Engineering Reviews* 7(4):1–39.

Du, Hongliang, Mengrui Liu, Xiaoye Yang, and Guangxi Zhai. 2015. The design of pH-sensitive chitosan-based formulations for gastrointestinal delivery. *Drug Discovery Today* 20(8):1004–1011.

EFSA. 2011. Guidance on the risk assessment of the application of nanoscience and nanotechnologies in the food and feed chain. *EFSA Journal* 9(5):2140.

El Gharras, Hasna. 2009. Polyphenols: Food sources, properties and applications—A review. *International Journal of Food Science & Technology* 44(12):2512–2518.

Ensign, Laura M., Richard Cone, and Justin Hanes. 2012. Oral drug delivery with polymeric nanoparticles: The gastrointestinal mucus barriers. *Advanced Drug Delivery Reviews* 64(6):557–570.

Faisant, Nathalie, Juergen Siepmann, and Jean-Pierre Benoit. 2002. PLGA-based microparticles: Elucidation of mechanisms and a new, simple mathematical model quantifying drug release. *European Journal of Pharmaceutical Sciences* 15(4):355–366.

Fang, Zhongxiang and Bhesh Bhandari. 2010. Encapsulation of polyphenols—A review. *Trends in Food Science & Technology* 21(10):510–523.

Feng, Chao, Zhiguo Wang, Changqing Jiang, Ming Kong, Xuan Zhou, Yang Li, Xiaojie Cheng, and Xiguang Chen. 2013. Chitosan/o-carboxymethyl chitosan nanoparticles for efficient and safe oral anticancer drug delivery: In vitro and in vivo evaluation. *International Journal of Pharmaceutics* 457(1):158–167.

Fischer, Hans C. and Warren C. W. Chan. 2007. Nanotoxicity: The growing need for in vivo study. *Current Opinion in Biotechnology* 18(6):565–571.

Flores, Silvia, Amalia Conte, Carmen Campos, Lía N. Gerschenson, and Matteo A. Del Nobile. 2007. Mass transport properties of tapioca-based active edible films. *Journal of Food Engineering* 81(3):580–586.

Fukui, Yuuka, and Keiji Fujimoto. 2009. The preparation of sugar polymer-coated nanocapsules by the layer-by-layer deposition on the liposome. *Langmuir* 25(17):10020–10025.

Garduño-Balderas, Luis Guillermo, Ismael M. Urrutia-Ortega, Estefany I. Medina-Reyes, and Yolanda I. Chirino. 2015. Difficulties in establishing regulations for engineered nanomaterials and considerations for policy makers: Avoiding an unbalance between benefits and risks. *Journal of Applied Toxicology* 35(10):1073–1085.

Garti, Nissim, Aviram Spernath, Abraham Aserin, and Rachel Lutz. 2005. Nano-sized self-assemblies of nonionic surfactants as solubilization reservoirs and microreactors for food systems. *Soft Matter* 1(3):206–218.

Gnanadhas, Divya P., Midhun B. Thomas, Monalisha Elango, Ashok M. Raichur, and Dipshikha Chakravortty. 2013a. Chitosan–dextran sulphate nanocapsule drug delivery system as an effective therapeutic against intraphagosomal pathogen Salmonella. *Journal of Antimicrobial Chemotherapy* 68(11):2576–2586.

Gnanadhas, Divya P., Midhun B. Thomas, Monalisha Elango, Ashok M. Raichur, and Dipshikha Chakravortty. 2013b. Chitosan–dextran sulphate nanocapsule drug delivery system as an effective therapeutic against intraphagosomal pathogen Salmonella. *Journal of Antimicrobial Chemotherapy* 68(11):2576–2586.

Grigoriev, Dmitry O. and Reinhard Miller. 2009. Mono- and multilayer covered drops as carriers. *Current Opinion in Colloid & Interface Science* 14(1):48–59.

Groo, Anne-Claire and Frederic Lagarce. 2014. Mucus models to evaluate nanomedicines for diffusion. *Drug Discovery Today* 19(8):1097–1108.

Guzey, Demet and David J. McClements. 2006. Formation, stability and properties of multilayer emulsions for application in the food industry. *Advances in Colloid and Interface Science* 128–130:227–248.

Hatanaka, Junya, Hina Chikamori, Hideyuki Sato, Shinya Uchida, Kazuhiro Debari, Satomi Onoue, and Shizuo Yamada. 2010. Physicochemical and pharmacological characterization of α-tocopherol-loaded nano-emulsion system. *International Journal of Pharmaceutics* 396(1–2):188–193.

Ilyasoglu, Hulya and Sedef N. El. 2014. Nanoencapsulation of EPA/DHA with sodium caseinate–gum arabic complex and its usage in the enrichment of fruit juice. *LWT—Food Science and Technology* 56(2):461–468.

Jain, Sanyog, Dinesh Kumar, Nitin K. Swarnakar, and Kaushik Thanki. 2012. Polyelectrolyte stabilized multilayered liposomes for oral delivery of paclitaxel. *Biomaterials* 33(28):6758–6768.

Jeong, Byeongmoon, You H. Bae, and Sung. W. Kim. 1999. Thermoreversible Gelation of PEG–PLGA–PEG Triblock Copolymer Aqueous Solutions. *Macromolecules* 32:7064.

Jiang, B., and B. Li. 2009. Tunable drug loading and release from polypeptide multilayer nanofilms. *International Journal of Nanomedicine* 4:37–53.

Johnston, Angus P. R., Christina Cortez, Alexandra S. Angelatos, and Frank Caruso. 2006. Layer-by-layer engineered capsules and their applications. *Current Opinion in Colloid & Interface Science* 11(4):203–209.

Joye, Iris J., Gabriel Davidov-Pardo, and David J. McClements. 2014. Nanotechnology for increased micronutrient bioavailability. *Trends in Food Science & Technology* 40(2):168–182.

Kantak, Chaitanya, Sebastian Beyer, Levent Yobas, Tushar Bansal, and Dieter Trau. 2011. A 'microfluidic pinball' for on-chip generation of Layer-by-Layer polyelectrolyte microcapsules. *Lab on a Chip* 11(6):1030–1035.

Khare, Atul R. and Niraj Vasisht. 2014. Nanoencapsulation in the food industry: Technology of the future. In *Microencapsulation in the Food Industry*, edited by Anilkumar G. Gaonkar, Niraj Vasisht, Atul R. Khare, and Robert Sobel, pp. 151–155 (Chapter 14). San Diego, CA: Academic Press.

Klinkesorn, Utai, Pairat Sophanodora, Pavinee Chinachoti, David J. McClements, and Eric A. Decker. 2005. Increasing the oxidative stability of liquid and dried tuna oil-in-water emulsions with electrostatic layer-by-layer deposition technology. *Journal of Agricultural and Food Chemistry* 53(11):4561–4566.

Kostewicz, Edmund S., Bertil Abrahamsson, Marcus Brewster, Joachim Brouwers, James Butler, Sara Carlert, Paul A. Dickinson et al. 2014. In vitro models for the prediction of in vivo performance of oral dosage forms. *European Journal of Pharmaceutical Sciences* 57:342–366.

Kovatcheva-Datchary, Petia, and Tulika Arora. 2013. Nutrition, the gut microbiome and the metabolic syndrome. *Best Practice & Research Clinical Gastroenterology* 27(1):59–72.

Lai, David Y. 2012. Toward toxicity testing of nanomaterials in the 21st century: A paradigm for moving forward. *Wiley Interdisciplinary Reviews: Nanomedicine and Nanobiotechnology* 4(1):1–15.

Li, Yan and David J. McClements. 2014. Modulating lipid droplet intestinal lipolysis by electrostatic complexation with anionic polysaccharides: Influence of cosurfactants. *Food Hydrocolloids* 35:367–374.

Li, Yan, Min Hu, Hang Xiao, Yumin Du, Eric A. Decker, and David J. McClements. 2010. Controlling the functional performance of emulsion-based delivery systems using multi-component biopolymer coatings. *European Journal of Pharmaceutics and Biopharmaceutics* 76(1):38–47.

Liu, Weilin, Jianhua Liu, Wei Liu, Ti Li, and Chengmei Liu. 2013. Improved physical and in vitro digestion stability of a polyelectrolyte delivery system based on layer-by-layer self-assembly Alginate–Chitosan-coated nanoliposomes. *Journal of Agricultural and Food Chemistry* 61(17):4133–4144.

Łukasiewicz, Sylwia and Krzysztof Szczepanowicz. 2014. In vitro interaction of polyelectrolyte nanocapsules with model cells. *Langmuir* 30(4):1100–1107.

Madrigal-Carballo, Sergio, Seokwon Lim, Gerardo Rodriguez, Amparo O. Vila, Christian G. Krueger, Sundaram Gunasekaran, and Jess D. Reed. 2010. Biopolymer coating of soybean lecithin liposomes via layer-by-layer self-assembly as novel delivery system for ellagic acid. *Journal of Functional Foods* 2(2):99–106.

Marsanasco, Marina, Andrés L. Márquez, Jorge R. Wagner, Silvia del V. Alonso, and Nadia S. Chiaramoni. 2011. Liposomes as vehicles for vitamins E and C: An alternative to fortify orange juice and offer vitamin C protection after heat treatment. *Food Research International* 44(9):3039–3046.

Martins, Joana T., Óscar L. Ramos, Ana C. Pinheiro, Ana I. Bourbon, Hélder D. Silva, Melissa C. Rivera, Miguel A. Cerqueira et al. 2015. Edible bio-based nanostructures: Delivery, absorption and potential toxicity. *Food Engineering Reviews* 7(4):1–23.

McClements, David J., ed. 2005. *Food Emulsions: Principles, Practice, and Techniques*, 2nd ed. Boca Raton, FL: CRC Press.

McClements, David J. 2010. Design of nano-laminated coatings to control bioavailability of lipophilic food components. *Journal of Food Science* 75(1):R30–R42.

McClements, David J. 2013. Edible lipid nanoparticles: Digestion, absorption, and potential toxicity. *Progress in Lipid Research* 52(4):409–423.

McClements, David J. and Eric A. Decker. 2000. Lipid oxidation in oil-in-water emulsions: Impact of molecular environment on chemical reactions in heterogeneous food systems. *Journal of Food Science* 65(8):1270–1282.

McClements, David J., Eric A. Decker, and J. Weiss. 2007. Emulsion-based delivery systems for lipophilic bioactive components. *Journal of Food Science* 72(8):R109–R124.

McClements, David J. and Yan Li. 2010. Structured emulsion-based delivery systems: Controlling the digestion and release of lipophilic food components. *Advances in Colloid and Interface Science* 159(2):213–228.

McClements, David J. and Hang Xiao. 2012. Potential biological fate of ingested nanoemulsions: Influence of particle characteristics. *Food & Function* 3(3):202–220.

Menrad, Klaus. 2003. Market and marketing of functional food in Europe. *Journal of Food Engineering* 56(2–3):181–188.

Mora-Huertas, Claudia E., Hatem Fessi, and Abdelhamid Elaissari. 2010. Polymer-based nanocapsules for drug delivery. *International Journal of Pharmaceutics* 385(1–2):113–142.

Mrsny, Randall J. 2012. Oral drug delivery research in Europe. *Journal of Controlled Release* 161(2):247–253.

Muheem, Abdul, Faiyaz Shakeel, Mohammad A. Jahangir, Mohammed Anwar, Neha Mallick, Gaurav K. Jain, Musarrat H. Warsi, and Farhan J. Ahmad. 2014. A review on the strategies for oral delivery of proteins and peptides and their clinical perspectives. *Saudi Pharmaceutical Journal* (in press).

Neal, Andrew L. 2008. What can be inferred from bacterium–nanoparticle interactions about the potential consequences of environmental exposure to nanoparticles? *Ecotoxicology* 17(5):362–371.

Oberdörster, Günter, Andrew Maynard, Ken Donaldson, Vincent Castranova, Julie Fitzpatrick, Kevin Ausman, Janet Carter et al. 2005. Principles for characterizing the potential human health effects from exposure to nanomaterials: Elements of a screening strategy. *Particle and Fibre Toxicology* 2(1):8.

Paini, Marco, Bahar Aliakbarian, Alessandro A. Casazza, Patrizia Perego, Carmelina Ruggiero, and Laura Pastorino. 2015. Chitosan/dextran multilayer microcapsules for polyphenol co-delivery. *Materials Science and Engineering: C* 46:374–380.

Pawar, Vivek K., Jaya G. Meher, Yuvraj Singh, Mohini Chaurasia, B. Surendar Reddy, and Manish K. Chourasia. 2014. Targeting of gastrointestinal tract for amended delivery of protein/peptide therapeutics: Strategies and industrial perspectives. *Journal of Controlled Release* 196:168–183.

Peppas, Nicholas A. and Balaji Narasimhan. 2014. Mathematical models in drug delivery: How modeling has shaped the way we design new drug delivery systems. *Journal of Controlled Release* 190:75–81.

Pinheiro, Ana C., Ana I. Bourbon, Miguel A. Cerqueira, Élia Maricato, Cláudia Nunes, Manuel A. Coimbra, and António A. Vicente. 2015. Chitosan/fucoidan multilayer nanocapsules as a vehicle for controlled release of bioactive compounds. *Carbohydrate Polymers* 115(0):1–9.

Pinheiro, Ana C., Ana I. Bourbon, Mafalda A. C. Quintas, Manuel A. Coimbra, and António A. Vicente. 2012. K-carrageenan/chitosan nanolayered coating for controlled release of a model bioactive compound. *Innovative Food Science & Emerging Technologies* 16:227–232.

Pinheiro, Ana C., Ana I. Bourbon, António A. Vicente, and Mafalda A. C. Quintas. 2013a. Transport mechanism of macromolecules on hydrophilic bio-polymeric matrices—Diffusion of protein-based compounds from chitosan films. *Journal of Food Engineering* 116(3):633–638.

Pinheiro, Ana C., Manuel A. Coimbra, and António A. Vicente. 2016. In vitro behaviour of curcumin nanoemulsions stabilized by biopolymer emulsifiers—Effect of interfacial composition. *Food Hydrocolloids* 52: 460–467.

Pinheiro, Ana C., Mita Lad, Hélder D. Silva, Manuel A. Coimbra, Michael Boland, and António A. Vicente. 2013b. Unravelling the behaviour of curcumin nanoemulsions during in vitro digestion: Effect of the surface charge. *Soft Matter* 9(11):3147.

Plapied, Laurence, Nicolas Duhem, Anne des Rieux, and Véronique Préat. 2011. Fate of polymeric nanocarriers for oral drug delivery. *Current Opinion in Colloid & Interface Science* 16(3):228–237.

Preetz, Claudia, Andrea Rübe, Ines Reiche, Gerd Hause, and Karsten Mäder. 2008. Preparation and characterization of biocompatible oil-loaded polyelectrolyte nanocapsules. *Nanomedicine: Nanotechnology, Biology, and Medicine* 4(2):106–114.

Priest, Craig, Anthony Quinn, Almar Postma, Alexander N. Zelikin, John Ralston, and Frank Caruso. 2008. Microfluidic polymer multilayer adsorption on liquid crystal droplets for microcapsule synthesis. *Lab on a Chip* 8(12):2182–2187.

Qiu, Xingping, Stefano Leporatti, Edwin Donath, and Helmuth Möhwald. 2001. Studies on the drug release properties of polysaccharide multilayers encapsulated ibuprofen microparticles. *Langmuir* 17(17):5375–5380.

Rivera, Melissa C., Ana C. Pinheiro, Ana I. Bourbon, Miguel A. Cerqueira, and António A. Vicente. 2015. Hollow chitosan/alginate nanocapsules for bioactive compound delivery. *International Journal of Biological Macromolecules* 79(0):95–102.

Sajid, Muhammad, Muhammad Ilyas, Chanbasha Basheer, Madiha Tariq, Muhammad Daud, Nadeem Baig, and Farrukh Shehzad. 2014. Impact of nanoparticles on human and environment: Review of toxicity factors, exposures, control strategies, and future prospects. *Environmental Science and Pollution Research* 22(6):4122–4143.

Salvia-Trujillo, Laura, Cheng Qian, Olga Martín-Belloso, and David J. McClements. 2013. Influence of particle size on lipid digestion and β-carotene bioaccessibility in emulsions and nanoemulsions. *Food Chemistry* 141(2):1472–1480.

Sharifi, Shahriar, Shahed Behzadi, Sophie Laurent, M. Laird Forrest, Pieter Stroeve, and Morteza Mahmoudi. 2012. Toxicity of nanomaterials. *Chem. Soc. Rev.* 41(6):2323–2343.

Shchukina, Elena M. and Dmitry G. Shchukin. 2011. LbL coated microcapsules for delivering lipid-based drugs. *Advanced Drug Delivery Reviews* 63(9):837–846.

Shu, Shujun, Chunyang Sun, Xinge Zhang, Zhongming Wu, Zhen Wang, and Chaoxing Li. 2010a. Hollow and degradable polyelectrolyte nanocapsules for protein drug delivery. *Acta Biomaterialia* 6(1):210–217.

Shu, Shujun, Xinge Zhang, Zhongming Wu, Zhen Wang, and Chaoxing Li. 2010b. Gradient cross-linked biodegradable polyelectrolyte nanocapsules for intracellular protein drug delivery. *Biomaterials* 31(23):6039–6049.

Siepmann, Juergen and Nicholas A. Peppas. 2001. Modeling of drug release from delivery systems based on hydroxypropyl methylcellulose (HPMC). *Advanced Drug Delivery Reviews* 48(2–3):139–157.

Siepmann, Juergen, Alexander Streubel, and Nicholas A. Peppas. 2002. Understanding and predicting drug delivery from hydrophilic matrix tablets using the "sequential layer" model. *Pharmaceutical Research* 19(3):306–314.

Silva, Hélder D., Miguel A. Cerqueira, Bartolomeu W. S. Souza, Clara Ribeiro, Maria C. Avides, Mafalda A. C. Quintas, Jane S. R. Coimbra, Maria G. Carneiro-da-Cunha, and António A. Vicente. 2011. Nanoemulsions of β-carotene using a high-energy emulsification–evaporation technique. *Journal of Food Engineering* 102(2):130–135.

Silva, Hélder, Miguel Cerqueira, and António Vicente. 2012. Nanoemulsions for food applications: Development and characterization. *Food and Bioprocess Technology* 5(3):854–867.

Singh, Amritpal, Ankush Sharma, Pooja, and Anju. 2014. Novel approaches for colon targeted drug delivery system. *International Journal of Research and Development in Pharmacy and Life Sciences* 3(2):877–886.

Song, Lei, Zheng L. Zhi, and John J. Pickup. 2014. Nanolayer encapsulation of insulin-chitosan complexes improves efficiency of oral insulin delivery. *International Journal of Nanomedicine* 9:2127–2136.

Soppimath, Kumaresh S., Tejraj M. Aminabhavi, Anandrao R. Kulkarni, and Walter E. Rudzinski. 2001. Biodegradable polymeric nanoparticles as drug delivery devices. *Journal of Controlled Release* 70(1–2):1–20.

Sosnik, Alejandro, José das Neves, and Bruno Sarmento. 2014. Mucoadhesive polymers in the design of nano-drug delivery systems for administration by non-parenteral routes: A review. *Progress in Polymer Science* 39(12):2030–2075.

Sosnik, Alejandro and Maya M. Raskin. 2015. Polymeric micelles in mucosal drug delivery: Challenges towards clinical translation. *Biotechnology Advances*. 33(6 Pt 3):1380–1392.

Städler, Brigitte, Rona Chandrawati, Kenneth Goldie, and Frank Caruso. 2009. Capsosomes: Subcompartmentalizing polyelectrolyte capsules using liposomes. *Langmuir* 25(12):6725–6732.

Sukhorukov, Gleb B., Alexei A. Antipov, Andreas Voigt, Edwin Donath, and Helmuth Möhwald. 2001. pH-controlled macromolecule encapsulation in and release from polyelectrolyte multilayer nanocapsules. *Macromolecular Rapid Communications* 22(1):44–46.

Sukhorukov, Gleb B., Edwin Donath, Heinz Lichtenfeld, Eberhard Knippel, Monika Knippel, Axel Budde, and Helmuth Möhwald. 1998. Layer-by-layer self assembly of polyelectrolytes on colloidal particles. *Colloids and Surfaces A: Physicochemical and Engineering Aspects* 137(1–3):253–266.

Szczepanowicz, Krzysztof, Urszula Bazylińska, Jadwiga Pietkiewicz, Lilianna Szyk-Warszyńska, Kazimiera A. Wilk, and Piotr Warszyński. 2015. Biocompatible long-sustained release oil-core polyelectrolyte nanocarriers: From controlling physical state and stability to biological impact. *Advances in Colloid and Interface Science* 222:678–691.

Szczepanowicz, Krzysztof, Dorota Dronka-Góra, Grażyna Para, and Piotr Warszyński. 2010a. Encapsulation of liquid cores by layer-by-layer adsorption of polyelectrolytes. *Journal of Microencapsulation* 27(3):198–204.

Szczepanowicz, Krzysztof, Hanna J. Hoel, Lilianna Szyk-Warszynska, Elżbieta Bielańska, Aud M. Bouzga, Gustav Gaudernack, Christian Simon, and Piotr Warszynski. 2010b. Formation of Biocompatible Nanocapsules with Emulsion Core and Pegylated Shell by Polyelectrolyte Multilayer Adsorption. *Langmuir* 26(15):12592–12597.

Szczepanowicz, Krzysztof, Karolina Podgórna, Lilianna Szyk-Warszyńska, and Piotr Warszyński. 2014. Formation of oil filled nanocapsules with silica shells modified by sequential adsorption of polyelectrolytes. *Colloids and Surfaces A: Physicochemical and Engineering Aspects* 441:885–889.

Truong-Le, Vu, Phillip M. Lovalenti, and Ahmad M. Abdul-Fattah. 2015. Stabilization challenges and formulation strategies associated with oral biologic drug delivery systems. *Advanced Drug Delivery Reviews* 93:95–108

Vander, Arthur J., James Sherman, and Dorothy S. Luciano. 2001. The digestion and absorption of food. In *Human Physiology: The Mechanism of Body Function*, edited by Eric P. Widmaier, Hershel Raff, and Hershal Strang, pp. 553–591. New York: McGraw-Hill.

Weiss, Jochen, Paul Takhistov, and David J. McClements. 2006. Functional materials in food nanotechnology. *Journal of Food Science* 71(9):R107–R116.

Woitiski, Camile B., Ronald J. Neufeld, Francisco Veiga, Rui A. Carvalho, and Isabel V. Figueiredo. 2010. Pharmacological effect of orally delivered insulin facilitated by multilayered stable nanoparticles. *European Journal of Pharmaceutical Sciences* 41(3–4):556–563.

Woitiski, Camile B., Bruno Sarmento, Rui A. Carvalho, Ronald J. Neufeld, and Francisco Veiga. 2011. Facilitated nanoscale delivery of insulin across intestinal membrane models. *International Journal of Pharmaceutics* 412(1–2):123–131.

Yada, Rickey Y., Neil Buck, Richard Canady, Chris DeMerlis, Timothy Duncan, Gemma Janer, Lekh Juneja et al. 2014. Engineered nanoscale food ingredients: Evaluation of current knowledge on material characteristics relevant to uptake from the gastrointestinal tract. *Comprehensive Reviews in Food Science and Food Safety* 13(4):730–744.

Yan, Shifeng, Jie Zhu, Zhichun Wang, Jingbo Yin, Yanzhen Zheng, and Xuesi Chen. 2011. Layer-by-layer assembly of poly(L-glutamic acid)/chitosan microcapsules for high loading and sustained release of 5-fluorouracil. *European Journal of Pharmaceutics and Biopharmaceutics* 78(3):336–345.

Ye Shiqu, Wang Chaoyang, Xinxing Liu, and Zhen Tong 2005. Multilayer nanocapsules of polysaccharide chitosan and alginate through layer-by-layer assembly directly on PS nanoparticles for release. *J Biomater Sci Polym Ed* 16(7):909–23.

Yu, Hailong and Qingrong Huang. 2012. Improving the oral bioavailability of curcumin using novel organogel-based nanoemulsions. *Journal of Agricultural and Food Chemistry* 60(21):5373–5379.

Zhang, Zipei, Ruojie Zhang, Long Chen, Qunyi Tong, and David J. McClements. 2015. Designing hydrogel particles for controlled or targeted release of lipophilic bioactive agents in the gastrointestinal tract. *European Polymer Journal* 72:698–716.

Zolnik, Banu S., África González-Fernández, Nakissa Sadrieh, and Marina A. Dobrovolskaia. 2010. Minireview: Nanoparticles and the immune system. *Endocrinology* 151(2):458–465.

8 Freeze Concentration as a Technique to Protect Valuable Heat-Labile Components of Foods

Guillermo Petzold, Patricio Orellana, Jorge J. Moreno, Julio Junod, and Graciela Bugueño

CONTENTS

8.1 Introduction 183
8.2 Fundamentals of Freeze Concentration 184
8.3 Methods of Freeze Concentration 186
    8.3.1 Suspension Crystallization 186
    8.3.2 Film Freeze Concentration 186
    8.3.3 Partial Block Cryoconcentration 187
    8.3.4 Complete Block Cryoconcentration 187
    8.3.5 Progressive Freeze Concentration 188
8.4 Advantages of Freeze Concentration over Heat Treatments to Concentrate Liquid Foods 189
8.5 Freeze Concentration Protection of Valuable Heat-Labile Components: Examples and Potential Applications 190
8.6 Conclusions and Future Trends 192
Acknowledgment 192
References 192

8.1 INTRODUCTION

Freeze concentration is an important concentration technique useful to concentrate liquid foods with high nutritional and organoleptic property. This is a valuable technique for preserving heat-labile components such as vitamins, polyphenols, and volatile compounds, among other properties or components. This is possible because freeze concentration process occurs at low temperatures where no vapor/liquid interface exists. However, the practical maximum concentrations are only between 45% and 55% ranges, but the energy consumption of freeze concentration is lower than in the vaporization method due to the lower latent heat of water solidification.

Therefore, the priority of freeze concentration over other concentration methods (in particular the evaporation) is explained for the quality of concentrate products. In general, the quality of the concentrated final products is excellent, because the aroma, as well as the flavor, remains in the liquid foods. In addition to this organoleptic property, the protection of heat-labile components of foods is an important advantage of the freeze concentration, as described in some studies, which can help increase the use of this technique in the future.

8.2 FUNDAMENTALS OF FREEZE CONCENTRATION

Concentration of liquid foods is an important unit operation, because concentrated products occupy less space and weight less, and food manufacturers can potentially save on transportation costs, warehousing costs, and handling costs for materials required for its products (Ramaswamy and Marcotte 2006).

The concentration of liquid foods can be made primarily through three mechanisms (Figure 8.1), high temperature (evaporation), freeze concentration, and reverse osmosis (or other membrane separation). Each of the above method has advantages and disadvantages relative to the solute concentration achieved and nutrient retention, among other features.

When freezing a solution, at the same form of most materials, important redistribution of solutes occurs during the crystallization (in this case crystallization of water) (Bartels-Rausch et al. 2012). When an ice nucleus begins to grow in a solution, solutes are rejected from the ice phase and accumulate at the solid–liquid interphase during freezing process. Thus, the impurities or solutes become segregated on the frozen interphase to increase their concentration compared to that in the original solution. This exclusion phenomenon of ice is not only common to a solution (including both inorganic and organic) but also to suspensions and that is a major principle of the freeze concentration techniques (Nakagawa et al. 2010). The ice growth in a solution is controlled by the rate of latent heat released during the phase change as well as by the rate of mass transfer (diffusion of water molecules from the solution to the crystal lattice and counterdiffusion of solutes away from

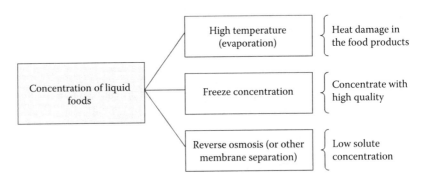

FIGURE 8.1 Mechanisms for concentration liquid foods and specific features. Freeze concentration generates high quality concentrates.

the growing crystal surface). The rate of crystal growth (G) is also a function of the supercooling (ΔT_S) reached by the specimen according to the phenomenological expression (Zaritzky 2006):

$$G = \beta(\Delta T_S)^n \tag{8.1}$$

where β and n are experimental constants.

The general claim of the freeze concentration process is essentially that it is capable of removing water by freezing it out from a solution as ice crystals. Ideally, the ice formed should be free of solutes. First, the solution is partially frozen, the ice crystals are physically separated from the residual solution (concentrated solution), and the ice is melted to form the product water. Ice crystals formed under the appropriate conditions can be very pure (Rahman et al. 2007).

Freeze concentration is a concentration technique that involves lowering the temperature of an aqueous solution sufficiently to partially freeze the water, resulting in (the traditional method of freeze concentration) a slurry of ice crystals dispersed in a concentrated solution (Hartel 1992). The freeze concentration process is schematically shown in Figure 8.2. The basic system consists of two stages: freezing (crystallization) and separation, obtaining ice and concentrate (a concentrate fruit juice, for example).

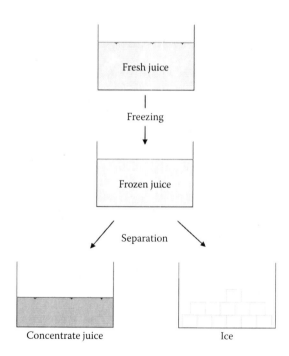

FIGURE 8.2 Stages of the basic freeze concentration process. In this case, a fresh juice is freezing and separate in concentrate juice and ice. The stages of this unit operation are freezing and separation.

Freeze concentration is not a new process. Before the 1800s, native North Americans made maple syrup from sap taken from maple trees. They used shallow vessels made of bark. They filled them with water and maple sap. Vessels were left to freeze overnight, and in the morning the ice was broken off and thrown away, concentrating the unfrozen sugar. The process was repeated several times until the end product became the concentrated maple syrup (Ramaswamy and Marcotte 2006).

8.3 METHODS OF FREEZE CONCENTRATION

8.3.1 SUSPENSION CRYSTALLIZATION

In the conventional method of suspension crystallization (folded chain single crystal), individual ice crystals are formed that are enlarged in size by Ostwald ripening (Huige 1972). Nevertheless, the separation between the ice crystals and the mother liquid is quite complicated and results in losses of concentrate trapped in a mash of solution and crystals (Deshpande et al. 1982; Miyawaki 2001). In this method, the size of ice crystals is still limited, so that the concentration based on this method needs very complicated system composed of surface-scraper heat exchange for generation of seed ice, recrystallization vessel for ice crystal growth, and washing tower for separation of ice crystals (Miyawaki et al. 2005). It has been reported that this complicated system makes cryoconcentration process the most expensive among all methods of liquid foods concentration. Therefore, the practical application of freeze concentration by suspension crystallization is still limited in the food industry (Aider and Halleux 2009).

A typical freeze concentration by suspension crystallization consists of three fundamental components: (a) a crystallizer or freezer, (b) an ice–liquid separator, a melter–condenser, and (c) a refrigeration unit. In the freeze concentration system, the solution is usually first chilled to a prefreezing temperature in a cooler, and then the solution enters the crystallizer where part of the water crystallizes. Cooling causes ice crystal growth and an increase in solute concentration. The resulting mixture of ice crystals and concentrated solution is pumped through a separator where crystals are separated and the concentrated solution is drained off. Ice crystals are removed and melted by hot refrigerant gas. The final products are cold water and concentrated solution, which flow separately (Welti-Chanes 2004).

8.3.2 FILM FREEZE CONCENTRATION

Another method is the crystallization of water present in the solution in the form of an ice on a cold surface. This method (film freeze concentration) consists of the formation of a single ice, which grows layer by layer from the solution to be concentrated. The crystal growth (dendrites) tends to be parallel and opposite to the direction of heat transfer (Flesland 1995). In this condition, the crystals adhere to the cold surface during the process, facilitating separation of the two phases. This concentration system is based on directional freezing, and the heat transfer rates are normally greater than mass transfer, due to the high thermal conductivity of ice and low mass diffusion coefficients. Therefore, solute diffusion will be the limiting factor for ice growth, and supercooling (constitutional supercooling) in the tip region will be observed.

However, the solute inclusion in ice is difficult to avoid in practical applications, especially for solute concentrations of commercial interest for freeze concentration, which means between 20% and 50% of dissolved solids (Raventós et al. 2012).

8.3.3 Partial Block Cryoconcentration

This technique consists of the following sequence: a liquid food or solution is introduced in a crystallizer chamber and partially frozen from the center by introducing a pipe in which a cooling agent is circulated. As the ice block increases, concentration of the remaining solution increases. In this cryoconcentration technology, the important parameter, influencing the meaning of the separation factor, is the density of a packing of crystals in the formed ice block.

The higher the density of the packing the better the solution separates (Aider and Halleux 2009). The authors reported that the ideal separation requires extremely small meanings of the moving forces that have low unit efficiency. Therefore, compromise regimes must be looked for, and the necessary meaning of the separation factor must be achieved by a separate operation—separation. Ratio of thermal and diffusion resistances in the volume of the boundary layer is an important parameter to evaluate the process efficiency (Aider and Halleux 2009).

Prevailing of the heat conduction mechanism will be characterized by the growth of porous ice structure because of lowering of the separation factor value. In the limit one can achieve the regimes, when the separations would not occur at all, and the concentration in the block of ice will be equal to the concentration of the initial solution. Another limiting regime can be the situation, when intensive convection currents prove to be able to prevent the formation of the ice block. The result of such regime will be not directed, but volumetric crystallization (Aider and Halleux 2009).

8.3.4 Complete Block Cryoconcentration

Complete block cryoconcentration technology was reported as a promising and effective concentration technology to produce concentrated liquid foods having high nutritive value and organoleptic properties (Aider et al. 2007; Aider and Halleux 2008a; Aider et al. 2009). The basis of complete block cryoconcentration is as follows: a food liquid the product is largely below the freezing point. After that, the whole frozen solution is thawed and the concentrated fraction is separated from the ice fraction by means of gravitational thawing assisted or not by other techniques to enhance the separation efficiency (Aider and Halleux 2008a, b). Under these conditions, the ice block acts as a solid carcass through which the concentrated fraction passes. By controlling the thawing temperature, it is possible to reach process efficiency, higher than 90%, meaning that the amount of the entrapped solute in the ice crystal is reduced to a minimal level (Aider and Halleux 2009).

In general, the efficiency of freeze concentration operation may be improved by the implementation of some freezing assisting techniques (Petzold and Aguilera 2009). Thus, an alternative for separating the concentrated solution from the ice fraction (in the particular case of complete block freeze concentration) is the use of different assisted techniques or external forces such as vacuum or centrifugation. In this way, vacuum (suction by a pump) has been proposed by Hsieh (2008) to get drinkable water from seawater to separate salt, converting the ice of seawater into freshwater, and Petzold et al. (2013) applying a vacuum (80 kPa) improved the efficiency by more than the atmospheric conditions in freeze concentration of sucrose solutions. Centrifugation has been proposed by Bonilla-Zavaleta et al. (2006) in frozen pineapple juice to separate ice from concentrated juice, while Luo et al. (2010) obtained ice crystals of high purity during the freeze concentration of brackish water. Virgen-Ortíz et al. (2012, 2013) proposed simple freeze centrifugation methods to concentrate dilute protein solutions, Petzold and Aguilera (2013) presented an effective centrifugal freeze concentration method with sucrose solutions, and recently Petzold et al. (2015) proposed this assisted technique applied to block freeze concentration of blueberry and pineapple juices.

The complete block freeze concentration process takes advantage of the hydraulic system existing in the frozen matrix formed by veins (or channels) between the ice crystals occluding the concentrated solution. This matrix in a frozen system is responsible for differences in the concentration of impurities in ancient polar ice, where solutes migrated through the microchannels between the ice crystals under the pressure of upper ice layers (Rempel et al. 2001).

8.3.5 Progressive Freeze Concentration

Progressive freeze concentration (PFC) is based on a completely different concept because a large ice mass is formed and grown on the cooling surface so that the separation of the mother solution is relatively easy (Bae et al. 1994; Liu et al. 1997).

At PFC, the effective partition coefficient of a solute between the ice and the liquid phases at the ice–liquid interface is the most important parameter. The value of the partition coefficient K changes between 0 (ideal freeze concentration) and 1 (no concentration) and decreases (giving higher ice purity) with a decrease in the advance rate of the ice front (u) and/or an increase in the stirring rate N (see Equations 8.2 through 8.4) (Miyawaki 2001; Gu et al. 2005):

$$K = K_0 / \left[K_0 + (1 + K_0) \exp(u/k) \right] \quad (8.2)$$

$$K = C_S / C_L \quad (8.3)$$

$$k = aN^b \quad (8.4)$$

where:
C_S is the solute concentration in ice–liquid interface
C_L the solute concentration in solution phase
K_0 is limiting partition coefficient at the ice–liquid interface
u is the advance rate of ice front
k is mass transfer coefficient at the interface
a and b are constants experimentally determined

Note that these equations (Equations 8.2 through 8.4) are valid only if K is constant during the concentration process (i.e., quasi-steady state), a condition reached if the volume of solution is sufficiently large (Butler 2002).

8.4 ADVANTAGES OF FREEZE CONCENTRATION OVER HEAT TREATMENTS TO CONCENTRATE LIQUID FOODS

The growing demand for fruit juices of high organoleptic and nutritional quality has led to the search for new and improved food processing technologies. Among the techniques for concentration of liquid foodstuffs, freeze concentration is of particular interest due to the low temperatures used in the process (Raventós et al. 2012).

The concentration of liquid foods (like fruit juices) is a delicate process, since they are sensitive to thermal treatments. Even at moderate temperatures, many of their components are unstable. At temperature between 40°C and 70°C, enzyme catalyzed reactions can alter juice properties within a few minutes. In order to inactivate the enzymes, juices must be heat treated. Moreover, the quality is strongly dependent on the concentration and composition of odorous compounds. Most flavor and aroma components are volatile and can be lost by evaporation (Deshpande et al. 1982).

Freeze concentration has long been recognized as one of the best concentration techniques. As compared to evaporation and membrane technology, freeze concentration has some significant potential advantages for producing a concentrate with high quality because the process occurs at low temperatures where no vapor/liquid interface exists, resulting in minimal loss of volatiles (Morison and Hartel 2007). Additionally, the energy consumption of freeze concentration is lower than in the vaporization method due to the lower latent heat of water solidification (0.33 kJ/g) than that of vaporization (2.26 kJ/g) (Ramaswamy and Marcotte 2006).

Freeze concentration process is particularly suited for the concentration of heat-labile liquid foods. In evaporators, volatile aromas in the feed are almost quantitatively lost with the water vapor. Normally, the quality can partly be restored by separating the aromas from the vapor leaving the evaporator in a distillation column and by feeding them back to the concentrated liquid. Very volatile aromas, however, are lost with the inert gasses and aromas with a volatile equal to—or less than—the volatile of water in the solution cannot be recovered from the vapor (Thijssen 1974).

Conversely, an important advantage of evaporation over freeze concentration is the level of concentration attainable. Evaporation can concentrate most liquid feeds of any dilution up to 50% solids easily, while, in the extreme, sugar solutions for the production of hard candies (toffees) are evaporated to about 98% solids (Schwartzberg 1989). On the other hand, in freeze concentration the practical maximum concentrations are only between 45% and 55% range, because in general, an increase in solute concentration produces an increase in nucleation and a decrease in the growth velocity of the ice crystals and in the mean diameter of the crystal, and at critical concentration, solutes may solidify along with ice and are difficult to separate (Welti-Chanes et al. 2006).

Evaporation has historically been the primary technology for liquid concentration in the food industry. However, the evaporation proceeds at higher temperature and promotes heat damage in the food products: the evaporation darkens the color of food, for example, milk, partly not only because of the increase in concentration of solids, but also because the reduction in water activity promotes chemical changes, such as caramelization of sugars, Maillard browning reactions, or lipid–protein interactions. The evaporation temperature also destroys some types of heat-labile vitamins, reduces the biological value of proteins, and promotes lipid oxidation. In addition, the effect of heat on natural antioxidants during evaporation is also considerable, reducing antioxidant efficiency through thermal decomposition (Pokorny and Schmidt 2001).

Therefore, the priority of freeze concentration over the evaporation is explained for the quality of concentrate products: freeze concentration has been known to be the best among the methods of concentration giving the highest retention in flavors and thermally fragile compounds (Deshpande et al. 1982).

8.5 FREEZE CONCENTRATION PROTECTION OF VALUABLE HEAT-LABILE COMPONENTS: EXAMPLES AND POTENTIAL APPLICATIONS

In general, freeze concentration technology is very suitable for the concentration of the sensitive juices (i.e., valuable heat-labile components). The quality of the concentrated juices is excellent, and the taste has to be measurably better than evaporation concentrates, because the aroma, as well as the flavor, remains in the juice. Ultimately, there is no loss of volatiles, making this technique very suitable for the concentration of thermosensitive fluids (Raventós et al. 2012).

Table 8.1 shows some examples of protection of valuable heat-labile components or valuable characteristics in different food matrix, indicating the freeze concentration techniques used. Thus, Liu et al. (1999) using PFC described that freeze concentrated and reconstituted tomato juice showed no differences in vitamin C and color in comparison with unconcentrated tomato juice. Also PFC was applied to Andes berry (*Rubus glaucus* Benth) pulp preserving the flavor (with a total volatiles loss near 20%), and sensorial analyzes show that PFC did not change the sensorial properties (appearance, color, taste, aroma, and overall quality) of fresh pulp (Ramos et al. 2005). On the other hand, Aider and Halleux (2008b), using block freeze concentration applied in cherry and apricot juices, showed that the best parameters were those of the cryoconcentrated juice in comparison with evaporated juice. Indeed, by using a subzero temperature to concentrate the juice, no significant degradation of ascorbic acid has occurred, and aromatic compounds were kept in the juice because the cryoconcentration technology did not allow their evaporation. Moreno et al. (2014) using block freeze concentration applied to coffee extract were found to show significant correlations between the antioxidant activity and chlorogenic acid and caffeine contents in the freeze concentrated extract. In addition, these authors concluded that the freeze concentration method increased the bioactive compound concentration and the antioxidant activity of the coffee extract, preserving volatile compounds and some sensory attributes (Moreno et al. 2015). Boaventura et al. (2013) demonstrated that it is possible to enhance the content of bioactive components and the *in vitro* antioxidant activity of yerba mate (an important infusion largely consumed in several South American countries) aqueous extract through PFC. Also, Boaventura et al. (2015) showed that yerba mate infusion, obtained by freeze concentration technology, was able to promote additional improvement on the antioxidant status of healthy individuals when compared to the traditional mate infusion.

TABLE 8.1
Examples of Protection of Valuable Heat-Labile Components or Valuable Characteristics

Food Matrix	Freeze Concentration Technique	Heat-Labile Component or Characteristic	Reference
Tomato juice	Progressive freeze concentration	Vitamin C Color	Liu et al. (1999)
Andes berry	Progressive freeze concentration	Flavor Sensory attributes	Ramos et al. (2005)
Cherry and apricot juices	Block cryoconcentration	Ascorbic acid Aroma	Aider and Halleux (2008b)
Tofu whey	Falling-film freeze concentration	Protein content Total isoflavones content	Belén et al. (2013)
Sour cherry and orange juices	Block freeze concentration	Vitamin C Aroma number Total antioxidant activity	Normohamadpor et al. (2013)
Pomegranate juice	Complete block cryoconcentration	Color	Khajehei et al. (2015)
Coffee extract and pear juice	Progressive freeze concentration	Flavor	Gunathilake et al. (2014)
Coffee extract	Block freeze concentration	Chlorogenic acid Cryptochlorogenic acid Caffeine Antioxidant activity (ABTS, DPPH)	Moreno et al. (2014)
Tofu whey	Block freeze concentration	Total isoflavones content Antioxidant activity (ABTS, FRAP)	Benedetti et al. (2015)
Yerba mate extract	Progressive freeze concentration Block freeze concentration	Phenolic compounds Methylxanthines Antioxidant activity (FRAP, DPPH)	Boaventura et al. (2013)
Yerba mate extract	Progressive freeze concentration	Total phenolic compounds Serum antioxidant capacity (FRAP)	Boaventura et al. (2015)
Potato juices	Progressive freeze concentration	Total phenolic compounds Antioxidant activity (FRAP)	Kowalczewski et al. (2012)
Apple juice and apple wine	Progressive freeze concentration	Organic acid Flavor components	Miyawaki et al. (2016)
Coffee extract	Falling-film freeze concentration Block freeze concentration	Volatile compounds Sensory attributes	Moreno et al. (2015)
Yerba mate extract	Block freeze concentration	Total phenolic content Antioxidant activity (FRAP, DPPH)	Nunes et al. (2015)
Black currant juice	Progressive freeze concentration	Aroma Flavor Color	Dette and Hansen (2010)

8.6 CONCLUSIONS AND FUTURE TRENDS

The freeze concentration process is a technique to concentrate liquid food and other raw materials, and in the case of liquid foods, the advantage over evaporation is the high quality of the concentrate products in terms of nutritional and organoleptic properties. In addition, in some cases the freeze concentration operation may be improved by the implementation of some freezing assisting techniques such as centrifugation or vacuum. No doubt that freeze concentration should focus on high value-added liquid foods, where preservation or retention of its functional or nutritional properties are valued by consumers and certainly has also interesting organoleptic qualities that can be retained.

ACKNOWLEDGMENT

Author Guillermo Petzold is grateful for the financial support provided by CONICYT through FONDECYT Project No. 11140747.

REFERENCES

Aider, M. and D. Halleux. Passive and microwave-assisted thawing in maple sap cryoconcentration technology. *Journal of Food Engineering* 85(1) (2008a): 65–72.

Aider, M. and D. Halleux. Production of concentrated cherry and apricot juices by cryoconcentration technology. *LWT—Food Science and Technology* 41(10) (2008b): 1768–1775.

Aider, M. and D. Halleux. Cryoconcentration technology in the bio-food industry: Principles and applications. *LWT—Food Science and Technology* 42(3) (2009): 679–685.

Aider, M., D. Halleux, and A. Akbache. Whey cryoconcentration and impact on its composition. *Journal of Food Engineering* 82(1) (2007): 92–102.

Aider, M., D. Halleux, and I. Melnikova. Skim milk whey cryoconcentration and impact on the composition of the concentrated and ice fractions. *Food and Bioprocess Technology* 2(1) (2009): 80–88.

Bae, S.K., O. Miyawaki, and S. Arai. Control of freezing front structure and its effect on the concentration-efficiency in the progressive freeze-concentration. *Cryobiology and Cryotechnology* 40 (1994): 29–32.

Bartels-Rausch, T., V. Bergeron, J.H.E. Cartwright et al. Ice structures, patterns, and processes: A view across the ice fields. *Reviews of Modern Physics* 84(2) (2012): 885–944.

Belén, F., S. Benedetti, J. Sánchez et al. Behavior of functional compounds during freeze concentration of tofu whey. *Journal of Food Engineering* 116(3) (2013): 681–688.

Benedetti, S., E.S. Prudêncio, G.L. Nunes et al. Antioxidant properties of tofu whey concentrate by freeze concentration and nanofiltration processes. *Journal of Food Engineering* 160 (2015): 49–55.

Boaventura, B.C.B., E.L. Da Silva, R.H. Liu et al. Effect of yerba mate (*Ilex paraguariensis* A. St. Hil.) infusion obtained by freeze concentration technology on antioxidant status of healthy individuals. *LWT—Food Science and Technology* 62(2) (2015): 948–954.

Boaventura, B.C.B., A.N.N. Murakami, E.S. Prudêncio et al. Enhancement of bioactive compounds content and antioxidant activity of aqueous extract of mate (*Ilex paraguariensis* A. St. Hil.) through freeze concentration technology. *Food Research International* 53(2) (2013): 686–692.

Bonilla-Zavaleta, E., E.J. Vernon-Carter, and C.I. Beristain. Thermophyscial properties of freeze-concentrated pineapple juice. *Italian Journal of Food Science* 18(4) (2006): 367.

Butler, Michael. Growth of solutal ice dendrites studied by optical interferometry. *Crystal Growth & Design* 2(1) (2002): 59–66.
Deshpande, S.S., H.R. Bolin, and D.K. Salunkhe. Freeze concentration of fruit juices. *Food Technology* 36(5) (1982): 68–82.
Dette, S. and H. Jansen. Freeze concentration of black currant juice. *Chemical Engineering & Technology* 33(5) (2010): 762–766.
Flesland, O.L.A. Freeze concentration by layer crystallization. *Drying Technology* 13(8–9) (1995): 1713–1739.
Gu, X., T. Suzuki, and O. Miyawaki. Limiting partition coefficient in progressive freeze-concentration. *Journal of Food Science* 70(9) (2005): 546–551.
Gunathilake, M., K. Shimmura, M. Dozen, and O. Miyawaki. Flavor retention in progressive freeze-concentration of coffee extract and pear (La France) juice flavor condensate. *Food Science and Technology Research* 20(3) (2014): 547–554.
Hartel, R.W. Evaporation and freeze concentration. In *Handbook of Food Engineering*, D.R. Heldman and D.B. Lund (eds.), pp. 341–392. New York: Marcel Dekker (1992).
Hsieh, H.C. Desalinating process. (2008). US Patent 7,467,526.
Huige, N.J.J.. Nucleation and growth of ice crystals from water and sugar solutions in continuous stirred tank crystallizers. PhD dissertation (1972). TU Eindhoven, Eindhoven, the Netherlands.
Khajehei, F., M. Niakousari, M.H. Eskandari, and M. Sarshar. Production of pomegranate juice concentrate by complete block cryoconcentration process. *Journal of Food Process Engineering* 38(5) (2015): 488–498.
Kowalczewski, P., K. Celka, W. Bialas, and G. Lewandowicz. Antioxidant activity of potato juice. *Acta scientiarum polonorum. Technologia alimentaria* 11(2) (2012): 175–181.
Liu, L., O. Miyawaki, and K. Hayakawa. Progressive freeze-concentration of model liquid food. *Food Science and Technology International, Tokyo* 3(4) (1997): 348–352.
Liu, L., O. Miyawaki, and K. Hayakawa. Progressive freeze-concentration of tomato juice. *Food Science and Technology Research* 5(1) (1999): 108–112.
Luo, C., W. Chen, and W. Han. Experimental study on factors affecting the quality of ice crystal during the freezing concentration for the brackish water. *Desalination* 260(1) (2010): 231–238.
Miyawaki, Osato. Analysis and control of ice crystal structure in frozen food and their application to food processing. *Food Science and Technology Research* 7(1) (2001): 1–7.
Miyawaki, O., M. Gunathilake, C. Omote et al. Progressive freeze-concentration of apple juice and its application to produce a new type apple wine. *Journal of Food Engineering* 171 (2016): 153–158.
Miyawaki, O., L. Liu, Y. Shirai et al. Tubular ice system for scale-up of progressive freeze concentration. *Journal of Food Engineering* 69(1) (2005): 107–113.
Moreno, F.L., M.X. Quintanilla-Carvajal, L.I. Sotelo et al. Volatile compounds, sensory quality and ice morphology in falling-film and block freeze concentration of coffee extract. *Journal of Food Engineering* 166 (2015): 64–71.
Moreno, F.L., M. Raventós, E. Hernández et al. Block freeze-concentration of coffee extract: Effect of freezing and thawing stages on solute recovery and bioactive compounds. *Journal of Food Engineering* 120 (2014): 158–166.
Morison, K.R. and R.W. Hartel. Evaporation and freeze concentration. In *Handbook of Food Engineering*, D.R. Heldman and D.B. Lund (eds.), pp. 495–552. New York: CRC Press (2007).
Nakagawa K., H. Nagahama, S. Maebashi, and K. Maeda. Usefulness of solute elution from frozen matrix for freeze-concentration technique. *Chemical Engineering Research and Design* 88(5) (2010): 718–724.

Normohamadpor Omran, M., M. Pirouzifard, P. Aryaey, and M. Hasan. Cryoconcentration of sour cherry and orange juices with novel clarification method; Comparison of thermal concentration with freeze concentration in liquid foods. *Journal of Agricultural Science and Technology* 15(5) (2013): 941–950.

Nunes, G.L., B.C.B. Boaventura, S.S. Pinto et al. Microencapsulation of freeze concentrated *Ilex paraguariensis* extract by spray drying. *Journal of Food Engineering* 151 (2015): 60–68.

Petzold, G. and J.M. Aguilera. Centrifugal freeze concentration. *Innovative Food Science and Emerging Technologies* 20 (2013): 253–258.

Petzold, G. and J.M. Aguilera. Ice morphology: Fundamentals and technological applications in foods. *Food Biophysics* 4 (2009): 378–396.

Petzold, G., K. Niranjan, and J.M. Aguilera. Vacuum-assisted freeze concentration of sucrose solutions. *Journal of Food Engineering* 115(3) (2013): 357–361.

Petzold, G., J. Moreno, P. Lastra et al. Block freeze concentration assisted by centrifugation applied to blueberry and pineapple juices. *Innovative Food Science and Emerging Technologies* 30 (2015): 192–197.

Pokorny, J. and S. Schmidt, Natural antioxidant functionality during food processing. In *Antioxidants in Food. Practical Applications*, J. Pokorny, N. Yanislieva, and M. Gordon (eds.) pp. 331–351. Boca Raton, FL: CRC Press (2001).

Rahman, M.S., M. Ahmed, and X.D. Chen. Freezing–melting process in liquid food concentration. In *Handbook of Food Preservation*, M.S. Rahman (ed.), 2nd edition, pp. 667–686. Boca Raton, FL: CRC Press (2007).

Ramaswamy, H. and M. Marcotte. *Food Processing. Principles and Applications*. Boca Raton, FL: Taylor & Francis (2006).

Ramos, F.A., J.L. Delgado, E. Bautista et al. Changes in volatiles with the application of progressive freeze-concentration to Andes berry (*Rubus glaucus* Benth). *Journal of Food Engineering* 69(3) (2005): 291–297.

Rempel, A.W., E.D. Waddington, J.S. Wettaufer, and M.G. Worster. Possible displacement of the climate signal in ancient ice by premelting and anomalous diffusion. *Nature* 411(6837) (2001): 568–571.

Raventós, M., E. Hernández, and J.M. Auleda. Freeze concentration applications in fruit processing. In *Advances in Fruit Processing Technologies*, S. Rodriguez and F.A.N. Fernandes (eds.), pp. 263–286. Boca Raton, FL: CRC Press (2012).

Schwartzberg, H.G. Food property effects in evaporation. In *Food Properties and Computer-Aided Engineering of Food Processing Systems*, R.P. Singh and A.G. Medina (eds.), pp. 443–470. Boston, MA: NATO ASI Series (1989).

Thijssen, H.A.C. Freeze-concentration. In *Advances in Preconcentration and Dehydration of Foods*, A. Spicer (ed.), pp. 115–116. Boston, MA: Applied Science Publishers (1974).

Virgen-Ortíz, J.J., V. Ibarra-Junquera, P. Escalante-Minakata et al. Improving sodium dodecyl sulfate polyacrylamide gel electrophoresis detection of low-abundance protein samples by rapid freeze centrifugation. *Analytical Biochemistry* 443(2) (2013): 249–251.

Virgen-Ortíz, J.J., V. Ibarra-Junquera, J.A. Osuna-Castro et al. Method to concentrate protein solutions based on dialysis-freezing-centrifugation: Enzyme applications. *Analytical Biochemistry* 426(1) (2012): 4–12.

Welti-Chanes, J., D. Bermúdez, A. Valdes-Fragoso et al. Principles of freeze-concentration and freeze-drying. In *Handbook of Frozen Foods*, Y.H. Hui, P. Cornillon, I.G. Legarette et al. (eds.), Chapter 2, pp. 13–17. New York: Marcel Dekker (2004).

Welti-Chanes, J., D. Bermúdez, A. Valdes-Fragoso et al. Principles and applications of freeze concentration and freeze-drying. In *Handbook of Food Science, Technology, and Engineering*, Y.H. Hui (ed.), vol. 3, pp. 106–102. Boca Raton, FL: CRC Press (2006).

Zaritzky, N. (2006). Physical–chemical principles in freezing. In *Handbook of Frozen Food Processing and Packaging*, D.-W. Sun (ed.), pp. 3–31. Boca Raton, FL: CRC Press (2006).

9 Influence of Osmotic Pretreatment and Drying Air Properties on Bioactive Compounds of Fruits

Patchimaporn Udomkun, Marcus Nagle, Dimitrios Argyropoulos, and Joachim Müller

CONTENTS

9.1 Introduction 195
9.2 Bioactive Compounds in Fruits 196
    9.2.1 Vitamin C 196
    9.2.2 Carotenoids 197
    9.2.3 Phenolic Compounds 198
9.3 Effect of Osmotic Pretreatment on Bioactive Compounds 199
    9.3.1 Concentration and Type of Osmotic Agent 199
    9.3.2 Agitation and Material Geometry 200
    9.3.3 Temperature 201
    9.3.4 Additional Influences on Osmotic Treatment 201
9.4 Effect of Drying Air Properties 202
    9.4.1 Drying Temperature 204
    9.4.2 Air Velocity 207
    9.4.3 Humidity 208
9.5 Conclusions 208
References 208

9.1 INTRODUCTION

Preservation of fruits can reduce postharvest losses and increase availability of fruit products globally. Convective air drying is widely used for preserving many fruit products because of its easy operation and low investment and running costs. Drying of fruits helps stabilize production income at a reasonable level, because seasonality and overproduction tend to cause market surplus, resulting in low wholesale prices.

However, drying processes can cause a considerable decrease in the nutritive value of fruit products. In order to preserve essential fresh fruit characteristics and to gain advantages from an economic perspective, combined processing methods involving osmotic dehydration and convective air drying are worth consideration (Sagar and Suresh Kumar, 2010).

Osmotic dehydration is a common treatment applied to fruits before drying, mainly because of two principal advantages: reduced energy requirement and improved final quality (Chandra and Kumari, 2013). The reduction in energy requirement is due to moisture removal, which happens by (1) the induction of water loss through cell membranes because of osmotic pressure and (2) subsequent flow along the intercellular space before diffusion into the hypertonic solution (Sereno et al., 2001). Improved quality through osmotic dehydration is demonstrated by diminished color changes, enhanced flavor, and decreased shrinkage and structural collapse (Reppa et al., 1999; Riva et al., 2005). Although the final osmodehydrated fruit products generally have enhanced quality after hot air drying as compared to those without pretreatment, other natural bioactive solutes present in the cells, such as organic acids, minerals, and salts, can also be leached into the osmotic solution because the cell membrane is not highly selective (Lerici et al., 1985; Lazarides et al., 1997; Shi and Le Maguer, 2002). Meanwhile, increasing demand for high-quality fruit products requires the refinement of processing, with the purpose of improving the efficiency of dehydration procedures and optimizing physicochemical and nutritional properties (Fernandes et al., 2011). Accordingly, the application of pretreatments and the control of drying parameters—airflow direction, temperature, air velocity, and humidity—are critical aspects in the production of dried fruits.

9.2 BIOACTIVE COMPOUNDS IN FRUITS

Recently, emphasis on the importance of fruit consumption has increased considerably, because fruits are naturally rich in various beneficial antioxidant compounds, such as polyphenols, carotenoids, and vitamins (Kaur and Kapoor, 2001; Proteggente et al., 2002). Many experimental, clinical, and epidemiologic studies have exemplified the health potentials of bioactive compounds in protection against cancer, as well as neurologic diseases and cardiovascular diseases (CVDs). This effect is due to their biochemical properties, such as free radical scavengers, hydrogen donors, singlet oxygen quenchers, and metal ion chelators (Ikram et al., 2009; Vega-Gálvez et al., 2009). In fruits, bioactive compounds can be present bound to the plant cell membranes or exist as free compounds. Therefore, the amount of these compounds retained in fruits depends on their stability during food preparation and processing before consumption (Rawson et al., 2011), which mainly relate to their sensitivity toward thermal degradation and oxidation processes (Kalt, 2005; Amarowicz et al., 2009; Leong and Oey, 2012).

9.2.1 VITAMIN C

Vitamin C, also known as ascorbic acid, is a water-soluble ketolactone with two ionizable hydroxyl groups, meaning it dissolves in water and is not stored in the body. It is commonly found in citrus, bell pepper, guava, tomato, kiwi, and papaya. In human health, vitamin C acts as a natural antioxidant, which can prevent damage of macromolecules in the body due to free radicals, reactive oxygen species, and nitrogen species (Padayatty et al., 2003). Vitamin C intake has been associated with the reduction of risks in many diseases, for instance, cancer, CVD, age-related macular degeneration, common cold, and asthma (Bendich and Langseth, 1995). Vitamin C has also been found to be crucial in the synthesis of collagen for efficient healing of wound and growth of bones, teeth, ligaments, and tendons; in the biosynthesis of neurotransmitters and L-carnitine; in the prevention of scurvy; and in the metabolism of cholesterol (Naidu, 2003). In fruit species, vitamin C can act as an antioxidant that inhibits enzymatic browning and other quality changes that occur by oxidative reactions. Vitamin C loss in fruit products is used as a chemical indicator during handling to evaluate the applied processing. Santos and Silva (2008) reviewed the dynamics of vitamin C during drying. Degradation depends on factors such as temperature, light, pH, and moisture content, as well as the presence of oxygen and metal ion catalysts. Furthermore, vitamin C in dried products is also considered a general indicator of the preservation of other less labile nutrients. In order to avoid excessive losses of vitamin C during processing, retention kinetics have been documented, with authors reporting first-order models to describe the degradation reaction in several processed fruits (Goula and Adamopoulos, 2006; Nicoleti et al., 2007).

9.2.2 CAROTENOIDS

Carotenoids are natural pigments that are responsible for the yellow, orange, and red colors in various fruits and vegetables. On the basis of functional groups, carotenoids can be classified into two groups. Xanthophylls contain oxygen as a functional group and include lutein and zeaxanthin, whereas carotenes contain only the parent hydrocarbon chain without any functional group and include α-carotene, β-carotene, and lycopene. The addition of polar groups (epoxy, hydroxyl, and keto) alters the polarity of carotenoids and affects their biological functions (Saini et al., 2015). In addition, bioavailability of certain carotenoids may depend on their deposition in cellular chromatoplasts (Schweiggert et al., 2011) and the presence of triglycerides (Yeum and Russell, 2002).

Carotenoids are known to be substantial intermediates in the biosynthesis of vitamin A (Haskell, 2013). Carotene is the most efficient provitamin A carotenoid that can be converted to retinol. Other provitamin A carotenoids include α-carotene, β-carotene, and β-cryptoxanthin. Vitamin A deficiency is a public health problem among young children, pregnant women, and nursing mothers in many low-income and some middle-income countries (West and Darnton-Hill, 2008). Vitamin A deficiency leading to xerophthalmia is an important cause of childhood blindness and is a major underlying cause of mortality and morbidity in children (Sherwin et al., 2012). Lutein and zeaxanthin (having no provitamin A activity) are the only carotenoids found in the retinal and lens tissues of the human eye. A diet rich in these particular carotenoids is believed to reduce age-related macular degeneration and cataract formation (Meyers et al., 2014). Lycopene has been found to be a powerful antioxidant commonly responsible for the protection of cellular system from a variety of reactive oxygen species and reactive nitrogen species. It also helps reduce CVDs risk in humans (Müller et al., 2015; Saini et al., 2015). Overall, epidemiological studies have demonstrated that the consumption of carotenoids-rich food is associated with a lower incidence of certain cancers and CVDs (Sharoni et al., 2012).

Fruits that are typically rich in carotenoids include tomato, papaya, melons, squash, pumpkin, bell pepper, sweet corn, and citrus. During drying, loss of carotenoids due to heat, low pH, dissolved oxygen, and light exposure have considerable consequences not only on the product color, which is an important attribute in the acceptability of dried fruits, but also on the health-promoting effects of these bioactive compounds. Although many factors have detrimental effects on carotenoids, including destruction and isomerization (with the loss of vitamin A activity), other effects may be beneficial, for example, increase in bioavailability (Thane and Reddy, 1997; Shi and Le Maguer, 2000). Overall, it is certain that the type and extent of processing can have a considerable effect on the type and content of carotenoids in dried fruit products.

9.2.3 Phenolic Compounds

Phenolic compounds are secondary metabolites widely found in fruits. These can be classified into water-soluble compounds, such as phenolic acids, phenylpropanoids, flavonoids, and quinones, or water-insoluble compounds, such as condensed tannins, lignins, and cell-wall-bound hydroxycinnamic acids (Haminiuk et al., 2012). The developing interest in these substances is not only due to their major influence on sensory qualities (such as color, flavor, and taste) of the fruit products but also due to their antioxidant property and free-radical scavenging capacity. Phenolic compounds have attracted increasing attention as potential agents for the prevention and treatment of some diseases associated with oxidative stress in humans, for example, cancer, multiple sclerosis, autoimmune diseases, and Parkinson disease (Kähkönen et al., 2001; Miletić et al., 2012). These compounds also have biological uses as antimicrobial, anti-inflammatory, antimutagenic, anticarcinogenic, antiallergic, antiplatelet, and vasodilatory agents (Boyer and Liu, 2004; Scalbert et al., 2005; Henríquez et al., 2010). Polyphenols are also known to protect against the harmful actions of reactive oxygen species (Robards et al., 1999; Kähkönen et al., 2001).

The composition, content, and distribution of phenolic phytochemicals depend on fruit ripeness, cultivar specificities, cultivation practices, geographic origin, growing season, and postharvest conditions (Robards et al., 1999; Kayano et al., 2003; Bureau et al., 2009). Moreover, Haminiuk et al. (2012) claimed several factors that can interfere with the release and/or absorption of polyphenols from fruits, such as (1) chemical structure and interactions with different macromolecules of proteins and fibers, (2) processing methods, and (3) biological interactions between polyphenols and cells and enzymes and proteins in the gastrointestinal system. Phenolic compounds are sensitive to heat and oxidation, and therefore, environmental variables, including temperature, oxygen concentration, and exposure time, are critical factors that affect the final content of phenolic compounds in processed fruit products. Additional elements that may affect their reactivity include water activity, enzymes, pH, the variety of phenolics, and food composition and structure (McSweeney and Seetharaman, 2013).

9.3 EFFECT OF OSMOTIC PRETREATMENT ON BIOACTIVE COMPOUNDS

Osmotic dehydration is commonly used for the partial removal of water from fruit tissues. This is done by immersing the fruit in a hypertonic aqueous solution, especially sugar and salt solutions. This process has received much attention as a drying pretreatment, in order to reduce the total processing time and energy consumption. It also inhibits enzymatic browning and thus helps in retaining the natural color of fruits, without the addition of sulfites or preservatives, and in preserving the volatile aromas during subsequent drying. Osmotic dehydration promotes the diffusion of water from fruit tissues into the solution and the diffusion of solutes from the osmotic solution into the product. However, the osmotic process induces changes that affect the physiologic properties of fruit tissues, including undesirable effects on turgor pressure, splitting and degradation of the middle lamella, alterations in cell wall resistance, deformation and failure of cell walls, lysis of plasmalemmas and tonoplasts, and tissue shrinkage (Chiralt et al., 2001). Together with tissue alteration, the penetration of solutes can additionally change the nutritional and functional qualities of fruits. During osmotic pretreatment, a loss of vitamins, polysaccharides, and minerals, which flow from the fruit to the osmotic solution, has been observed (Torreggiani and Bertolo, 2001; García-Martínez et al., 2002; Peiró et al., 2006). Giovanelli et al. (2013) manifested that osmotic pretreatment resulted in a great loss of total phenolics, total anthocyanins, and the antioxidant activity in blueberries, especially during sucrose pretreatment, when compared to fructose and glucose treatments. On the contrary, it was found that osmotic treatment did not reduce the contents of polyphenols and flavonoids, as well as the antioxidant properties, in button mushrooms (Singla et al., 2010). Kinetics of the osmotic dehydration process depends upon factors such as the concentration and temperature of the osmotic solution, agitation, food-to-osmotic-solution ratio, food structure, shape and size, molecular weight of the osmotic solute, and hydrostatic pressure (Rastogi et al., 2002; Telis et al., 2004; Silva et al., 2012, 2014).

9.3.1 Concentration and Type of Osmotic Agent

It is well established that increased concentration of the osmotic solution results in corresponding increases in water loss to equilibrium level and drying rate (Gekas and Mavroudis, 1998; Gekas et al., 1998; Shi and Le Maguer, 2002; Li and Ramaswamy, 2010; Nahimana et al., 2010). Studies by Saurel et al. (1994) demonstrated that a dense solute-barrier layer formed at the surface of the food material when the concentration of the solution increased, which enhanced the dewatering effect and reduced the loss of nutrients during the process. A similar solute-barrier layer also formed in the osmotic solution with higher-molecular-weight solutes, even at low concentration. Djendoubi Mrad et al. (2013) investigated the effect of sucrose concentration on the quality attributes of pears. The result showed that increasing the concentration of sucrose had a noticeable effect on the loss of ascorbic acid and total phenolic content of pear pieces. The loss of ascorbic compounds could be explained by the fact that the mass flux inside the fruits and at the surface depends on the concentration of the osmotic agent. In addition, the loss of phenolic compounds during osmodehydration of fruits may be attributed to the diffusion of hydrophilic phenolic compounds with water into the osmotic solution (Renard, 2005; Devic et al., 2010). However, Djendoubi Mrad et al. (2013) commented that this explanation is not sufficient because the essential phenolic compounds of pear are procyanidins, which are polymerized compounds and cannot leach easily in water. Beserra-Almeida et al. (2011) studied the effect of sucrose concentration on the retention of polyphenols and the antioxidant activity in osmotically dehydrated bananas. It was observed that the higher the solution's concentration, the greater the retention of the antioxidant activity. This is likely the result of the protective effect on the fruit surface promoted by the incorporation of sucrose in the product, thus preventing the outflow of compounds that confer the antioxidant activity. The same concentration also enhanced the retention of tannins, which compose one of the groups of substances that contribute to the antioxidant activity in fruits.

The specific effect of the osmotic solution is of great importance when choosing the osmotic agent. Several solutes, alone or in combination, have been used in hypertonic solutions for osmotic dehydration (Ispir and Tòrul, 2009; Mahayothee et al., 2009; Vicente et al., 2012). The common solute types used as an osmotic agent are sucrose, glucose, sorbitol, glycerol, glucose syrup, corn syrup, and fructooligosaccharide. Generally, osmotic agents with low molecular weight penetrate easily into fruit cells as compared to osmotic agents with higher molecular weight (Phisut et al., 2013). Nonetheless, cost, organoleptic compatibility with the end product and additional preservation action by the solute are important factors considered while selecting osmotic agents (Torreggiani, 1995); however, their effects on retention of bioactive compounds should be mainly considered. Giovanelli et al. (2013) reported that total phenolics and the antioxidant activity of blueberries were better preserved in glucose and fructose solutions, whereas anthocyanin compounds were better preserved in sucrose solution. They attributed this result to the formation of a peripheral layer of sucrose on the berry surface, which slows down the diffusion phenomena, as previously described in osmodehydrated strawberry (Riva et al., 2005).

9.3.2 Agitation and Material Geometry

Osmotic dehydration can be enhanced by agitation or circulation of the osmotic solution around the sample. Agitation ensures a continuous contact of the sample surface with concentrated osmotic solution, securing a large gradient at the product–solution interface. For short process periods, agitation has no effect on the solids gain; however, over longer holding times, solids gain decreases drastically with agitation. The agitation-induced decrease in the rate of solids gain over longer osmosis periods can be an indirect effect of higher water loss (due to agitation), altering the solute concentration gradient inside the food particle. Since diffusion of solutes into the natural tissue is slow, most of the solute accumulates in a thin subsurface layer.

The geometry of fruit pieces affects the behavior of the osmotic concentration due to the variation of the surface area per unit volume (or mass) and diffusion length of water and solutes involved in mass transport. According to Lerici et al. (1985), up to a certain total surface area/half thickness (A/L) ratio, higher-specific-surface-area sample shape, such as rings, showed higher water loss and sugar gain value compared to lower-surface-area-samples, such as slices and sticks.

9.3.3 Temperature

Temperature is the most important variable that affects the mass transfer during osmotic dehydration. Beristain et al. (1990) stated that an increase in temperature of osmotic solution results in increases in water loss, whereas solids gain is less affected by temperature. Considering the effect of temperature on the bioactive compounds in fruits, Almeida et al. (2014) found that increasing drying temperature led to a marked decrease of phenolic compounds in banana. Similar results were observed by Djendoubi Mrad et al. (2013) for osmodehydrated pears. They reported that an increase in temperature of the sucrose solution from 20°C to 60°C resulted in a loss of ascorbic acid and total phenolic content in the samples. This finding corroborates the results of Devic et al. (2010), who observed that an increase in temperature favored the collapse of cells, causing a reduction of the permeability and selectivity of the membrane and favoring the leaching of phenolic compounds toward the osmotic solution, together with the diffusion of water. Another mechanism that should be considered is hydrolysis of the molecules, which reduces the polymerization degree of some phenolic compounds and favors the leaching of compounds with lower molecular weight. This mechanism appears to be more dependent on time than on temperature (Devic et al., 2010). The increase in temperature of the osmotic dehydration process can also enhance the thermal degradation of cell membrane integrity, promoting oxidation reactions of the phenolic compounds (Djendoubi Mrad et al., 2013). Furthermore, Phisut et al. (2013) reported that less 1,1-Diphenyl-2-picryl-hydrazyl (DPPH) antioxidant activity and lower contents of phenolic compounds and vitamin C were observed in cantaloupe slices produced by slow osmotic dehydration (30°Brix sucrose solution for 24 h, then 40° and 50°Brix sucrose solution for 24 h) compared to fast osmotic dehydration (50°Brix sucrose solution for 24 h). This could be attributed to the idea that extended exchange in slow osmotic dehydration environments may induce a greater loss of natural solutes such as acids, vitamins, and small molecules of phenolic compounds into the osmotic solution (Vijayakumari et al., 2007; Devic et al., 2010).

Alternatively, cold or freezing conditions have also been applied before or during osmotic treatment. Freezing fruit materials before submersion in the osmotic solution enhances the dehydration effects and uptake of solutes while subsequently preserving the structure of the tissues (Siramard and Charoenrein, 2014). In addition, it has been found that freezing pretreatment and osmotic drying at low temperatures (5°C) help preserve bioactive compounds in fruits such as strawberries and sour cherries (Taiwo et al., 2003; Blanda et al., 2009; Nowicka et al., 2015).

9.3.4 Additional Influences on Osmotic Treatment

Application of additional mechanisms such as vacuum conditions, increased hydrostatic pressure, ultrasonic energy, and electrical pulses to osmotic dehydration can help mitigate intrinsic problems typically associated with slow mass-transfer rates during osmotic drying (Rastogi et al., 2005). Shi and Fito Maupoey (1993) were among the first to document the benefits of applying vacuum conditions during osmotic drying. Subsequently, much work has been done to investigate the benefits of vacuum treatments to intensify capillary flow and water transfer and document the basic effects on moisture reduction, solute uptake, and color retention (Ito et al., 2007; Lombard et al., 2008; Corrêa et al., 2010; An et al., 2013; Ramallo et al., 2013; Corrêa et al., 2014). Although much of the previous work has focused on the impregnation of the product with the osmotic agents (sugars in most cases), much less investigation has been carried out to determine the effect of this treatment on the bioactive compound aspect of the fruit products. Some studies note the preservation of volatile compounds (Chiralt and Talens, 2005; Torres et al., 2007; Pino et al., 2008). Interestingly, the application of vacuum conditions during osmotic treatment can also be used for fortification of fruit products (Fito et al., 2001), most commonly with calcium (Torres et al., 2008; Barrera et al., 2009), but also with β-carotene (Santacruz-Vázquez et al., 2008) and phenolic compounds (Bellary and Rastogi, 2012). In contrast, Bórquez et al. (2010) reported high losses of vitamin C during osmotic treatment of raspberries under vacuum conditions. A number of recent studies have examined the benefits of applying ultrasound and electric impulses during osmotic treatment (frequently in conjunction with vacuum conditions) to facilitate moisture removal and solute uptake (Deng and Zhao, 2008; Moreno et al., 2011, 2012; Kek et al., 2013; Corrêa et al., 2015). However, the effect of these approaches on retention or leaching of bioactive substances remains unclear.

Studies have shown that an increase of atmospheric pressure and consequently hydrostatic pressure in osmotic dehydration systems results in changes of mass transfer (Nuñez-Mancilla et al., 2011; Verma et al., 2014). Nuñez-Mancilla et al. (2013, 2014) exhibited that the radical scavenging activity of strawberries showed higher antioxidant activity at 400 MPa rather than at low pressure (100, 200, and 300 MPa). The total phenolic content also increased with pressure, being maximum at 400 MPa, whereas vitamin C content did not show changes with increasing pressure. Similar result reported by Patras et al. (2009), who worked with strawberry purées, indicated that pressurization preserved 94% of vitamin C in the samples.

9.4 EFFECT OF DRYING AIR PROPERTIES

Convection drying is one of the most common and cost-effective preservation methods used to extend the storability of fruits in ambient temperatures. The major objective of the drying process for fresh fruits is to reduce moisture content to a safe storage level, usually from approximately 70–85 g/100 g to approximately 10–15 g/100 g wet basis. The removal of moisture prevents the development of microorganisms that cause decay and minimizes many of the moisture-intermediated deteriorative reactions such as enzymatic reactions, nonenzymatic browning, and oxidation of lipids and pigments. It also brings about a substantial reduction in weight and volume, thus minimizing packing, storage, and transportation costs.

Drying is a simultaneous heat and mass transfer process, accompanied by phase change. Two phases, a constant rate period and a falling rate period, can be observed with respect to the rate of moisture removal during the hot-air drying process. During drying, evaporation occurs at regions near the surface, and the surface water is removed first, which is signified by the constant rate period. The factors influencing the rate of drying in this period include temperature, air velocity, and relative humidity (Rizvi, 1995). As drying progresses, the drying rate decreases with the reduction of moisture content. The falling rate period is demonstrative of an increased resistance to both heat and mass transfer via the inner cells by mechanisms of capillary flow, liquid diffusion, and vapor diffusion. The falling rate is influenced by drying temperature, as well as chemical composition, physical structure, and thickness of the product (Schultz et al., 2007).

Moisture migration and heat transfer induce stress inside fruit materials, which consequently leads to numerous physical, chemical, and nutritional changes (Krokida et al., 2001; Aguilera, 2003; Lewicki and Jakubczyk, 2004; Prachayawarakorn et al., 2008). The physical properties of fruits change primarily because the viscoelastic matrix contracts into the space previously occupied by the water removed from the cellular structure to the surface and surrounding air. Several studies have attempted to characterize the physical changes in terms of parameters such as changes in volume, area, size, and shape (Ochoa et al., 2002; Aguilera, 2003; Kerdpiboon et al., 2007; Toğrul and İspir, 2007). However, these external changes are directly related to the microstructure of materials being processed.

Another complex phenomenon observed during drying that needs to be controlled is discoloration. It is well documented that enzymatic and nonenzymatic reactions and pigment degradation can lead to color changes in dried fruits. Enzymatic browning takes place in early stages of processing through release of polyphenols and activation of polyphenoloxidase (PPO). Normally, the primary oxidation products, named *o*-quinones, are highly reactive and prompt further reactions with phenolic and nonphenolic compounds, such as ascorbic acid and primary and secondary amines, in fruit flesh (Nicolas et al., 2003). Enzymatic browning can be controlled by reducing phenoxyl radicals and quinone forms of phenolics to precursor phenols in a coupled oxidation reaction (Rojas-Graü et al., 2007). The application of osmotic dehydration together with acid drips and thermal treatments is a strategic way to reduce this browning, even though this may be detrimental to the flavor and nutritional qualities of the products (Chow et al., 2011). Nonenzymatic browning, called the *Maillard reaction*, is also a cause of color changes that usually occur during drying and storage. Owing to the presence of sugars and amino acids with intermediate moisture content, temperatures more than 50°C, pH 5–7, and long processing times, brown pigments, such as Amadori compounds, and their derivatives are formed.

Several studies have reported that drying under different conditions may cause the loss of bioactive compounds, such as vitamin C, tocopherols, polyphenols, and carotenoids, in fruits (Zepka and Mercadante, 2009; Bennett et al., 2011; Kurozawa et al., 2014; Megías-Pérez et al., 2014; Sogi et al., 2014). Nonetheless, drying process was also reported to increase the concentration of bioactive compounds. Drying can lead to the solubilization of compounds bounded to insoluble fiber portions in the product (Toor and Savage, 2005) or freed by cell wall breakage (Omoni and Aluko, 2005). Thus, this complex impact of drying air properties opens interesting prospects for finding optimal process adjustments to fulfill the expectation and demands of consumers in terms of high-quality dried products that are both safe and nutritious (Sablani, 2006).

9.4.1 Drying Temperature

Generally, the loss of bioactive compounds depends on the properties of drying air, particularly drying temperature. Bennett et al. (2011) stated that high drying temperature can lead to a complete loss of phenolic compounds in several fruits. Vega-Gálvez et al. (2012) demonstrated that an increase in drying temperature caused degradation of total phenolics in dried apples. Specifically, the mean values of total phenolic contents diminished by 14%–30% when temperature was raised from 40°C to 80°C. A similar result regarding the influence of temperature on the degradation of these phenolic compounds was also observed in grapes (Adiletta et al., 2015) and plums (Del Caro et al., 2004). However, some contrasting results have been reported pertaining to the influence of drying temperature on these functional compounds. As mentioned by Udomkun et al. (2015), total phenolic content in dried papayas greatly increased by approximately 27%–40% when temperature was raised from 50°C to 80°C. Djendoubi Mrad et al. (2012) reported that a maximum reduction of 30% total phenolic contents was observed in pear during drying at 30°C for 10 h, whereas drying at 70°C for 2 h resulted in a reduction of only 3%. Madrau et al. (2009) also found that apricots dried at 75°C were considerably higher in hydroxycinnamic acids than those dried at 55°C.

On the contrary, the decrease in flavonols, another type of phenolic compound, was considerably more marked in apricots dried at 75°C. The lower drying temperature is believed to be responsible for the difference in phenolic compounds. The degradation of hydroxycinnamic acids under the experimental conditions (high presence of air and thus oxygen) could have been influenced by the PPO activity. During the dehydration process, the PPO activity remains high for longer periods when the drying temperature is approximately 55°C–60°C, whereas the enzyme becomes inactivated after short exposure periods at temperatures of 75°C–80°C (Arslan et al., 1998). Moreover, unlike hydroxycinnamic acids, the degradation of flavonols is not directly correlated to the PPO activity, because these compounds disappear proportionally with the increase in temperature. Flavonoids are therefore not degraded by the same mechanism as that of phenolic acids. This means they are not direct or poor substrates for oxidases, because PPO does not act directly on glycosides (Amiot et al., 1995).

Garau et al. (2007), Devic et al. (2010), and Da Silva et al. (2013) claimed that continuing drying at low temperature caused higher degradation of total phenolic contents in orange byproducts, apple, and pineapple, respectively. Que et al. (2008) suggested that the generation of phenolic compounds at high temperatures might be due to the availability of phenolic precursors from the nonenzymatic interconversion between phenolic molecules. Likewise, during thermal processing, their release might occur due to the breakdown of cellular constituents and covalent bonds, because phenolic acids are mainly bonded to carbohydrate and proteinaceous moieties. As a result, the released polyphenolic constituents become more responsive for extraction (Hartley et al., 1990). Although the disruption of cells during thermal processing may also result in the release of oxidative and hydrolytic enzymes, these enzymes are capable of oxidizing endogenous polyphenolics. However, exposure to high temperature (even for a short time) can inactivate these enzymes and protect polyphenolics from further decomposition (Wojdyło et al., 2009). Michalczyk et al. (2009) found that polyphenolic and anthocyanin contents in bilberries were considerably decreased because of long exposure time and high temperatures. De Ancos et al. (2000) suggested that the degradation of anthocyanins and other polyphenolic compounds may depend on many factors, such as PPO activity, organic acid content, sugar concentration, and pH, rather than only heat treatment.

Udomkun et al. (2015) also stated that the antioxidant activity of papaya obtained by using DPPH and ferric reducing antioxidant power (FRAP) assays were distinctly affected by thermal processing at high temperature, whereas 2,2-azinobis-3-ethyl-benzothiazoline-6-sulfonic acid (ABTS) antioxidant activity was not influenced by temperature. A similar result was reported by López et al. (2012), who claimed that a maximum antioxidant capacity in goldenberry was observed when the sample was dried at 90°C for both FRAP and DPPH methods when compared to the sample dried at 50°C. Although some endogenous antioxidants in biological materials can be destroyed during drying, the application of higher temperature may not negatively affect the antioxidant activity. Miranda et al. (2009) and Vega-Gálvez et al. (2009) have described this behavior and speculated that it might be related with the formation and accumulation of melanoidins and/or the Maillard reaction products, which also function as antioxidant substances through a chain-breaking mechanism. Several studies of dried fruits mentioned that in cases where strong browning takes place, strong antioxidant capacity is also observed (Krokida et al., 2003, Que et al., 2008; Rigi et al., 2014). On the other hand, drying at lower temperature over prolonged times can cause a decrease of antioxidant capacity because of oxidation process (Garau et al., 2007).

Many authors have correlated the presence of phenolics with antioxidant activity in several dried fruits. Frequently, an electron-transfer-based antioxidant capacity assay, particularly DPPH assay, shows more linear correlation with the total phenolic contents. However, it does not imply that the antioxidant activity is influenced by phenolic compounds alone. Other constituents, such as reducing carbohydrates, tocopherols, carotenoids, terpenes, and pigments (and also the synergistic effect among them), surely contribute to the total antioxidant activity also (Babbar et al., 2011). However, data on the effects of drying process on total phenolic compounds and antioxidant activity of fruits are rather conflicting because of several factors, likely explained by the difference in drying methods, type of extraction solvents, antioxidant assays, and interactions of several antioxidant reactions (Manzocco et al., 2000; Sulaiman et al., 2011).

The nutritional importance of vitamin C (ascorbic acid) as an essential vitamin is well-established. Since vitamin C is water soluble, it is the least stable of all vitamins and can be easily destroyed during processing and storage. In the literature, it is stated that vitamin C is heat sensitive and prone to degradation under different influences of pH, temperature, oxygen concentration, metal catalysts, and enzymes (Santos and Silva, 2008; Leong and Oey, 2012). The increase of vitamin C loss with air temperature has been reported during drying of several fruits, such as pear (Djendoubi Mrad et al., 2012), pineapple (Ramallo and Mascheroni, 2012), salak (Ong and Law, 2011), kiwi (Kaya et al., 2010), and papaya (Kurozawa et al., 2014). Djendoubi Mrad et al. (2012) evaluated the vitamin C degradation during drying of pear at temperatures from 30°C to 70°C. According to the results, the highest loss of approximately 70% was observed in the samples dried at 70°C. Similarly, Ramallo and Mascheroni (2012) verified that pineapple samples dried at 45°C had better vitamin C retention than those obtained by air drying at 75°C. Gamboa-Santos et al. (2014) also found that higher retention of vitamin C in dried strawberries was particularly evident at the mildest temperatures of 40°C–50°C than at other temperatures.

In contrast, Yılmaz et al. (2015) reported that the degradation of vitamin C in gooseberry was high after being subjected to low drying temperature. This phenomenon may be due to the presence of oxygen and the enzyme L-ascorbate oxidase during drying at low temperatures, which results in rapid oxidation of ascorbic acid to dehydroascorbic acid (DHAA) and subsequent hydrolysis to diketogulonic acid. Consequently, the oxygen-scavenging ability of ascorbic acid is compromised by the opening of lactone ring structure, thus destroying the antioxidant capacity of vitamin C (Leong and Oey, 2012). It is known that degradation of ascorbic acid leads to several products through DHAA, such as ketogulonic acid and furfural (Mertz et al., 2010). Nevertheless, Santos and Silva (2008) confirmed that at the beginning of the drying process, the effect of moisture content seems to be predominant, whereas the effect of temperature becomes more dominant as the process proceeds.

Several factors such as high temperature, light, oxygen, pH, catalysts, and metal ions have been found to affect the carotenoid contents in fruit products through the degradation of all-*trans* and *cis*-isomer and/or isomerization from all-*trans* to *cis*-isomer (Shi et al., 2002). In addition, Kuti and Konuru (2005) and García-Valverde et al. (2013) stated that contents of total carotenoids, lycopene, and β-carotene in tomatoes are significantly affected, not only by the degree of maturity but also by the variety and locality of cultivation. Vadivambal and Jayson (2007) mentioned that hot-air drying of agricultural products causes decomposition of carotenoids because of extended drying time and high temperature. This has been corroborated by many studies, for example, as reported by Pott et al. (2003) in mangoes from Kent and Tommy Atkins varieties, which were dehydrated at 75°C. Studies have revealed that in addition to the natural occurrence in plants, additional *cis*-isomers can be formed during thermal processing, resulting in the loss of provitamin A activity and reduction of bioavailability and bioconvertibility (Rodriguez-Amaya et al., 2008).

With respect to the effect of drying temperature on the degradation of carotenoids, Karabulut et al. (2007) demonstrated that convective hot-air drying at 50°C for both sulfurated and nonsulfurated apricots caused the greatest losses of β-carotene, whereas elevated temperatures at 75°C provided the lowest decrease in β-carotene content of the samples. This result is probably because the solubility of β-carotene increases at high temperatures. Thermal processing of tomatoes has been reported to increase the extractable lycopene content in processed products when compared to fresh samples (Dewanto et al., 2002). Another degrading effect on β-carotene content can be attributed to oxidation. However, Udomkun et al. (2015) found that drying temperature did not manifest any negative effects on quantitative changes in total lycopene, β-carotene, and β-cryptoxanthin contents of dried papayas. A slight variation was found for total lycopene and β-carotene changes during drying at different temperatures. It was unclear whether extended drying time or higher temperatures led to reduced carotenoid retention in dried papayas. Although several studies have shown that higher temperature can cause substantial loss of carotenoids, the extraction yield may also be improved, resulting in an apparent increase in the carotenoid contents. This can be attributed to the fact that most bioactive compounds are bound to other molecules or cell structures. Drying processes have a strong potential to damage the cell membrane, thus generating an opportunity for the bound bioactive compounds to be released from chromotoplasts into the cytoplasmic medium. Consequently, they are more easily extracted by solvents. In addition, Rodriguez-Amaya and Kimuru (2004) suggested that the susceptibility of the carotenoid compounds to degradation could increase by raising their exposure to oxidation process during drying at low temperatures for extensive time.

9.4.2 Air Velocity

Besides temperature, other drying air properties can also affect the activity and stability of bioactive compounds by chemical and enzymatic degradation and by volatilization and/or thermal decomposition (Dorta et al., 2012). With regard to the velocity of drying air, Vega-Gálvez et al. (2012) showed that total phenolic content and DPPH radical scavenging activity of apple samples increased with increasing drying air velocity. This was probably due to high convective forces acting at the air–solid interface, thus retarding heat diffusion into solid apples. The glycosides of phenolics, localized in the hydrophilic regions of the cell, such as vacuoles and apoplasts, or other soluble phenols in the cytoplasm and in the cell nuclei (Sakihama et al., 2002) apparently are protectively shielded from heat by the cell walls. Thus, internal resistance to heat diffusion is an important parameter to be considered when quality is at risk during heat treatment in the drying process. However, Udomkun et al. (2015) exhibited that air velocity did not clearly influence total phenolic content, antioxidant activities, and carotenoid contents in dried papayas.

As reported by Da Silva et al. (2013), vitamin C of pineapple was more preserved at an air velocity of 1.5 m s^{-1} than for the experimental runs carried out at 1.0 m s^{-1}. This behavior might be associated with the low interfacial accumulation of heat and moisture under high convective forces. Moreover, the degradation of vitamin C occurred probably because of the oxidation of ascorbic acid to DHAA over a longer-time drying. The oxidation of ascorbic acid is followed by the hydrolysis of DHAA to 2,3-diketogulonic acid. This substance may further undergo polymerization with amino acids, such as lysine or glutamic acid, leading to the active formation of brown pigments (Dewanto et al., 2002). However, the mechanism of ascorbic acid degradation is specific for each food system and depends on many factors, including other components in the food, pH, and reducing or oxidizing conditions (Serpen et al., 2007).

9.4.3 Humidity

Currently, limited work has been performed to determine the effect of humidity on retention of bioactive compounds in dried fruits. Sigge et al. (1999) reported that high humidity negatively affected vitamin C retention in green bell pepper only at high drying temperatures (65°C–75°C). Udomkun et al. (2015) mentioned that specific humidity did not show a significant effect on any chemical quality parameter of dried papaya samples. Furthermore, it was reported that some interaction effects were found to considerably influence quality alterations of air-dried papayas. For example, the synergies of drying temperature, specific humidity, and air velocity led to significant effects on antioxidant activities, total phenolic contents, and total carotenoids. These results suggested that all drying parameters have a significant relationship to the quality changes of papaya during drying. Therefore, the optimization of drying process, particularly combined drying conditions, could provide cost-effective solutions to produce dried fruits with high content of bioactive compounds and improve nutritional properties of these products.

9.5 CONCLUSIONS

Fruits are an important part of a healthy diet, because they are good sources of bioactive compounds. From a biological point of view, most fruits crops experience considerable postharvest losses due to rapid senescence, which causes high perishability. Consequently, postharvest procedures are required for generating fruit products with extended shelf life. Drying is one of the most widely used preservation methods that allows for greater flexibility in the availability and marketability of products, regardless of high production volume. Nowadays, there is an increasing demand for natural products, including high-quality dried fruits, in which nutritional and sensory properties have minimal alteration. Among the drying methods available for fruit processing industries, conventive air drying is considered a low-input conservation process, which involves the reduction of moisture in the product in order to slow down microbial and chemical deterioration. However, drying provokes numerous changes in materials, especially in terms of color, texture, and nutritional properties. These changes are can be mitigated by osmotic treatment and also by application of subsequent drying parameters. Thus, the optimization of osmotic treatment and drying process variables is recommended for consideration in order to improve nutritional properties of convectively dried fruits.

REFERENCES

Adiletta, G., Russo, P., Senadeera, W., and Di Matteo, M. Drying characteristics and quality of grape under physical pretreatment. *Journal of Food Engineering* 172: (2016) 9–18.

Aguilera, J.M. Drying and dried products under the microscope. *Food Science and Technology International* 9(3): (2003) 137–143.

Amarowicz, R., Carle, R., Dongowski, G., Durazzo, A., Galensa, R., Kammerer, D., Maiani, G., and Piskula, M.K. Influence of postharvest processing and storage on the content of phenolic acids and flavonoids in foods. *Molecular Nutrition & Food Research* 53(S2): (2009) S151–S183.

Amiot, M.J., Tacchini, M., Aubert, S.Y., and Oleszek, W. Influence of cultivar, maturity stage, and storage conditions on phenolic composition and enzymatic browning of pear fruits. *Journal of Agricultural and Food Chemistry* 43, (1995) 1132–1137.

An, K., Li, H., Zhao, D., Ding, S., Tao, H., and Wang, Z. Effect of osmotic dehydration with pulsed vacuum on hot-air drying kinetics and quality attributes of cherry tomatoes. *Drying Technology* 31(6): (2013) 698–706.

Arslan, O., Temur, A., and Tozlu, I. Polyphenol oxidase from Malatya apricot. *Journal of Agricultural and Food Chemistry* 46: (1998) 1239–1241.

Babbar, N., Oberoi, H.S., Uppal, D.S., and Patil, R.T. Total phenolic content and antioxidant capacity of extracts obtained from six important fruit residues. *Food Research International* 44: (2011) 391–396.

Barrera, C., Betoret, N., Corell, P., and Fito, P. Effect of osmotic dehydration on the stabilization of calcium-fortified apple slices (var. Granny Smith): Influence of operating variables on process kinetics and compositional changes. *Journal of Food Engineering* 92(4): (2009) 416–424.

Beserra-Almeida, M.M., Machado de Sousa, P.H., Campos Arriaga, A., Matias do Prado, G., Carvalho Magalhães, C., Arraes Maia, G., and Gomes de Lemos, T. Bioactive compounds and antioxidant activity of fresh exotic fruits from Northeastern Brazil. *Food Research International* 44: (2011) 2155–2159.

Bellary, A.N. and Rastogi, N.K. Effect of hypotonic and hypertonic solutions on impregnation of curcuminoids in coconut slices. *Innovative Food Science & Emerging Technologies* 16: (2013) 33–40.

Bendich, A. and Langseth, L. The health effects of vitamin C supplementation: A review. *Journal of the American College of Nutrition* 14(2): (1995) 124–136.

Bennett, L.E., Jegasothy, H., Konczak, I., Frank, D., Sudharmarajan, S., and Clingeleffer, P.R. Total polyphenolics and anti-oxidant properties of selected dried fruits and relationships to drying conditions. *Journal of Functional Foods* 3(2): (2011) 115–124.

Beristain, C.I., Azuara, E., Cortes, R., and Garcia, H.S. Mass transfer during osmoti dehydration of pineapple. *International Journal of Food Science Technology* 8: (1990) 122–130.

Blanda, G., Cerretani, L., Cardinali, A., Barbieri, S., Bendini, A., and Lercker, G. Osmotic dehydrofreezing of strawberries: Polyphenolic content, volatile profile and consumer acceptance. *LWT—Food Science and Technology* 42(1): (2009) 30–36.

Bórquez, R.M., Canales, E.R., and Redon, J.P. Osmotic dehydration of raspberries with vacuum pretreatment followed by microwave-vacuum drying. *Journal of Food Engineering* 99(2): (2010) 121–127.

Boyer, J. and Liu, R.H. Apple phytochemicals and their health benefits. *Nutrition Journal* 3(5): (2004) 5–19.

Bureau S., Renard, C.M.G.C., Reich, M., Ginies, C., Jean-Marc, A. Change in anthocyanin concentrations in red apricot fruits during ripening. *LWT-Food Science and Technology* 42: (2009) 372–377.

Chandra, S. and Kumari, D. Recent development in osmotic dehydration of fruit and vegetables: A review. *Critical Reviews in Food Science and Nutrition* 55(4): (2013) 552–561.

Chiralt, A., Martínez-Navarrete, N., Martínez-Monzó, J., Talens, P., Moraga, G., Ayala, A., and Fito, P. Changes in mechanical properties throughout osmotic process: Cryoprotectant effect. *Journal of Food Engineering* 49(2–3): (2001) 129–135.

Chiralt, A. and Talens, P. Physical and chemical changes induced by osmotic dehydration in plant tissues. *Journal of Food Engineering* 67(1–2): (2005) 167–177.

Chow, Y.N., Louarme, L., Bonazzi, C., Nicolas, J., and Billaud, C. Apple polyphenoloxidase inactivation during heating in the presence of ascorbic acid and chlorogenic acid. *Food Chemistry* 129(3): (2011) 761–767.

Corrêa, J., Justus, A., De Oliveira, L., and Alves, G.E. Osmotic dehydration of tomato assisted by ultrasound: Evaluation of the liquid media on mass transfer and product quality. *International Journal of Food Engineering* 11(4): (2015) 505–516.

Corrêa, J.L.G., Ernesto, D.B., Alves, J.G.L.F., and Andrade, R.S. Optimization of vacuum pulse osmotic dehydration of blanched pumpkin. *International Journal of Food Science & Technology* 49(9): (2014) 2008–2014.

Corrêa, J.L.G., Pereira, L.M., Vieira, G.S., and Hubinger, M.D. Mass transfer kinetics of pulsed vacuum osmotic dehydration of guavas. *Journal of Food Engineering* 96(4): (2010) 498–504.

Da Silva, D.I.S., Nogueira, G.D.R., Duzzioni, A.G., and Barrozo, M.A.S. Changes of antioxidant constituents in pineapple (*Ananas comosus*) residue during drying process. *Industrial Crops and Products* 50: (2013) 557–562.

De Ancos, B., González, E.M., and Cano, M.P. Ellagic acid, vitamin C and total phenolic contents and radical scavenging capacity affected by freezing and frozen storage in raspberry fruit. *Journal of Agricultural and Food Chemistry* 48: (2000) 4565–4570.

Del Caro, A., Piga, A., Pinna, I., Fenu, P. M., and Agabbio, M. Effect of drying conditions and storage period on polyphenolic content, antioxidant capacity and ascorbic acid of prunes. *Journal of Agricultural and Food Chemistry* 52: (2004) 4780–4784.

Deng, Y. and Zhao, Y. Effects of pulsed-vacuum and ultrasound on the osmodehydration kinetics and microstructure of apples (Fuji). *Journal of Food Engineering* 85(1): (2008) 84–93.

Devic, E., Guyot, S., Daudin, J., and Bonazzi, C. Effect of temperature and cultivar on polyphenol retention and mass transfer during osmotic dehydration of apples. *Journal of Agricultural and Food Chemistry* 58: (2010) 606–616.

Dewanto, V., Wu, X., Adom, K.K., and Liu, R.H. Thermal processing enhances the nutritional values of tomatoes by increasing total antioxidant activity. *Journal of Agricultural and Food Chemistry* 50: (2002) 3010–3014.

Djendoubi Mrad, D., Bonazzi, C., Courtois, C., Kechaou, N., Mihoubi· N.B. Moisture desorption isotherms and glass transition temperatures of osmo-dehydrated apple and pear. *Food and Bioproducts Processing* 91(2): (2013) 121–128.

Djendoubi Mrad, N., Boudhrioua, N., Kechaou, N., Courtois, F., and Bonazzi· C. Influence of air drying temperature on kinetics, physicochemical properties, total phenolic content and ascorbic acid of pears. *Food and Bioproducts Processing* 90(3): (2012) 433–441.

Dorta, E., Lobo, M.G., and González, G. Using drying treatments to stabilise mango peel and seed: Effect on antioxidant activity. *LWT—Food Science and Technology* 45: (2012) 261–268.

Fernandes, F.N., Rodrigues, S., Law, C., and Mujumdar, A. Drying of exotic tropical fruits: A comprehensive review. *Food and Bioprocess Technology* 4(2): (2011) 163–185.

Fito, P., Chiralt, A., Betoret, N., Gras, M., Cháfer, M., Martínez-Monzó, J., Andrés, A., and Vidal, D. Vacuum impregnation and osmotic dehydration in matrix engineering: Application in functional fresh food development. *Journal of Food Engineering* 49(2–3): (2001) 175–183.

Gamboa-Santos, J., Megías-Pérez, R., Cristina Soria, A., Olano, A., Montilla, A., Villamiel, M. Impact of processing conditions on the kinetic of vitamin C degradation and 2-furoylmethyl amino acid formation in dried strawberries. *Food Chemistry* 153: (2014) 164–170.

Garau, M.C., Simal, S., Roselló, C., and Femenia, A. Effect of air-drying temperature on physico-chemical properties of dietary fibre and antioxidant capacity of orange (*Citrus aurantium* v. Canoneta) by-products. *Food Chemistry* 104: (2007) 1014–1024.

García-Martínez, E., Martínez-Monzó, J., Camacho, M. M., and Martínez-Navarrete, N. Characterisation of reused osmotic solution as ingredient in new product formulation. *Food Research International* 35: (2002) 307–313.

García-Valverde, V., Navarro-Gonzáles, I., García-Alonso, J., and Periago, J. M. Antioxidant bioactive compounds in selected industrial processing and fresh consumption tomato cultivars. *Food Bioprocess Technology* 6: (2013) 391–402.

Gekas, V., Gonzalez, C., Sereno, A., Chiralt, A., and Fito, P. Mass transfer properties of osmotic solutions. I. Water activity and osmotic pressure. *International Journal of Food Properties* 1(2): (1998) 95–112.

Gekas, V. and Mavroudis, N. Mass transfer properties of osmotic solutions. II. Diffusivities. *International Journal of Food Properties* 1(2): (1998) 181–195.

Giovanelli, G., Brambilla, A., Sinelli, N. Effects of osmo-air dehydration treatments on chemical, antioxidant and morphological characteristics of blueberries. *LWT—Food Science and Technology* 54: (2013) 577–584.

Goula, A.M. and Adamopoulos, K.G. Retention of ascorbic acid during drying of tomato halves and tomato pulp. *Drying Technology* 24(1): (2006) 57–64.

Haminiuk, C.W.I., Maciel, G.M., Plata-Oviedo, M.S.V., and Peralta, R.M. Phenolic compounds in fruits: An overview. *International Journal of Food Science and Technology* 47(10): (2012) 2023–2044.

Hartley, R.D., Morrison, III W.H., Himmelsbach, D.S., and Borneman, W.S. Cross-linking of cell wall phenolics to arabinoxylans in graminaceous plants. *Phytochemistry* 29: (1990) 3705–3709.

Haskell, M.J. Provitamin A carotenoids as a dietary source of vitamin A. In S.A. Tanumihardjo, ed. *Carotenoids and Human Health*. Humana Press, New York (2013) pp. 249–260.

Henríquez, C., Speisky, H., Chiffelle, I., Valenzuela, T., Araya, M., Simpson, R., and Almonacid, S. Development of an ingredient containing apple peel, as a source of polyphenols and dietary fiber. *Journal of Food Science* 75: (2010) H172–H181.

Ikram, E.H.K., Khoo, H.E., Jalil, A.M.M., Ismail, A., Idris, S., Azlan, A., Nazri, H.S.M., Diton, N.A.M., and Mokhtar, R.A.M. Antioxidant capacity and total phenolic content of Malaysian underutilized fruits. *Journal of Food Composition and Analysis* 22(5): (2009) 388–393.

Ispir, A. and Tòrul, I.T. The influence of application of pretreatment on the osmotic dehydration of apricots. *Journal of Food Processing and Preservation* 33: (2009) 58–74.

Ito, A.P., Tonon, R.V., Park, K.J., and Hubinger, M.D. Influence of process conditions on the mass transfer kinetics of pulsed vacuum osmotically dehydrated mango slices. *Drying Technology* 25(10): (2007) 1769–1777.

Kähkönen, M., Hopia, A., and Heinonen, M. Berry phenolics and their antioxidant activity. *Journal of Agricultural and Food Chemistry* 49: (2001) 4076–4082.

Kalt, W. Effects of production and processing factors on major fruit and vegetable antioxidants. *Journal of Food Science* 70(1): (2005) R11–R19.

Karabulut, I., Topcu, A., Duran, A., Turan, S., and Ozturk, B. Effect of hot air drying and sun drying on color values and β-carotene content of apricot (*Prunus armenica* L.). *LWT-Food Science and Technology* 40: (2007) 753–758.

Kaur, C. and Kapoor, H.C. Antioxidants in fruits and vegetables—The millennium's health. *International Journal of Food Science & Technology* 36(7): (2001) 703–725.

Kaya, A., Aydin, O., and Kolayli, S. Effect of different drying conditions on the vitamin C (ascorbic acid) content of Hayward kiwifruits (*Actinidiadeliciosa Planch*). *Food and Bioproducts Processing* 88(2–3): (2010) 165–173.

Kayano, S., Yamada, N.F., Suzuki, T., Ikami, T., Shioaki, K., Kikuzaki, H., Mitani, T., and Nakatani, N. Quantitative evaluation of antioxidative components in prunes (*Prunus domestica* L.). *Journal of Agricultural Food Chemistry* 51: (2003) 1480–1485.

Kek, S.P., Chin, N.L., and Yusof, Y.A. Direct and indirect power ultrasound assisted preosmotic treatments in convective drying of guava slices. *Food and Bioproducts Processing* 91(4): (2013) 495–506.

Kerdpiboon, S., Devahastin, S., and Kerr, W.L. Comparative fractal characterization of physical changes of different food products during drying. *Journal of Food Engineering* 83(4): (2007) 570–580.

Krokida, M.K., Karanthanos, V.T., Maroulis, Z.B., and Marinos-Kouris, D. Drying kinetics of some vegetables. *Journal of Food Engineering* 59(4): (2003) 391–403.

Krokida, M.K., Oreopoulou, V., Maroulis, Z.B., and Marinos-Kouris, D. Effect of pre-treatment on viscoelastic behaviour of potato strips. *Journal of Food Engineering* 50(1): (2001) 11–17.

Kurozawa, L.E., Terng, I., Hubinger, M.D., and Park, K.J. Ascorbic acid degradation of papaya during drying: Effect of process conditions and glass transition phenomenon. *Journal of Food Engineering* 123: (2014) 157–164.

Kuti, J.O. and Konuru, B. Effects of genotype and cultivation environment on lycopene content in red-ripe tomatoes. *Journal of the Science of Food and Agriculture* 85(12): (2005) 2021–2026.

Lazarides, H.N., Gekas, V., and Mavroudis, N. Apparent mass diffusivities in fruit and vegetable tissues undergoing osmotic process. *Journal of Food Engineering* 31: (1997) 315–324.

Leong, S.Y. and Oey, I. Effects of processing on anthocyanins, carotenoids and vitamin C in summer fruits and vegetables. *Food Chemistry* 133: (2012) 1577–1587.

Lerici, C.R., Pinnavia, G., Rosa, M.D., and Bartolucci, L. Osmotic dehydration of fruit: Influence of osmotic agents on drying behaviour and product quality process. *Journal of Food Technology* 7: (1985) 147–155.

Lewicki, P.P. and Jakubczyk, E. Effect of hot air temperature on mechanical properties of dried apples. *Journal of Food Engineering* 64(3): (2004) 307–314.

Li, H. and Ramaswamy, H.S. Osmotic dehydration: Dynamics of equilibrium and pseudo-equilibrium kinetics. *International Journal of Food Properties* 13(2): (2010) 234–250.

Lombard, G.E., Oliveira, J.C., Fito, P., and Andrés, A. Osmotic dehydration of pineapple as a pre-treatment for further drying. *Journal of Food Engineering* 85(2): (2008) 277–284.

López, J., Vega-Galvez, A., Torres, M.J., Lemus-Mondaca, R., Quispe-Fuentes, I., Di Scala, K. Effect of dehydration temperature on physico-chemical properties and antioxidant capacity of goldenberry (*Physalis peruviana* L.). *Chilean Journal of Agricultural Research* 73(3): (2012)293–300.

Madrau, M., Piscopo, A., Sanguinetti, A., Del Caro, A., Poiana, M., and Romeo, F. Effect of drying temperatura on polyphenolic content and antioxidant activity of apricots. *European Food Research Technology* 228: (2009) 441–448.

Mahayothee, B., Udomkun, P., Nagle, M., Haewsungcharoen, M., Janjai, S., and Müller, J. Effect of pretreatments on colour alterations of litchi during drying and storage. *European Food Research and Technology* 229: (2009) 329–337.

Manzocco, L., Calligaris, S., Mastrocola, D., Nicoli, M.C., and Lerici, C.R. Review of non-enzymatic browning and antioxidant capacity in processed foods. *Trends in Food Science and Technology* 11(9–10): (2000) 340–346.

McSweeney, M. and Seetharaman, K. State of polyphenols in the drying process of fruits and vegetables. *Critical Reviews in Food Science and Nutrition* 55(5): (2013) 660–669.

Megías-Pérez, R., Gamboa-Santos, J., Soria, A.C., Villamiel, M., and Montilla, A. Survey of quality indicators in commercial dehhydrated fruits. *Food Chemistry* 150: (2014) 41–48.

Mertz, C., Brat, P., Caris-Veyrat, C., and Gunata, Z. Characterization and thermal lability of carotenoids and vitamin C of tamarillo fruit (*Solanum betaceum* Cav.). *Food Chemistry* 119(2): (2010) 653–659.

Meyers, K.J., Mares, J.A., Igo, R.P., Truitt, B., Liu, Z., Millen, A.E. Genetic evidence for role of carotenoids in age-related macular degeneration in the carotenoids in age-related eye disease study (CAREDS). *Investigative Ophthalmology and Visual Science* 55(1): (2014) 587–599.

Michalczyk, M., Macura, R., and Matuszak, I. The effect of air-drying, freeze-drying and storage on the quality and antioxidant activity of some selected berries. *Journal of Food Processing and Preservation* 33: (2009) 11–21.

Miletić, N., Popović, B., Mitrović, O., and Kandić, M. Phenolic content and antioxidant capacity of fruits of plum cv. Stanley (*Prunus domestica* L.) as influenced by maturity stage and on-tree ripening. *Australian Journal of Crop Science* 6: (2012) 681–687.

Moreno, J., Simpson, R., Estrada, D., Lorenzen, S., Moraga, D., and Almonacid, S. Effect of pulsed-vacuum and ohmic heating on the osmodehydration kinetics, physical properties and microstructure of apples (cv. Granny Smith). *Innovative Food Science & Emerging Technologies* 12(4): (2011) 562–568.

Moreno, J., Simpson, R., Pizarro, N., Parada, K., Pinilla, N., Reyes, J.E., and Almonacid, S. Effect of ohmic heating and vacuum impregnation on the quality and microbial stability of osmotically dehydrated strawberries (cv. Camarosa). *Journal of Food Engineering* 110(2): (2012) 310–316.

Müller, L., Caris-Veyrat, C., Lowe, G., and Böhm, V. Lycopene and its antioxidant role in the prevention of cardiovascular diseases—A critical review. *Critical Reviews in Food Science and Nutrition* (2015), doi: 10.1080/10408398.2013.801827.

Nahimana, H., Zhang, M., Mujumdar, A.S., and Ding, Z. Mass transfer modeling and shrinkage consideration during osmotic dehydration of fruits and vegetables. *Food Reviews International* 27(4): (2010) 331–356.

Naidu, K.A. Vitamin C in human health and disease is still a mystery? An overview. *Nutrition Journal* 2(1): (2003) 7.

Nicolas, J., Billaud, C., Philippon, J., and Rouet-Mayer, M.A. Enzymatic browning-biochemical aspects. In B. Caballero, L. Trugo, and P.M. Finglas, eds. *Encyclopaedia of Food Sciences and Nutrition*, 2nd ed. Academic Press, London, UK (2003) pp. 678–686.

Nicoleti, J.F., Silveira, V., Telis-Romero, J., and Telis, V.R.N. Influence of drying conditions on ascorbic acid during convective drying of whole persimmons. *Drying Technology* 25(5): (2007) 891–899.

Nowicka, P., Wojdyło, A., Lech, K., and Figiel, A. Influence of osmodehydration pretreatment and combined drying method on the bioactive potential of sour cherry fruits. *Food and Bioprocess Technology* 8(4): (2015) 824–836.

Nuñez-Mancilla, Y., Pérez-Won, M., Uribe, E., Vega-Gálvez, A., Di Scala, K. Osmotic dehydration under high hydrostatic pressure: Effect on antioxidant activity, total phenolics compounds, vitamin C and colour of strawberry (*Fragaria vesca*). *LWT-Food Science and Technology* 52: (2013) 151–156.

Nuñez-Mancilla, Y., Perez-Won, M., Vega-Gálvez, A., Arias, V., Tabilo-Munizaga, G., Briones-Labarca, V., Lemus-Mondaca, R., and Di Scala, K. Modeling mass transfer during osmotic dehydration of strawberries under high hydrostatic pressure conditions. *Innovative Food Science & Emerging Technologies* 12(3): (2011) 338–343.

Núñez-Mancilla, Y., Vega-Gálvez, A., Pérez-Won, M., Zura, L., García-Segovia, P., and Di Scala, K. Effect of osmotic dehydration under high hydrostatic pressure on microstructure, functional properties and bioactive compounds of strawberry (*Fragaria Vesca*). *Food and Bioprocess Technology* 7(2): (2014) 516–524.

Ochoa, M.R., Kesseler, A.G., Pirone, B.N., Márquez, C.A., and De Michelis, A. Shrinakge during convective drying of whole rose hip (*Rosa Rubiginosa* L.) fruits. *LWT-Food Science and Technology* 35(5): (2002) 400–406.

Omoni, A.O. and Aluko, R.E. The anti-carcinogenic and anti-atherogenic effects of lycopene: A review. *Trends in Food Science and Technology* 16(8): (2005) 344–350.

Ong, S.P. and Law, C.L. Drying kinetics and antioxidant phytochemicals retention of salak fruit under different drying and pretreatment conditions. *Drying Technology* 29(4): (2011) 429–441.

Padayatty, S.J., Katz, A., Wang, Y., Eck, P., Kwon, O., Lee, J.-H., Chen, S., Corpe, C., Dutta, A., Dutta, S.K., and Levine, M. Vitamin C as an antioxidant: Evaluation of its role in disease prevention. *Journal of the American College of Nutrition* 22(1): (2003) 18–35.

Patras, A., Brunton, N. P., Da Pieve, S., and Butler, F. Impact of high pressure processing on total antioxidant activity, phenolic, ascorbic acid, anthocyanin content and colour of strawberry and blackberry purées. *Innovative Food Science and Emerging Technologies*, 10: (2009) 308–313.

Peiró, R., Dias, V.M.C., Camacho, M.M, and Martĺnez-Navarrete, N. Micronutrient flow to the osmotic solution during grapefruit osmotic dehydration. *Journal of Food Engineering* 74: (2006) 299–307.

Phisut, N., Rattanawadee, M., and Aekkasak, K. Effect of osmotic dehydration process on the physical, chemical and sensory properties of osmo-dried cantaloupe. *International Food Research Journal* 20(1): (2013) 189–196.

Pino, J.A., PanadÉS, G., Fito, P., Chiralt, A., and Ortega, A. Influence of osmotic dehydration on the volatile profile of guava fruits. *Journal of Food Quality* 31(3): (2008) 281–294.

Pott, I., Marx, M., Neidhart, S., Mühlbauer, W., and Carle, R. Quantitative determination of β-carotene stereoisomers in fresh, dried, and solar-dried mangoes (*Mangifera indica* L.). *Journal of Agricultural and Food Chemistry* 51(16): (2003) 4527–4531.

Prachayawarakorn, S., Tia, W., Plyto, N., and Soponronnarit, S. Drying kinetics and quality attributes of low-fat banana slices dried at high temperature. *Journal of Food Engineering* 85(4): (2008) 509–517.

Proteggente, A.R., Pannala, A.S., Paganga, G., Buren, L.V., Wagner, E., Wiseman, S., Put, F.v.d., Dacombe, C., and Rice-Evans, C.A. The antioxidant activity of regularly consumed fruit and vegetables reflects their phenolic and vitamin C composition. *Free Radical Research* 36(2): (2002) 217–233.

Que, F., Mao, L., Fang, X., and Wu, T. Comparison of hot air-drying and freeze drying on the physicochemical properties and antioxidant activities of pumpkin (*Cucurbita moschata* Duch.) flours. *International Journal of Food Science and Technology* 43(7): (2008) 1195–1201.

Ramallo, L., Hubinger, M., and Mascheroni, R. Effect of pulsed vacuum treatment on mass transfer and mechanical properties during osmotic dehydration of pineapple slices. *International Journal of Food Engineering* 9(4): (2013) 403–412.

Ramallo, L.A. and Mascheroni, R.H. Quality evaluation of pineapple fruit during drying process. *Food and Bioproducts Processing* 90: (2012) 275–283.

Rastogi, N.K., Raghavarao, K.S.M.S., Niranjan, K., and Knorr, D. Recent developments in osmotic dehydration: Method to enhance mass transfer. *Trends in Food Science and Technology* 13(2): (2002) 48–59.

Rastogi, N.K., Raghavarao, K., and Niranjan, K. Developments in osmotic dehydration. In D.-W. Sun, ed. *Emerging Technologies for Food Processing*. Academic Press, London, UK (2005) pp. 221–250.

Rawson, A., Patras, A., Tiwari, B.K., Noci, F., Koutchma, T., and Brunton, N. Effect of thermal and non-thermal processing technologies on the bioactive content of exotic fruits and their products: Review of recent advances. *Food Research International* 44(7): (2011) 1875–1887.

Renard, C.M.G.C. Effects of conventional boiling on the polyphenols and cell walls of pears. *Journal of the Science of Food and Agriculture* 85: (2005) 310–318.

Reppa, A., Mandala, J., Kostaropoulos, A.E., and Saravacos, G.D. Influence of solute temperature and concentration on the combined osmotic and air drying. *Drying Technology* 17: (1999) 1449–1458.

Rigi, S., Kamani, M.H., and Atash, M.M.S. Effect of temperature on dying kinetics, antioxidant capacity and vitamin C content of papaya (*Carica papaya* Linn.). *International Journal of Plant, Animal and Environmental Sciences* 4(3): (2014) 413–417.

Riva, M., Campolongo, S., Leva, A.A., Maestrelli, A., and Torreggiani, D. Structure-property relationships in osmo-air-dehydrated apricot cubes. *Food Research International* 38: (2005) 533–542.

Rizvi, S.S.H. Thermodynamic properties of foods in dehydration. In A. Rao and S.S.H. Rizvi, eds. *Engineering Properties of Foods*. Marcel Dekker, New York (1995) pp. 223–309.

Robards K., Prenzler P.D., Tucke, G., Swatsitang, P., and Glover, W. Phenolic compounds and their role in oxidative processes in fruits. *Food Chemistry* 66: (1999) 401–436.

Rodriguez-Amaya, D.B. and Kimura, M. Carotenoids in foods. In: D.B. Rodriguez-Amaya and M., Kimura, eds. *HarvestPlus Handbook for Carotenoid Analysis*. HarvestPlus, Washington, DC (2004) pp. 2–7.

Rodriguez-Amaya, D. B., Kimura, M., Godoy, H. T., and Amaya-Farfan, J. Updated Brazilian database on food carotenoids: Factors affecting carotenoids composition. *Journal of Food Composition and Analysis* 21: (2008) 445–463.

Rojas-Graü, M.A., Tapia, M.S., Rodríguez, F.J., Carmona, A.J., and Martin-Belloso, O. Aliginate and gellan-based edible coatings as carriers of antibrowning agents applied on fresh-cut Fuji apples. *Food Hydrocolloids* 21(1): (2007) 118–127.

Sablani, S.S. Drying of fruits and vegetables: Retention of nutritional/functional quality. *Drying Technology* 24(2): (2006) 123–135.

Sagar, V.R. and Suresh Kumar, P. Recent advances in drying and dehydration of fruits and vegetables: A review. *Journal of Food Science and Technology* 47(1): (2010) 15–26.

Saini, R.K., Nile, S.H., and Park, S.W. Carotenoids from fruits and vegetables: Chemistry, analysis, occurrence, bioavailability and biological activities: A review. *Food Research International* 76: (2015) 735–750.

Sakihama, Y., Cohen, M., Grace, S., and Yamasaki, H. Plant phenolic antioxidant and pro-oxidant activities: Phenolics-induced oxidative damage mediated by metals in plants. *Toxicology* 177: (2002) 67–80.

Santacruz-Vázquez, C., Santacruz-Vázquez, V., Jaramillo-Flores, M.E., Chanona-Pérez, J., Welti-Chanes, J., and Gutiérrez-López, G.F. Application of osmotic dehydration processes to produce apple slices enriched with β-carotene. *Drying Technology* 26(10): (2008) 1265–1271.

Santos, P.H.S. and Silva, M.A. Retention of vitamin C in drying processes of fruits and vegetables: A review. *Drying Technology* 26: (2008) 1421–1437.

Saurel, R., Raoult-Wack, A., Rios, G., and Guilbert, S. Mass transfer phenomena during osmotic dehydration of apple. I. Fresh plant tissue. *International Journal of Food Science and Technology* 29: (1994) 531–537.

Scalbert, A., Manach, C., Morand, C., and Rémésy, C. Dietary polyphenols and the prevention of diseases. *Critical Reviews in Food Science and Nutrition* 45: (2005) 287–306.

Schultz, E.L., Mazzuco, M.M., Machado, R.A.F., Bolzan, A., Quadri, M.B., and Quadri, M.G.N. Effect of pre-treatments on drying, density and shrinkage of apple slices. *Journal of Food Engineering* 8(3): (2007) 1103–1110.

Schweiggert, R.M., Steingass, C.B., Heller, A., Esquivel, P., and Carle, R. Characterization of chromoplasts and carotenoid of red and yellow fleshed papaya (*Carica papaya* L.). *Planta* 234: (2011) 1031–1044.

Sereno, A.M., Moreira, R., and Martinez, E. Mass transfer coefficients during osmotic dehydration of apple in single and combined aqueous solutions of sugar and salt. *Journal of Food Engineering* 47(1): (2001) 43–49.

Serpen, A., Gokmen, V., Bahçeci, K.S, and Acar, J. Reversible degradation kinetics of vitamin C in peas during frozen storage. *European Food Research and Technology* 224: (2007) 749–753.

Sharoni, Y., Linnewiel-Hermoni, K., Khanin, M., Salman, H., Veprik, A., and Danilenka, M. Carotenoids and apocarotenoids in cellular signalling related to cancer: A review. *Molecular Nutrition and Food Research* 56(2): (2012) 259–269.

Sherwin, J.C., Reacher, M.H., Dean, W.H., and Ngondi, J. Epidemiology of vitamin A deficiency and xerophthalmia in at risk populations. *Transactions of the Royal Society of Tropical Medicine and Hygiene* 106(4): (2012) 205–214.

Shi, J. and Le Maguer, M. Lycopene in tomatoes: Chemical and physical properties affected by food processing. *Critical Reviews in Food Science and Nutrition* 40(1): (2000) 1–42.

Shi, J. and Le Maguer, M. Osmotic dehydration of foods: Mass transfer and modeling aspects. *Food Reviews International* 18(4): (2002) 305–335.

Shi, J., Le Maguer, M., and Bryan, M. Lycopene from tomatoes. In J. Shi, G. Mazza, and M. Le Maguer, eds. *Functional Foods—Biochemical and Processing Aspect.* CRC Press, Boca Raton, FL (2002) pp. 135–168.

Shi, X.Q. and Fito Maupoey, P. Vacuum osmotic dehydration of fruits. *Drying Technology* 11(6):1429–1442. (1993).

Sigge, G.O., Hansmann, C.F., and Joubert, E. Optimizing the dehydration conditions of green bell peppers (*Capsicum annuum* L.): Quality criteria. *Journal of Food Quality* 22(4): (1999) 439–452.

Silva, M.A.C., Da Silva, Z.E., Mariani, V.C., and Darche, A. Mass transfer during the osmotic dehydration of West Indian cherry. *LWT-Food Science and Technology* 45(2): (2012) 246–252.

Silva, K.S., Fernandes, M.A., and Mauro, M.A. Effect of calcium on the osmotic dehydration kinetics and quality of pineapple. *Journal of Food Engineering* 134: (2014) 37–44.

Singla, R., Ganguli, A., and Ghosh, M. Antioxidant activities and polyphenolic properties of raw and osmotically dehydrated dried mushroom (*Agaricus bisporous*) snack food. *International Journal of Food Properties* 13(6): (2010) 1290–1299.

Siramard, S. and Charoenrein, S. Effect of ripening stage and infusion with calcium lactate and sucrose on the quality and microstructure of frozen mango. *International Journal of Food Science & Technology* 49(9): (2014) 2136–2141.

Sogi, D.S., Siddiq, M., and Dolan, K.D. Total phenolics, carotenoids and antioxidant properties of Tommy Atkin mango cubes as affected by drying techniques. *LWT-Food Science and Technology* 62(1), Part 2: (2014) 564–568.

Sulaiman, S.F., Yusoff, N.A.Md., Eldeen, I.M., Seow, E.M., Sajak, A.A.B., Supriatno, and Ooi K.L. Correlation between total phenolic and mineral contents with antioxidant activity of eight Malaysian bananas (*Musa* sp.). *Journal of Food Composition and Analysis* 24: (2011) 1–10.

Taiwo, K.A., Eshtiaghi, M.N., Ade-Omowaye, B.I.O., and Knorr, D. Osmotic dehydration of strawberry halves: Influence of osmotic agents and pretreatment methods on mass transfer and product characteristics. *International Journal of Food Science & Technology* 38(6): (2003) 693–707.

Telis, V.R.N., Murari, R.C.B.D.L., and Yamashita, F. Diffusion coefficients during osmotic dehydration of tomatoes in ternary solutions. *Journal of Food Engineering* 61(2): (2004) 253–259.

Thane, C. and Reddy, S. Processing of fruit and vegetables: Effect on carotenoids. *Nutrition & Food Science* 97(2): (1997) 58–65.

Toğrul, İ.T. and İspir, A. Effect on effective diffusion coefficients and investigation of shrinkage during osmotic dehydration of apricot. *Energy Conversion and Management* 48(10): (2007) 2611–2621.

Toor, R.K. and Savage, G.P. Antioxidant activity in different fractions of tomatoes. *Food Research International* 38(5): (2005) 487–494.

Torreggiani, D. Technological aspects of osmotic dehydration in foods. In G.V. Barbosa-Canovas and J. Welti-Chanes, eds. *Food Preservation by Moisture Control: Fundamentals and Applications.* ISOPOW Practicum II. Technomic Publishing, Lancaster, PA (1995) pp. 281–304.

Torreggiani, D. and Bertolo, G. Osmotic pre-treatments in fruit processing: Chemical, physical and structural effects. *Journal of Food Engineering* 49: (2001) 247–253.

Torres, J.D., Castelló, M.L., Escriche, I., and Chiralt, A. Quality characteristics, respiration rates, and microbial stability of osmotically treated mango tissue (*Mangifera indica* L.) with or without calcium lactate. *Food Science and Technology International* 14(4): (2008) 355–365.

Torres, J.D., Talens, P., Carot, J.M., Chiralt, A., and Escriche, I. Volatile profile of mango (*Mangifera indica* L.), as affected by osmotic dehydration. *Food Chemistry* 101(1): (2007) 219–228.

Udomkun, P., Nagle, M., Mahayothee, B., Nohr, D., Koza, A., and Müller, J. Influence of air drying properties on non-enzymatic browning, major bio-active compounds and antioxidant capacity of osmotically pretreated papaya. *LWT-Food Science and Technology* 60(2): (2015) 914–922.

Vadivambal, R. and Jayson, D.S. Changes in quality of microwave-treated agricultural products—A review. *Biosystems Engineering* 98(1): (2007) 1–16.

Vega-Galvez, A., Ah-Hen, K., Chacana, M., Vergara, J., Martinez-Monzo, J., Garcia-Segovia, P., Lemus-Mondaca, R., Di Scala, K. Effect of temperature and air velocity on drying kinetics, antioxidant capacity, total phenolic content, colour, texture and microstructure of apple (var. *Granny Smith*) slices. *Food Chemistry* 132: (2012) 51–59.

Vega-Gálvez, A., Di Scala, K., Rodríguez, K., Lemus-Mondaca, R., Miranda, M., López, J., and Perez-Won, M. Effect of air-drying temperature on physico-chemical properties, antioxidant capacity, colour and total phenolic content of red pepper (*Capsicum annuum*, L., Var. Hungarian). *Food Chemistry* 117: (2009) 647–653.

Verma, D., Kaushik, N., and Rao, P.S. Application of high hydrostatic pressure as a pretreatment for osmotic dehydration of banana slices (*Musa cavendishii*) finish-dried by dehumidified air drying. *Food and Bioprocess Technology* 7(5): (2014) 1281–1297.

Vicente, S., Nieto, A., Hodara, K., Castro, M., and Alzamora, S. Changes in structure, rheology, and water mobility of apple tissue induced by osmotic dehydration with glucose or trehalose. *Food and Bioprocess Technology* 5(8): (2012) 3075–3089.

Vijayakumari, K., Pugalenthi, M., and Vadivel, V. Effect of soaking and hydrothermal processing methods on the levels of antinutrients and in vitro protein digestibility of *Bauhinia purpurea* L. seeds. *Food Chemistry* 103: (2007) 968–975.

West, K., Jr. and Darnton-Hill, I. Vitamin A deficiency. In R. Semba, M. Bloem and P. Piot, eds. *Nutrition and Health in Developing Countries*. Humana Press (2008) pp. 323–332.

Wojdyło, A., Figiel, A., and Oszmianski, J. Effect of drying methods with the application of vacuum microwaves on the bioactive compounds, color, and antioxidant activity of strawberry fruits. *Journal of Agricultural and Food Chemistry* 57: (2009) 1337–1343.

Yeum, K.-J. and Russell, R.M. Carotenoid bioavailability and bioconversion. *Annual Review of Nutrition* 22(1): (2002) 483–504.

Yilmaz, F.M., Karaaslan, M., and Vardin, H. Optimization of extraction parameters on the isolation of phenolic compounds from sour cherry (*Prunus cerasus* L.) pomace. *Journal of Food Science and Technology* 52(5): (2015) 2851–2859.

Zepka, L.Q. and Mercadante, A.Z. Degradation compounds of carotenoids formed during heating of a simulated cashew apple juice. *Food Chemistry* 117(1): (2009) 28–34.

10 Effect of Hydrostatic High Pressure Treatments on Bioactive Compounds of Vegetable Products

Francisco González-Cebrino, Jesús J. García-Parra, and Rosario Ramírez

CONTENTS

10.1 Introduction: General Background 219
10.2 Effect of HPP and HPT Processing on the Bioactive Compounds of Vegetables 222
    10.2.1 Case Study: Plum and Pumpkin as Model Matrices 222
10.3 Effect of Hydrostatic High Pressure Treatments on Anthocyanin Compounds 224
10.4 Effect of Hydrostatic High Pressure Treatments on Carotenoid Compounds 227
10.5 Effect of Hydrostatic High Pressure Treatments on Phenolic Compounds 229
10.6 Effect of Hydrostatic High Pressure Treatments on Compounds with Antioxidant Activity 231
10.7 Factors that Affect the Preservation of Bioactive Compounds in Vegetable Products 234
10.8 Concluding Remarks and Future Trends 235
Acknowledgments 236
References 236

10.1 INTRODUCTION: GENERAL BACKGROUND

There are several traditional methods for food preservation, with thermal treatments being well known and widely used. The usefulness of these treatments is primarily based on the inactivation of microorganisms and enzymes to avoid the degradation of food in subsequent storage. Degradation reaction foods are accelerated by temperature, so that heat treatments adversely affect the quality of food (Patras et al., 2009a).

A growing consumer demand for healthy, nutritious, varied, and convenient processed food has led to a continuous improvement in conventional technologies and the development of new processing alternatives. The goal of these efforts is the production of safer foods while enhancing nutritional and sensory quality, but challenges still remain. Prevailing technologies, such as thermal processing, cause extensive and undesirable chemical changes, while some minimal processing strategies cannot eliminate all microbial pathogens.

Currently, at the industrial level, high pressure processing (HPP) is applied at room or refrigeration temperatures, and this gives a pasteurization effect, as there is a non-complete elimination of microorganisms. High pressure combined with mild or high temperatures, known as pressure-assisted thermal processing (PATP) or high pressure thermal (HPT) treatment can produce a sterilization effect. Differences between both technologies are summarized in Table 10.1. Typical HPP conditions include holding pressures of 400–600 MPa, an initial low-medium temperature of 10°C–40°C, and a holding time of 1–15 min. These conditions will produce a pasteurization, rather than a sterilization effect.

The application of HPP is relatively simple; it involves injecting water into a steel cylinder using powerful pumps, so that a high volume of water enters into the tank than normal. These conditions are lethal to microorganisms; however, pressure minimally alters the nutritional, texture, color, and flavor of food compared to the effect of thermal technologies (Oey et al., 2008a, b). Therefore, HPP provides a way to maintain the quality and freshness of food without using chemical preservatives or high temperatures.

HPP offers several competitive advantages over existing technologies. The main purpose of the application of HPP is the extension of the shelf life of food and safe products obtained by the elimination of spoilage or pathogens microorganisms.

The lethal effect of high pressure exerted on the microorganisms lies primarily in inducing changes in biochemical reactions in the genetic mechanisms and cell wall and membranes (Hoover et al., 1989). The inactivation of microorganisms by high pressure may be due to various factors, such as increased membrane permeability (broken cell membranes), inhibition of energy-producing reactions, denaturation of proteins and essential enzymes for development and cell reproduction (several enzyme systems are inhibited or inactivated by pressure) (Pothakamury et al., 1995),

TABLE 10.1
Comparison between HPP or HPT Treatments

	Pressure	Temperature	Holding Time	Effect	State of Art
HPP—high pressure processing	400–600 MPa	10°C–40°C	1–15 min	Pasteurization	Present in the market
PATP—pressure assisted thermal processing or HPT—High pressure thermal processing	>600 MPa	60°C–120°C	Less than thermal treatment	Severe pasteurization or sterilization	Future in research

and also damage to the mechanisms of replication and transcription of the genetic material (decrease of DNA synthesis).

High pressure treatments can be applied to both liquid and solid foods with high moisture contents. Treatments are usually applied once food is packaged in their final package, which is another great advantage, because food can be pasteurized after being cut and packaged, thus avoiding the risk of contamination in the manufacturing environment. Packages used should have sufficient flexibility and elasticity (plastic) to recover the initial volume after compression and thus prevent irreversible deformations.

Many foods treated by HPP are currently available in the market, such as fruit juices, deli meats, or ready-to-eat meals and salsas. In contrast, the application of HPT treatment is being studied, but it is not yet available for consumers. In the case of vegetables products, HPP maintains the original qualities of products such as juices, smoothies, and other fruit-based drinks, processed through. Thus, the true taste of freshly squeezed juice is preserved. Furthermore, the nutritional properties remain almost intact compared to traditional methods, allowing the creation of a range of high-quality products. Besides, this technology allows the preservation of juices that were seriously degraded by a heat treatment, such as pomegranate, apple, carrot, and beet. Products treated by HPP generally receive good acceptance from consumers, mainly because they are perceived as being of high sensory quality with better retention of nutrients and other life-enhancing components.

HPP is already established as a successful food processing technology. One reason for this success is that, unlike thermal processing and other preservation technologies, HPP effects are uniform and nearly instantaneous throughout the food and thus independent of food geometry and equipment size. These observations have facilitated the scale-up of laboratory findings to full-scale production. However, HPT treatment presents new challenges for industrial scale.

HPT treatments consist of a combination of high pressure (500–900 MPa) and temperature (60°C–120°C) over a short holding time. The rapid temperature increase during compression (due to the adiabatic heating) and temperature decrease in the product upon decompression could help reduce the hardness of heating effects encountered in conventional thermal technologies. HPT treatments offer several advantages such as quick elimination of microorganisms, uniform and rapid increase and decrease in temperature of the product, reduced thermal impact and processing time, and extension in the shelf life of the processed food (Subramanian et al., 2006). Some authors have reported that HPT treatments provides better color, flavor, and aroma retention compared to traditional thermally treated products (Gupta et al., 2010). Although there are no HPT-processed shelf-stable foods commercially available, it would be interesting to provide to the food industry a viable process for heat-sensitive products, such as fruit-derived products, that would suffer a severe loss of quality by traditional thermal processing.

The effects of HPT treatment are widely studied in microorganisms and microbial spores (Nguyen et al., 2013). In general, combined pressure and temperature treatment is applied to inactivate the vegetative cells and bacterial spores that are resistant to the application of high pressures at low room temperatures (Mújica-Paz et al., 2011).

The study of phytochemicals with nutritious and healthy interest is an area of particular relevance today, the evidence linking supply with consumer health and consumer rejection of overly processed foods and food additives. Although the main advantage of the high-pressure technology against heat treatments is the nutritional and sensory quality of the products, there is little scientific information on changes in phytochemicals of some fruits and vegetables. In fact, benefits of HPT treatment in preserving the quality attributes of foods like color or health-related compounds are quite less studied than microbiological aspects.

10.2 EFFECT OF HPP AND HPT PROCESSING ON THE BIOACTIVE COMPOUNDS OF VEGETABLES

10.2.1 Case Study: Plum and Pumpkin as Model Matrices

In this chapter, we want to explore the effect of HPP and HPT treatment on the bioactive compounds of vegetables. With this propose, we will utilize two different vegetable matrices as models: pumpkin (var. *Butternut*) and red plum (var. *Crimson Globe*). Plums are one of the most important botanical genera among stone fruit, which are adapted to a board range of climatic and soil factors (Ertekin et al., 2006). Pumpkins (*Cucurbita moschata*, Duch) are grown worldwide for their sweet pulp and seeds, either for direct consumption or for producing different foods such as jams, jellies, and purées (Gliemmo et al., 2009). Both could serve to evaluate the effect of these technologies on the phytochemical compounds with nutritional interest. The advantage of studying the effect of two such different products is that it is possible to assess the effect of both technologies on matrices that can respond in different ways and with different processing needs. Pumpkin and plum were selected as vegetable models, because they are rich in different phytochemicals (carotenoids vs. anthocyanins), and they have different pH (pumpkin: 6.6 vs. plum: 3.4); they would need a different intensity of treatment for their microbiological stabilization.

HPT treatments are normally applied to preserve low-acid foods. However, HPP does not always allow the inactivation of enzymes such as the polyphenoloxidase, which could reduce the shelf life of processed products (González-Cebrino et al., 2013). The application of HPT treatments could also be interesting for the inactivation of enzymes that are not inactivated by HPP to extend the shelf life of those acid foods without the addition of chemicals.

It is essential to detail the conditions of application of the high-pressure treatments due to their importance in the understanding on the effect that they produce on both vegetables.

HPP was applied in a semi-industrial discontinuous hydrostatic unit (Hiperbaric Wave 6000/55, Hiperbaric, S.A., Burgos, Spain). Purée were pressurized at nine processing combinations of pressure level and processing time (400, 500, and 600 MPa for 1, 150, and 300 s) with water at 10°C as the pressure-transmitting medium. The time taken to reach the target pressure was 180–230 s and decompression took 1–2 s. The initial temperature of water in the high-pressure vessel was 10°C. When high-pressure treatments at 400, 500, and 600 MPa were applied, the temperature reached approximately 20°C, 22.5°C, and 25°C, respectively, due to the adiabatic compression, which was calculated as 2.5°C per 100 MPa (Balasubramaniam et al., 2004), because there was no temperature monitoring system in the pressure unit.

For applying HPT processing, a multivessel Resato unit (FPU-100-50, serial no. 14685/42798. Roden, the Netherlands) was used. The machine uses ethylene glycol as a pressure-transmitting medium, and is equipped with thermostatic jacket for the temperature control. The initial temperature of the assays were 60°C, 70°C, and 80°C and the pressures applied were 300, 600, and 900 MPa. The rate of pressure build-up was 10 MPa s^{-1} and the holding time was 1 min. The increases of pressure and temperature during the treatments applied are showed in Figure 10.1. Graphics shows the temperature of the pressurization media, and the pressure measured in the vessels during the treatments applied to purées.

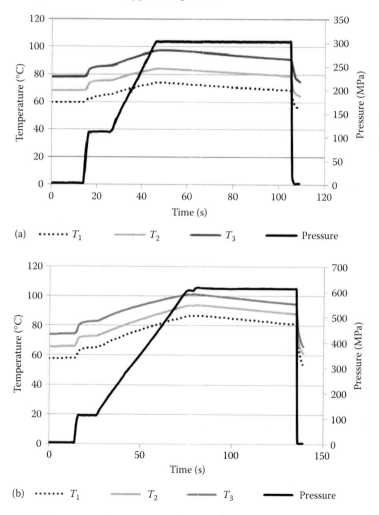

FIGURE 10.1 Pressure and temperature changes during the high pressure–high temperature treatments applied in plum and pumpkin purées for 60 s at 300 MPa (a), 600 MPa (b).

(Continued)

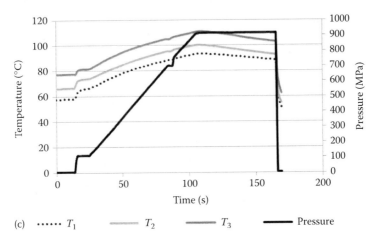

FIGURE 10.1 (Continued) Pressure and temperature changes during the high pressure–high temperature treatments applied in plum and pumpkin purées for 60 s at 900 MPa (c). T_1: initial temperature of treatment at 60°C; T_2: initial temperature of treatment at 70°C; and T_3: initial temperature of treatment at 80°C.

10.3 EFFECT OF HYDROSTATIC HIGH PRESSURE TREATMENTS ON ANTHOCYANIN COMPOUNDS

Almost all fruits and vegetables, and cereal grains contain appreciable amounts of natural polyphenols. The three major groups of dietary polyphenols are flavonoids, phenolic acids, and phenolic polymers. Flavonoids are the largest and the most studied group of plant phenols. The term *flavonoids* are assigned generically to a group of polyphenolic compounds that are secondary metabolites of plants. These compounds are synthesized from phenylalanine and malonyl-CoA. Flavonoids can be divided into various classes depending on the oxidation and unsaturation of the heterocyclic ring, and differences based on the nature and number of substituent bounded to the rings within each class can be established (Robards et al., 1999). Most plant tissues can synthesize flavonoids, which are structured as water-soluble glycosides in leaves and fruits. The most common linked sugars are glucose, galactose, rhamnose, arabinose, xylose, and glucuronic acid (Manach et al., 2004). These compounds have important metabolic functions, and they are responsible for the resistance of plants to photo-oxidation of ultraviolet sunlight; it is believed to function as a defense against herbivorous (Havsteen, 2002). The best-known functions of these compounds are enzyme inhibition (cyclooxygenase, lipoxygenase, etc.) and their antioxidant capacity, being able to act directly like reacting with free radicals to give more stable compounds, or have an additive effect on the endogenous antioxidant defense, increasing or maintaining such defense (Tomás-Barberán et al., 2000). Anthocyanidins exhibit the basic structure of flavylium cation and are usually linked to glycosides, resulting in anthocyanins, which are responsible for blues, purples, reds, and combinations of these colorations of red fruits such as cherries, plums, raspberries, and grapes (Clifford, 2000; Scalbert and Williamson, 2000).

Effect of Hydrostatic High Pressure Treatments on Vegetable Products

There are six important anthocyanins present in food: cyanidin, delphinidin, peonidin, pelargonidin, petunidin, and malvidin. The most common glucosides are glucose, galactose, rhamnose, and arabinose, normally attached to position 3 or positions 3 and 5, generating diglycosides (Clifford, 2000). Cyanidin glycosylated with glucose is the most abundant plum structure, responsible for the red color of skin and flesh, followed by glycosylated cyanidin with rutinoside (Tomás-Barberán et al., 2001) (Figures 10.2 and 10.3).

Anthocyanins are important bioactive compounds in red plum varieties. The preservation of bioactive compounds from vegetables has a great interest in the food industry. In this case, the effect of high pressure treatments (HPP and HPT) was evaluated for the preservation of anthocyanins from red plum (var. *Crimson Globe*). Anthocyanin changes in the plum purée after processing are shown in Table 10.2. In general, control purées showed higher anthocyanin content than purées treated by HPP or HPT treatment. Total anthocyanin contents were reduced after applying HPP in the plum purée compared to control purées. Treatments at 600 MPa/150 s

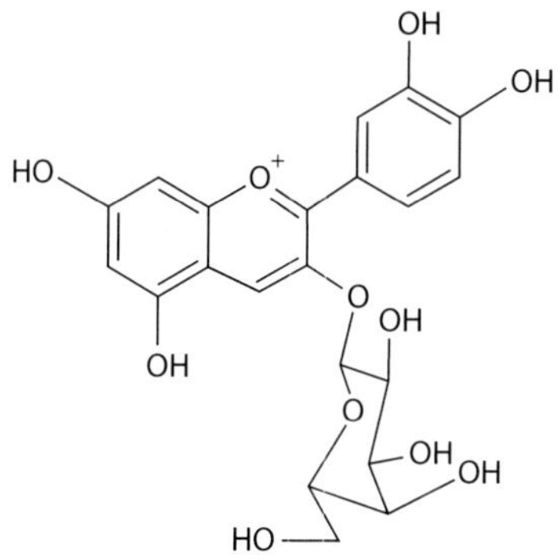

FIGURE 10.2 Chemical structure of cyanidin-3-rutinoside.

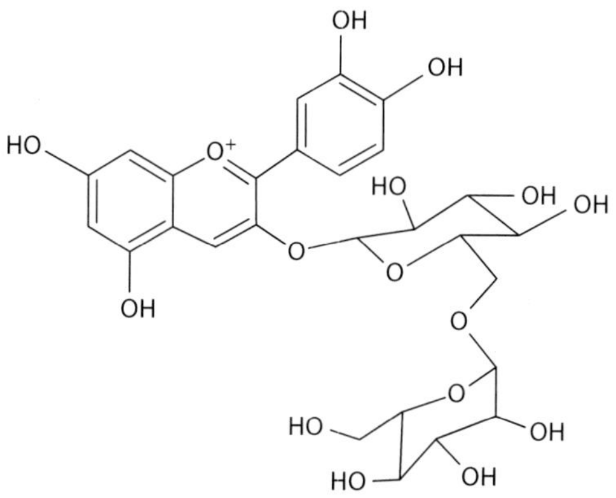

FIGURE 10.3 Chemical structure of cyanidin-3-glucoside.

TABLE 10.2
Percentage of Retention of Total Anthocyanin Contents (mg 3-O-Rut 100 g⁻¹ Fresh Weight) in Control Purée and Treated Purées

Control	High Pressure Processing									SEM	p_{value}
	400 MPa			500 MPa			600 MPa				
	1 s	150 s	300 s	1 s	150 s	300 s	1 s	150 s	300 s		
100.0[a]	82.9[ab]	80.5[bc]	85.4[bc]	85.4[bc]	85.4[bc]	85.4[bc]	85.4[bc]	78.0[c]	87.8[b]	1.16	.001

Control	High Pressure Thermal Processing									SEM	p_{value}
	300 MPa			600 MPa			900 MPa				
	60°C	70°C	80°C	60°C	70°C	80°C	60°C	70°C	80°C		
100.0[a]	79.9[c]	81.5[bc]	90.4[abc]	76.9[c]	101.0[a]	92.1[abc]	78.2[c]	93.1[abc]	89.4[abc]	1.84	.001

Sources: González-Cebrino, F. et al., *Innov. Food Sci. Emerg. Technol.*, 20, 34–41, 2013; García-Parra, J. et al., *Innov. Food Sci. Emerg. Technol.*, 26, 26–33, 2014.

[a–c] Different letters in the same row indicate significant statistical differences due to the treatment applied (Tukey's test, $p < .05$). SEM, standard error of mean.

achieved the lowest values of total anthocyanins, with 78% of the initial control purée, while the application of 600 MPa/300 s maintained the 87.8% of the original content. Regarding HPT treatments, a combination of 600 MPa/70°C maintained the levels of total anthocyanins to similar values as control purées. In contrast, treatments at lower initial temperatures (60°C and at 300, 600, and 900 MPa) retained the lowest anthocyanin content.

García-Parra et al. (2014) found high correlations between the decrease in the intensity of red color and the anthocyanins content in the plum purée. Significant positive correlations ($p < .01$) were found between the parameter CIE a* and the content of anthocyanins, associating the changes in CIE a* after processing with changes in the red pigment content in the plum purée. Therefore, the preservation of anthocyanins would not only be positive from a nutritional point of view, but also it would provide a better color of processed purées.

Total anthocyanin contents were reduced after applying HPP in the plum purée. Thus, the application of some HPT treatments could be more effective even to HPP to preserve these bioactive components. These different results between both high pressure treatments could be due to the low inactivation of the polyphenol oxidase (PPO) enzyme after HPP (González-Cebrino et al., 2013), which may limit the shelf life of the processed plum purée. Other factors such as the activity of other oxidative enzymes and effect of the processing conditions on the pigments stability could be also related to the observed changes in the anthocyanins after HPP (Ferrari et al., 2011).

10.4 EFFECT OF HYDROSTATIC HIGH PRESSURE TREATMENTS ON CAROTENOID COMPOUNDS

Carotenoids are natural pigments responsible for the yellow, orange and red fruits, roots, flowers, and inflorescences. Carotenoids are found in chloroplasts of higher plants, although the photosynthetic tissue color is masked by chlorophyll. Degeneration in chloroplast-to-chromoplast chlorophyll loss causes the appearance of yellow and orange coloration in many mature fruits. The color of these chromoplasts is due to its content of carotenoids and xanthophylls, adapting generally from yellow to orange colorations, depending on the intensity of maturation due to enrichment in carotenoids accompanied by the disappearance of chlorophyll. Carotenoids are lipophilic and therefore water insoluble substances. The basic structure of the carotenoids is tetraterpene 40 carbon atoms, and linear symmetric formed from eight isoprenoid units, containing five carbon atoms linked, such that the order is reversed in the center (Figure 10.4).

This basic skeleton can be modified in several ways such as by hydrogenation, cyclization, rearrangement, isomerization, by introducing oxygen function or by a combination of these processes, resulting in a variety of structures. The hydrocarbon carotenoids are collectively referred to as *carotenes*; those containing oxygen are known as *xanthophylls*. Over 600 carotenoids have been isolated and characterized naturally. However, only about 40 are present in a typical human diet. Of these 40, over 20 carotenoids have been identified in the human blood and tissues. About 90% of the carotenoids in the diet and human body are represented by β-carotene, α-carotene, lycopene, lutein, and cryptoxanthin (Gerster, 1997).

Certain beneficial properties in the prevention of cardiovascular diseases, cancer, and other chronic diseases are attributed to carotenoids (Astrog et al., 1997). Based on epidemiological studies, a positive association between a reduced risk of chronic diseases and greater dietary intake of carotenoids is suggested (Agarwal et al., 2000).

Pumpkin due to its pH should be processed by HPT to reach an adequate shelf life; for this reason, only the effect of HPT treatment has been evaluated on carotenoids. Carotenoids found in the pumpkin purée were lutein, α-carotene, and β-carotene. Two of them, lutein and β-carotene, were significantly altered after HPT processing ($p < .05$) (Figure 10.5). Lutein content showed the highest increase after treatments at 300 MPa/60°C; the other treatments assayed showed intermediate values. With respect to β-carotene content, the purée processed at 300 MPa at 60°C showed the lowest content, although purées treated at 600 and 900 MPa (at any temperature) had higher levels than those treated at 300 MPa.

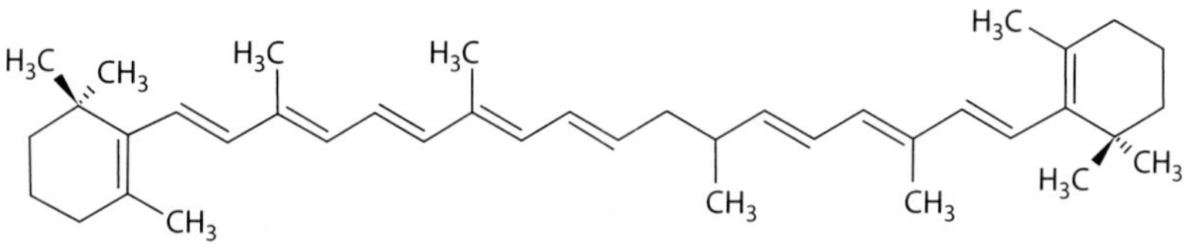

FIGURE 10.4 Chemical structure of a β-carotene molecule.

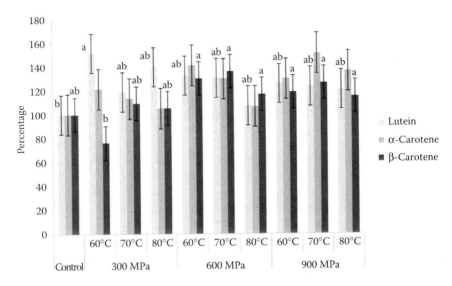

FIGURE 10.5 Percentage of retention with respect to the initial content of individual carotenes after applying different HPT processing to a pumpkin purée. (From García-Parra, J. et al., *Food Bioprod. Process.*, 98: (2016) 124–132. With permission.)

In general, in HPT treatments, pressures over 300 MPa (such as 600 or 900 MPa) were effective to maintain individual carotenoids or even increase them. Oey et al. (2008a) reported that HPP at room temperatures did not affect or eventually increased carotenoid contents in vegetables. High pressure could modify the extractability of carotenoids due to structural changes in a cell membrane induced by physicochemical changes through intense treatments. The extractability of carotenoids was dependent on the pressure applied and type of plant, as has been reported for different fruits and vegetables, such as orange juice, carrot, and tomato purée (Oey et al., 2008a). This effect was also corroborated by Sánchez-Moreno et al. (2004), who reported a denaturation of the carotenoid-binding protein during the thermal treatment, which enhanced the release of carotenoids in the tomato purée. Therefore, the combinations of both effects (temperature and pressure) would favor the release of carotenoids.

The application of HPT treatments would have a positive effect to preserve or even increase carotenoid contents. In this respect, Tauscher (1998) found a very low carotenes loss (less than 5%) in carrot based product at 600 MPa/75°C/40 min. Van Der Plancken et al. (2012) did not find major losses in α- or β-carotene in the carrot purée treated at 800 MPa and 40°C–74°C. Similarly, Sánchez et al. (2014) found that carotenoid contents of commonly consumed vegetables (carrot, tomato, or red pepper) were not significantly influenced by HPP (625 MPa, 5 min, 20°C) or HPT (625 MPa, 5 min, 70°C and 625 MPa, 5 min, 117°C). Moreover, at temperatures of sterilization treatments, Vervoort et al. (2012) did not find significant changes in α- and β-carotene in carrots after HPT treatment at 700 MPa and 124.8°C for 3 min. Nguyen et al. (2007) also reported a 92% carotene retention in carrots after 700 MPa and 121°C treatment for 1 min.

While, in general, HPP did not alter the initial levels of carotenoids in pumpkin (González Cebrino et al., in press), HPT processing at moderate temperatures may be a suitable alternative to preserve and even increase the bioavailability of carotenoids. This effect would be related to pressure intensity, and it would be an important factor about the retention of bioactive compounds on the vegetable matrix.

10.5 EFFECT OF HYDROSTATIC HIGH PRESSURE TREATMENTS ON PHENOLIC COMPOUNDS

Polyphenols are secondary metabolites of plants that possess one or more aromatic rings with one or more hydroxyl groups. They are the most abundant antioxidants on the dietary intake, their main dietary sources being fruits and plant-derived beverages (Scalbert et al., 2005).

The effect of HPP and HPT treatment on the total phenolic compounds content (TPC) of plum and pumpkin was evaluated (Table 10.3). TPC did not show significant modifications after HPP. Neither the pressure intensity nor the holding time did significantly modify the TPC in comparison to the fresh purée. Other authors have also reported that TPC was not affected by HPP in different vegetables. Patras et al. (2009a) reported that TPC in tomato and carrot purées were not affected by HPP (at 400, 500, or 600 MPa for 15 min at an initial temperature of 20°C). Wang et al. (2012) have also reported no changes in TPC content in purple sweet potato nectar after HPP at 400 MPa for 10 min, 500 MPa for 5 min, and 600 MPa for 2.5 min.

Indeed, other studies have even indicated that TPC was increased after HPP such as in purées based on yellow plum (var. *Songold*), pressurized at 400 and 600 MPa for 7 min (González-Cebrino et al., 2012), apple (Landl et al., 2010), nectarine (García-Parra et al., 2011), or strawberry (Patras et al., 2009b) pressurized at 600 MPa for 5, 10, and 15 min, respectively. Other authors studied have also reported an increase in TPC in vegetable juices from blueberry processed at 200 MPa/5, 9, and 15 s at maximum temperature of 42°C (Barba et al., 2013), cashew apple at 250 and 400 MPa for 3 and 5 s (25°C) (Queiroz et al., 2010), and pomegranate processed at 400 MPa/5 min (25°C) (Ferrari et al., 2011). This increase in TPC may be related to an increased extractability of some of the compounds from solid suspended particles following HPP (Chen et al., 2015). The increase would depend on the holding time, temperature, and the pressure level of treatments.

On the other hand, TPC were also reduced after HPP in fruit smoothies processed at 450 MPa for 5 min at 20°C (Keenan et al., 2010), asparagus juice pressurized at 600 MPa for 10 and 20 min at 29°C (Chen et al., 2015), and aloe vera gel treated at 400 MPa for 1, 3, and 5 min (Vega-Gálvez et al., 2012). Reductions in TPC could be caused by pressure-induced chemical oxidation of different phenolic compounds by enzymes (Clariana et al., 2011). PPO enzyme is responsible of browning by oxidizing polyphenols; therefore, losses could be attributed to the activity of this enzyme.

HPT treatment maintained or increased the TPC levels in the same red plum cultivar (Table 10.3). The highest increase (17%) was found in the purée at 600 MPa

TABLE 10.3
Percentage of Retention of Total Phenolic Compounds Content (mg Gallic Acid 100 g^{-1} Fresh Weight) in Control Purée and Treated Purées

High Pressure Processing

	Control	400 MPa			500 MPa			600 MPa			SEM	p_{value}
		1 s	150 s	300 s	1 s	150 s	300 s	1 s	150 s	300 s		
Plum	100.0	84.4	91.3	80.4	94.6	94.3	94.1	93.6	103.5	91.1	1.70	.080

High Pressure Thermal Processing

	Control	300 MPa			600 MPa			900 MPa			SEM	p_{value}
		60°C	70°C	80°C	60°C	70°C	80°C	60°C	70°C	80°C		
Plum	100.0[bc]	102.1[bc]	98.9[bc]	109.8[ab]	99.1[bc]	117.3[a]	105.9[abc]	92.6[c]	98.2[bc]	109.9[ab]	0.99	.001
Pumpkin	100.0[c]	146.8[ab]	148.9[ab]	147.9[ab]	149.8[ab]	165.1[a]	159.9[ab]	138.9[b]	146.0[ab]	142.6[ab]	3.34	.001

Sources: González-Cebrino, F. et al., *Innov. Food Sci. Emerg. Technol.*, 20, 34–41, 2013; García-Parra, J. et al., *Innov. Food Sci. Emerg. Technol.*, 26, 26–33, 2014; García-Parra, J. et al., *Food Bioprod. Process.*, in press.

[a–c] Different letters in the same row indicate significant statistical differences due to the treatment applied (Tukey's test, $p < .05$). SEM: standard error of mean.

and 70°C respect to the initial content. TPC did not change after applying the rest of the HPT treatments. In the pumpkin purée, TPC increased significantly between 39% and 65% following all applied treatments. In the same way as the plum purée, HPT treatment at 600 MPa and 70°C obtained the highest TPC in pumpkin. Corrales et al. (2008) reported that treatments at 600 MPa and 70°C increased phenolic compounds from grape by-product recovery approximately 1.5-fold higher than the initial. Huang et al. (2013) indicated that the increase of TPC induced by HPT processing is possibly due to the easier extraction of phenolic compounds in the pulp during thermal processing due to the cells disruption, which would release substrates and promote changes in TPC (Clariana et al., 2011). In contrast, Van der Plancken et al. (2012) commented that HPT treatment can induce losses in some compounds, although this strongly depends on the type of compound. Chakraborty et al. (2015) reported that TPC from pineapple were significantly reduced at all pressure levels at 70°C (the loss in TPC ranged from 3% to 11%). The maximum TPC was obtained at 600 MPa/50°C.

The effect of HPP and HPT treatment on the levels of phenolic compounds would be dependent on the food matrix (which would favor or not the release of TPC content) and its specific profile of phenolic compounds. In addition, the treatment conditions (i.e., temperature/pressure/time) and the inactivation of oxidative enzymes such as the PPO are also important factors that affect the final levels of TPC after processing. In this sense, according to García-Parra et al. (2014, in press) HPT treatments at 600 MPa at 70°C (which better preserved TPC) showed a high inactivation of the PPO after processing.

10.6 EFFECT OF HYDROSTATIC HIGH PRESSURE TREATMENTS ON COMPOUNDS WITH ANTIOXIDANT ACTIVITY

Dietary intake of antioxidants may help protect cells from oxidative damage caused by free radicals. Antioxidant molecules have been shown to counteract oxidative stress in *in vitro* assays (e.g., in cells or animal studies). Examples of antioxidants include anthocyanins, β-carotene, catechins, flavonoids, lutein, lycopene, selenium, and vitamins C and E (Carlsen et al., 2010).

The effect of HPP and HPT treatment was evaluated in plum and pumpkin purées (Table 10.4). Significant decreases of total antioxidant activity (TAA) were observed after HPP in plum, reaching losses between 9% (600 MPa/1 s) and 15% (500 MPa/150 s). Purée treated at 400 MPa for 150 s and 600 MPa for 600 s showed similar levels respect to the initial purée.

HPP is known as an efficient technology to preserve the levels of compounds with antioxidant activity (Ramírez et al., 2009). However, Barba et al. (2013) found that treatments at 400 MPa for 15 min and 600 MPa for 5–15 min reduced antioxidant values in blackberry juice. Patras et al. (2009b) observed a decrease in the TAA of strawberry purée treated at 400 MPa, while no variations were reported at 600 MPa. They observed that blackberry purée pressurized at 600 MPa obtained an increase in TAA. On the other hand, Patras et al. (2009a) and Wang et al. (2012) reported no changes in TAA of carrot and tomato purées and purple sweet potato nectar after HPP, respectively (as it was observed by TPC).

TABLE 10.4
Percentage of Retention of Total Antioxidant Activity (mg Trolox 100g^{-1} Fresh Weight) in Control and Treated Purées

High Pressure Processing

	Control	400 MPa			500 MPa			600 MPa			SEM	p_{value}
		1 s	150 s	300 s	1 s	150 s	300 s	1 s	150 s	300 s		
Plum	100.0^a	86.8^b	93.3^{ab}	87.1^b	89.1^b	84.9^b	89.1^b	91.0^b	86.5^b	93.4^{ab}	1.15	.029

High Pressure Thermal Processing

		300 MPa			600 MPa			900 MPa				
		60°C	70°C	80°C	60°C	70°C	80°C	60°C	70°C	80°C		
Plum	100.0^{de}	96.4^e	94.7^e	96.0^e	104.7^{cd}	102.6^{de}	96.1^e	110.9^{bc}	113.0^{ab}	119.2^a	1.55	.001
Pumpkin	100.0^c	81.5^d	83.7^d	83.1^d	95.4^c	115.0^a	103.3^{bc}	94.5^c	110.0^{ab}	113.1^a	2.25	.001

Sources: González-Cebrino, F. et al., *Innov. Food Sci. Emerg. Technol.*, 20, 34–41, 2013; García-Parra, J. et al., *Innov. Food Sci. Emerg. Technol.*, 26, 26–33, 2014; García-Parra, J. et al., *Food Bioprod. Process.*, in press.

$^{a–e}$ Different letters in the same row indicate significant statistical differences due to the applied treatment (Tukey's test, $p < .05$). SEM: standard error of mean.

Some of the HPT treatment applied increased the levels of TAA of both plum and pumpkin purées (Table 10.4), especially when they were processed at the most intense processing conditions of pressure (900 MPa) and temperature (70°C and 80°C). In contrast, the less intense treatments (concretely at 300 MPa) showed lower levels than control. The increase in the TAA after HPT treatments has been previously reported in carrot juice treated at 100–800 MPa from 30°C up to 65°C for 90 min (Oey et al., 2004).

This effect could be associated with an increase in the extraction of antioxidant components after the treatment due to cell membrane disruption as a result of high pressure, as it was previously explained, which also improves the bioavailability of antioxidant complexes present in vegetables by breaking down cell walls.

Jayachandran et al. (2015) reported a degradation of antioxidants in a litchi-based mixed fruit beverage with an increase in the temperature above 60°C. They indicated that the combined effect of pressure and temperature on the antioxidant activity is more marked at a longer treatment times (they assayed 5, 10, 15, and 20 min). Ludikhuyze and Hendrickx (2002) reported that the antioxidant capacity of orange juice decreased at temperatures ranging from 30°C to 70°C as a function of treatment time (up to 20 min).

In this case, the application of most HPP treatments assayed was not effective enough for maintaining TAA of the red plum purée, although antioxidant compounds were well preserved at the most intense processing conditions, because slight reductions with respect to initial levels were reported. Reductions in TAA of plum after HPP or HPT treatment could be related to lower levels of anthocyanins after the application of some treatments, because phenolic compounds were generally well preserved after both treatments. These compounds would provide an important antioxidant activity to plum. In fact, Lozano et al. (2009) found that red flesh and peel plum varieties with high total antioxidant activity showed strong correlations with the phenolic and anthocyanin contents.

In turn, in pumpkin, the major compounds with antioxidant activity would be phenolics and carotenoids, and these compounds were not affected by processing and may even increase (Oey et al., 2008a), which is consistent with the levels of TAA obtained in pumpkin after HPT treatment.

According to Barba et al. (2013), reductions in the TAA could be also caused by the activation of oxidative enzymes during pressurization. These authors suggested that the involvement of the residual PPO in the degradation of antioxidant compounds takes place during HPP. We consider that the lack of inactivation of PPO and/or other oxidative enzymes after HPP in plum would be responsible at some extent to the reduction of some bioactive compounds after processing.

Regarding the effect of processing conditions on the TAA, at higher pressures, more antioxidants are extracted from the tissues; nevertheless, at the same time, the increase of heating could also favor the deterioration of these compounds. The application of a short treatment time (1 min) could be favorable for a better retention of some antioxidants such as phenolic compounds, carotenoids in pumpkin, and anthocyanins in plum. This could explain differences with other studies that apply longer holding times. In addition, other antioxidants that have not been evaluated in this study, such as ascorbic acid, could also play an important role in these results.

Similar to TPC, the effect of high pressure on TAA depends not only on the pressurization conditions but also on the type of food matrix. The effect of treatments on the TAA would depend on the effect of processing in each vegetable matrix that has different proportion of antioxidants such as carotenes and phenolic compounds.

10.7 FACTORS THAT AFFECT THE PRESERVATION OF BIOACTIVE COMPOUNDS IN VEGETABLE PRODUCTS

The application of HPP or HPT treatment needs to be optimized to minimize the effect of processing conditions on the bioactive compounds present in vegetables. For example, Chauhan et al. (2011) suggested that the development of mathematical models could be successfully used for the prediction of antioxidant activity, total phenolics, and flavonoids compounds while processing derived food products. Using the optimized condition, they reported more retention of total antioxidants, phenolics, and flavonoids compounds in black grape juice. In this case, the studies carried out in plum and pumpkin indicate that in order to optimize treatments in each matrix, several individual bioactive compounds content need to be studied, because each antioxidant compound have different sensitivity to pressure, temperature, or holding time. The inactivation of degradative enzymes is also an important point to be taken into account to evaluate the processing effect on each matrix.

In principle, because of the pH conditions of pumpkin and plum, from a microbiological point of view, for the conservation of plum purée (pH ~ 3.4), a pasteurization treatment would be necessary to reach an adequate shelf life, whereas for the conservation of pumpkin (pH ~ 6.6), more intense treatments such as a severe pasteurization or sterilization would be required. For this reason, HPP would be more adequate to preserve acid products, such as plum, while pumpkin would reach longer shelf life by HPT treatments.

In low-acid foods such as pumpkin, from a microbiological point of view, a thermal treatment with a process value F0 of 3 min has been adopted as the minimum standard for a sterilization process, although at industrial level, generally an F0 of 5 min is applied. Nowadays, the limit to reach a sterilization effect is not clear. Vervoort et al. (2012) established that an initial temperature of 90°C was required to result in a maximum temperature of 124.8°C after pressure build up till 700 MPa. According to them, these conditions applied during a holding time of 3 min should correspond to a F0-value of at least 5 min. Therefore, the conditions applied for HPT treatment in pumpkin would be equivalent to a severe pasteurization treatment, because lower temperatures and shorter times than those required for sterilization were reached (Vervoort et al., 2012).

HPP is effective to preserve plum-derived products due to the very low pH of this fruit, which could maintain their microbiological stability better during storage (González Cebrino et al., 2013). HPP maintained the levels of bioactive compounds of the plum purée. This treatment was effective to maintain the content of phenolic compounds of the plum purée, but it slightly reduced the anthocyanin content and the antioxidant activity. Nevertheless, in general, it can be assumed that HPP did not adversely affect the nutritional quality of the plum purée. However, the poor inactivation of enzymes responsible for browning, such as PPO, may limit the shelf life of plum purées during refrigerated storage (González Cebrino et al., 2013). This effect could be overcome by adding food additives such as ascorbic acid.

HPT treatments are adequate to maintain the shelf life of low-acid products such as pumpkin. However, HPT treatments would be also interesting in very-acid foods, such as plum, because a significant reduction in the activity of PPO enzyme was achieved (García-Parra et al., 2014), in contrast to HPP (González-Cebrino et al., 2013). For plum, the effect of HPT processing is very interesting since this treatment would allow the inactivation of the PPO and maintain the content of bioactive compounds, such as anthocyanins and antioxidant compounds, without adding additives (García-Parra et al., 2014).

In products such as pumpkin, HPT treatments are also very effective to maintain and even increase the bioactive compounds, by the increase in the extractability of carotenoids and phenolic compounds, which could increase their bioavailability.

In both vegetables (plum and pumpkin), treatments of 600 MPa at an initial temperature of 70°C was the tested combination, which maintained bioactive compounds of plum and pumpkin purées better. At these conditions, the PPO enzyme was importantly inactivated (García-Parra et al., 2014, García-Parra et al., in press).

Therefore, the effect of HPP and HPT treatments on the levels of bioactive compounds would be dependent on three factors: (1) the food matrix (which would favor the release/degradation/preservation of bioactive compounds); (2) processing conditions such as pressure intensity, holding time, and temperature of the treatment; and (3) the effect of treatments on oxidative enzymes such as the PPO. In this sense, according to García-Parra et al. (2013, 2014, in press), HPT treatments that better-preserved bioactive compounds showed high inactivation of the oxidative enzymes after processing.

10.8 CONCLUDING REMARKS AND FUTURE TRENDS

HPP and HPT treatments would allow the preservation of the nutritional quality of vegetables models such as plum and pumpkin. To reach an adequate shelf life of the processed product, HPP would be more appropriate to acid products. However, the resistance of oxidative enzymes to HPP also needs to be taken into account to reach an adequate shelf life. On the other hand, HPT treatment would be adequate to preserve bioactive compounds content of low-acid products and also in acid products that does not reach a sufficient inactivation of the PPO enzyme.

HPT treatment could extend the shelf life of pumpkin-derived products. In plum, this technology would also be effective to increase the shelf life without needing additives, which is in line with the current demands of consumers, who prefer *natural* products. HPT treatment, despite being more intense treatment than HPP, maintained betters the bioactive compounds of both products.

The inactivation of enzymes such as the PPO could play an important role for the preservation of bioactive compounds in vegetables by HPP and HPT treatments. Research about the most adequate conditions to maintain the bioactive compounds and to reach the maximum inactivation of degradation enzymes and microorganisms need to carried out jointly to know in deep the effect of these technologies in each food matrix.

ACKNOWLEDGMENTS

The authors wish to acknowledge the financial support by the project INIA RTA2010-00079-C02 (Obtención de purés de frutos de alto valor funcional mediante tecnologías innovadoras de procesado). R. Ramírez thanks the Gobierno de Extremadura for her employment (DOE 22/07/14).

REFERENCES

Agarwal, S., Rao, A.V. Carotenoids and chronic diseases. *Drug Metabolism and Drug Interactions* 17(1–4): (2000) 189–210.

Astrog, P., Gradelet, S., Berges, R., Suschetet, M. Dietary lycopene decreases initiation of liver preneoplastic foci by diethylnitrosamine in rat. *Nutrition and Cancer* 29: (1997) 60–68.

Balasubramaniam, V.M., Ting, E.Y., Stewart, C.M., Robbins, J.A. Recommended laboratory practices for conducting high-pressure microbial inactivation experiments. *Innovative Food Science Emerging Technologies* 5: (2004) 299–306.

Barba, F. J., Esteve, M. J., Frigola, A. Physicochemical and nutritional characteristics of blueberry juice after high pressure processing. *Food Research International* 50: (2013) 545–549.

Carlsen, M.H., Halvorsen, B.L., Holte, K., Bøhn, S.K., Dragland, S. et al. The total antioxidant content of more than 3100 foods, beverages, spices, herbs and supplements used worldwide. *Nutrition Journal* 9: (2010) 3.

Chakraborty, S., Rao, P.S., Mishra, H.N. Effect of combined high pressure–temperature treatments on color and nutritional quality attributes of pineapple (*Ananas comosus* L.) puree. *Innovative Food Science and Emerging Technologies* 28: (2015) 10–21.

Chauhan, O.P., Raju, P.S., Ravi, N., Roopa, N. Bawa, A.S. Studies on retention of antioxidant activity, phenolics and flavonoids in high pressure processed black grape juice and their modelling. *International Journal of Food Science and Technology* 46: (2011) 2562–2568.

Chen, X., Qin, W., Ma, L., Xu, F., Jin, P., Zheng, Y. Effect of high pressure processing and thermal treatment on physicochemical parameters, antioxidant activity and volatile compounds of green asparagus juice. *LWT—Food Science and Technology* 62: (2015) 927–933.

Clariana, M., Valverde, J., Wijngaard, H., Mullen, A.M., Marcos, B. High pressure processing of swede (*Brassica napus*): Impact on quality properties. *Innovative Food Science and Emerging Technologies* 12: (2011) 85–92.

Clifford, M.N. Anthocyanins: Nature, occurrence and dietary burden. *Journal Science Food Agriculture* 80: (2000) 1063–1072.

Corrales, M., Toepfl, S., Butz, P., Knorr, D., Tauscher, B. Extraction of anthocyanins from grape by-products assisted by ultrasonics, high hydrostatic pressure or pulsed electric fields: A comparison. *Innovative Food Science and Emerging Technologies* 9: (2008) 85–91

Ertekin, C., Gozlekci, S., Kabas, O., Sonmez, S., and Akinci, I. Some physical, pomological and nutritional properties of two plum (*Prunus domestica* L.) cultivars. *Journal of Food Engineering* 75: (2006) 508–514.

Ferrari, G., Maresca, P., Ciccarone, R. The effects of high hydrostatic pressure on the polyphenols and anthocyanins in red fruit products. *Procedia Food Science* 1: (2011) 847–853.

García-Parra, J., Contador, R., Delgado-Adámez, J., González-Cebrino, F., and Ramírez, R. The applied pretreatment (blanching, ascorbic acid) at the manufacture process affects the quality of nectarine purée processed by hydrostatic high pressure. *International Journal of Food Science & Technology* 49: (2013) 1203–1214.

García-Parra, J., González-Cebrino, F., Cava, R., Ramírez, R. Effect of a different high pressure thermal processing compared to a traditional thermal treatment on a red flesh and peel plum purée. *Innovative Food Science and Emerging Technologies* 26: (2014) 26–33

García-Parra, J., González-Cebrino, F., Delgado-Adámez J., Cava, R., Ramírez, R. High pressure assisted thermal processing of pumpkin pureé: Effect on microbioal counts, color, Bioactive compounds and polyphenoloxidase enzyme. *Food and Bioproducts Processing* 98: (2016) 124–132.

García-Parra, J., González-Cebrino, F., Delgado, J., Lozano, M., Hernández, T., and Ramírez, R. Effect of thermal and high pressure processing on the nutritional value and quality attributes of a nectarine purée with industrial origin during the refrigerated storage. *Journal of Food Science* 76: (2011) C618–C625.

Gerster, H. The potential role of lycopene for human health. *The Journal of the American College of Nutrition* 16: (1997) 109–26.

Gliemmo, M.F., Latorre, M.E., Gerschenson, L.N., Campos, C.A. Color stability of pumpkin (*Cucurbita moschata, Duchesne ex Poiret*) puree during storage at room temperature: Effect of pH, potassium sorbate, ascorbic acid and packaging material. *LWT-Food Science and Technology* 42: (2009) 196–201.

González-Cebrino, F., Durán, R., Delgado-Adámez, J., Contador, R., Ramírez, R. Changes after high-pressure processing on physicochemical parameters, bioactive compounds, and polyphenol oxidase activity of red flesh and peel plum purée. *Innovative Food Science and Emerging Technologies* 20: (2013) 34–41.

González-Cebrino, F., Durán, R., Delgado-Adámez, J., Contador, R., Ramírez, R. Application of high pressure processing on pumpkin purée: Effects on the bioactive compounds and polyphenol oxidase enzyme stability. *Food Science and Technology International* 22(3): (2016) 235–245.

González-Cebrino, F., García-Parra, J., Contador, R., Tabla, R., Ramírez, R. Effect of high-pressure processing and thermal treatment on quality attributes and nutritional compounds of Songold plum purée. *Journal of Food Science* 77: (2012) C866–C873.

Gupta, R., Balasubramaniam, V.M., Schwartz, S.J., Francis, D.M. Storage stability of lycopene in tomato juice subjected to combined pressure-heat treatments. *Journal of Agricultural and Food Chemistry* 58(14): (2010) 8305–8313.

Havsteen, B.H. The biochemistry and medical significance of the flavonoids. *Pharmacology & Therapeutics* 96(2–3): (2002) 67–202.

Hoover, D.G., Metrick, C., Farkas, D.F. Effects of high hydrostatic pressure on milk. *Milchwssenschaft* 47(12): (1989) 760–763.

Huang, W., Bi, X., Zhang, X., Liao, X., Hu, X., Wu, J. Comparative study of enzymes, phenolics, carotenoids and color of apricot nectars treated by high hydrostatic pressure and high temperature short time. *Innovative Food Science and Emerging Technologies* 18: (2013) 74–82.

Jayachandran, L.E., Chakraborty, S., Rao, P.S. Effect of high pressure processing on physicochemical properties and bioactive compounds in litchi based mixed fruit beverage. *Innovative Food Science and Emerging Technologies* 28: (2015) 1–9.

Keenan, D.F., Brunton, N.P., Gormley, T.R., Butler, F., Tiwari, B.K., Patras, A. Effect of thermal and high hydrostatic pressure processing on antioxidant activity and colour of fruit smoothies. *Innovative Food Science and Emerging Technologies* 11: (2010) 551–556.

Landl, A., Abadias, M., Sárraga, C., Viñas, I., Picouet, P.A. Effect of high pressure processing on the quality of acidified Granny Smith apple purée product. *Innovative Food Science and Emerging Technologies* 11: (2010) 557–564.

Lozano, M., Vidal-Aragón, M.C., Hernández, M.T., Ayuso, M.C., Bernalte, M.J., García, J., Velardo, B. Physicochemical and nutritional properties and volatile constituents of six Japanese plum (*Prunus salicina Lindl.*) cultivars. *European Food Research and Technology* 469(228): (2009) 403–410.

Ludikhuyze, L., Hendrickx, M.E.G. Effects of high-pressure on chemical reactions related to food quality. In Hendrickx, M.E.G. and Knorr, D. (Eds.), *Ultra High Pressure Treatments of Foods Food Engineering Series*. Part II (pp. 167–188), New York: Kluwer Academic/Plenum Publishers (2002).

Manach, C., Scalbert, A., Morand, C., Rémésy, C., Jiménez, L. Polyphenols: Food sources and bioavailability. *American Journal of Clinical Nutrition* 79: (2004) 727–747.

Mújica-Paz, H., Valdez-Fragoso, A., Samson, C.T., Welti-Chanes, J., Torres, J.A. High-pressure processing technologies for the pasteurization and sterilization of foods. *Food and Bioprocess Technology* 4(6): (2011) 969–985.

Nguyen, L.T., Balasubramaniam, V.M., Ratphitagsanti, W. Estimation of accumulated lethality under pressure-assisted thermal processing. *Food and Bioprocess Technology* 7(3): (2013) 633–644.

Nguyen, T.L., Rastogi, N.K., Balasubramaniam, V.M. Evaluation of the instrumental quality of pressure-assisted thermally processed carrots. *Journal of Food Science* 72(5): (2007) E264–E270.

Oey, I., Lille, M., Van Loey A., Hendrickx, M. Effect of high-pressure processing on colour, texture and flavour of fruit and vegetable-based food products: A review. *Trends in Food Science & Technology* 19: (2008b) 320–328.

Oey, I., Van der Plancken, I., Van Loey, A., Hendrickx, M. Does high pressure processing influence nutritional aspects of plant based food systems? *Trends Food Science Technology* 19: (2008a) 300–3008.

Oey, I., Van Loey, A., Hendrick, M. Pressure and temperature stability of water-soluble antioxidants in orange and carrot juice: A kinetic study. *European Journal of Food Research and Technology* 219: (2004) 161–166.

Patras, A., Brunton, N.P., Da Pieve, S., Butler, F. Impact of high pressure processing on total antioxidant activity, phenolic, ascorbic acid, anthocyanin content and colour of strawberry and blackberry purées. *Innovative Food Science and Emerging Technologies* 10: (2009b) 308–313.

Patras, A., Brunton, N., Da Pieve, S., Butler, F., Downey, G. Effect of thermal and high pressure processing on antioxidant activity and instrumental colour of tomato and carrot purées. *Innovative Food Science and Emerging Technologies* 10: (2009a) 16–22.

Pothakamury, U., Barbosa-Cánovas, G.V. Fundamental aspects of controlled release in foods. *Trends in Food Science & Technology* 6: (1995) 387–406.

Queiroz, C., Moreira, C.F.F, Lavinas, F.C., Lopes, M.L.M., Fialho, E., Valente-Mesquita, V.L. Effect of high hydrostatic pressure on phenolic compounds, ascorbic acid, and antioxidant activity in cashew apple juice. *High Pressure Research: An International Journal* 30: (2010) 507–513.

Ramírez, R., Saravia, J., Pérez Lamela, C., Torres, A. Reaction kinetics analysis of chemical changes in pressure-assisted thermal processing. *Food Engineering Reviews* 603(1): (2009) 16–30.

Robards, K., Prentzler, P.D., Tucker, G., Swatsitang, P., Glover, W. Phenolic compounds and their role in oxidative processes in fruits. *Food Chemistry* 66: (1999) 401–436.

Sánchez, C., Baranda, A.B., Martínez de Marañón, I. The effect of high pressure and high temperature processing on carotenoids and chlorophylls content in some vegetables. *Food Chemistry* 163: (2014) 37–45.

Sánchez-Moreno, C., Plaza, L., De Ancos, B., Cano, MP. Effect of combined treatments of high-pressure and natural additives on carotenoid extractability and antioxidant activity of tomato puree (*Lycopersicum esculentum* Mill.). *European Food Research Technology* 219: (2004) 151–160.

Scalbert, A., Williamson, G. Dietary intake and bioavailability of polyphenols. *Journal of Nutrition* 130: (2000) 2073S–2085S.

Scalbert, A., Johnson, I.T., Saltmarsh, M. Polyphenols: Antioxidants and beyond. *American Journal of Clinical Nutrition* 81: (2005) 215S–217S.

Subramanian, A., Ahn, J., Balasubramaniam, V.M., Rodriguez-Saona, L. Determination of spore inactivation during thermal and pressure-assisted thermal processing using FT-IR spectroscopy. *Journal of Agricultural and Food Chemistry* 54(26): (2006) 10300–10306.

Tauscher, B. Effect of high pressure treatment to nutritive sub- stances and natural pigments. In Autio, K. (Ed.), *Fresh Novel Foods by High Pressure* (pp. 83–95), Espoo, Finland: Technical Research Centre of Finland, VTT Symposium 186 (1998).

Tomás-Barberán, F.A., Ferreres, F., Gil, M.I. Antioxidant phenolic metabolites from fruit and vegetables and changes during postharvest storage and processing. *Studies in Natural Products Chemistry* 23: (2000) 739–795.

Tomás-Barberán, F.A., Gil, M.I., Cremin, P., Waterhouse, A.L., Hess-Pierce, B., Kader, A. A. HPLD–DAD–ESI/MS analysis of phenolic compounds in nectarines, peaches, and plums. *Journal of Agricultural and Food Chemistry* 49: (2001) 4748–4760.

Van der Plancken, I., Verbeyst, L., De Vleeschouwer, K., Grauwet, T., Heiniö, R. et al. (Bio) chemical reactions during high pressure/high temperature processing affect safety and quality of plant-based foods. *Trends in Food Science and Technology* 23: (2012) 28–38.

Vega-Gálvez, A., Giovagnoli, C., Pérez-Won, P., Reyes, J. E., Vergara, J., Miranda, M., Uribe, E., Di Scala, K. Application of high hydrostatic pressure to aloe vera (*Aloe barbadensis* Miller) gel: Microbial inactivation and evaluation of quality parameters. *Innovative Food Science and Emerging Technologies* 13: (2012) 57–63.

Vervoort, L., Van der Plancken, I., Grauwet, T., Verlinde, P., Matser, A., Hendrickx, M., Van Loey, A. Thermal versus high pressure processing of carrots: A comparative pilot-scale study on equivalent basis. *Innovative Food Science and Emerging Technologies* 15: (2012) 1–13.

Wang, Y., Liu, F., Cao, X., Chen, F., Hu, X., Liao, X. Comparison of high hydrostatic pressure and high temperature short time processing on quality of purple sweet potato nectar. *Innovative Food Science and Emerging Technologies* 16: (2012) 326–334.

11 Bioactive Compound Encapsulation and Its Behavior during *In Vitro* Digestion

Cristian Ramírez, Helena Nuñez, and Ricardo Simpson

CONTENTS

11.1 Introduction 241
11.2 Importance of Encapsulation of Bioactive Compound during Digestion.........242
    11.2.1 Chewing as the Starting Point of Digestion.................................242
    11.2.2 Food Structure and Its Protective Role during Digestion.............244
    11.2.3 Study Case: Effect of Glucose Release from Sodium Alginate Gel–Starch Mixtures Based on *In Vitro* Digestion Studies..........246
    11.2.4 Microencapsulation of Food Nutrients 252
    11.2.5 Control Release of Nutrient during Digestion 253
    11.2.6 Inclusion of Microcapsules in Food Matrices 254
References..255

11.1 INTRODUCTION

Recently, interest has increased in understanding the interaction of functional nutrients with the structure of the food matrix and how an adequate food formulation can positively affect these interactions, allowing controlled or slow release of nutrients. For example, food processing plays an important role in food formulation, which modifies the food structure, transforming hard and solid structures into soft and porous structures that allow easy nutrient release or transforming liquid food into solid food by thermal or chemical processing, retaining the nutritional compound in the food matrix.

Independent of food structure or texture, the digestion of food starts at the mouth with chewing, which reduces the food into particles that can then be easily swallowed. This process is an important step that is often excluded in simulated digestion delivery studies. This step is important because the particle size after chewing could play an important role in the protection of labile compounds during digestion primarily because a structure (micro or nano) encloses the compounds, protecting them from mechanical damage, enzymatic action, and pH changes during digestion.

Therefore, the problem that we have identified is that the structural changes from hard to soft structures or vice versa could determine the behavior during chewing primarily due to changes in the structure, texture, and water content, which modify the behavior associated with the nutrient bioavailability at the end of digestion due to changes in the particle size. This change in the particle size could affect the delivery rate of nutrients because food structures contained in small particles can function as encapsulating agents that protect or retain the nutrient during digestion. Consequently, foods do not all have identical behavior as protective agents.

11.2 IMPORTANCE OF ENCAPSULATION OF BIOACTIVE COMPOUND DURING DIGESTION

11.2.1 CHEWING AS THE STARTING POINT OF DIGESTION

Chewing plays an important role in digestion because it is the point at which food is reduced into particles whose sizes must be small enough to form a bolus that can be swallowed (van der Bilt et al., 2006). However, the particle size of chewed food depends on many factors that are related to the food texture, water content and food size, which influence the human perception of the food and, therefore, the number of chews and the force and frequency of chewing, which directly affect the particle size (Turgeon and Rioux, 2011).

The particle size obtained during chewing has been studied to understand the effect of the particle size on nutrient release during the later steps in digestion. As the particle size decreases, the surface area that is exposed to the digestive enzymes is increased, causing an increase in the digestion rate (Ranawana et al., 2014). This situation was clearly observed in rice where *in vitro* digestion of whole and chewed rice resulted in significant differences in the glucose release rate, with a lower glucose release for whole grain and a higher release when the particle size was smaller than 1000 µm. Similar effects have been reported for other type of nutrients such as β-carotene, as demonstrated by the study performed by Lemmens et al. (2010), who evaluated the effect of the particle size on β-carotene bioaccessibility from raw and thermally processed carrots. These authors found that the particle size of raw carrots had a relevant effect on β-carotene release for particle fractions smaller than 125 µm; however, when the carrot was cooked, the chewing process was not relevant because the bioaccessibility was more related to cell disruption because of the thermal treatment. This research is interesting primarily because it is independent of the particle size in the raw state, where the carrot structure (cell wall) played a protective role in β-carotene bioaccessibility because the cell wall prevented it from being released. In fact, the particle size obtained during chewing affects nutrient release during digestion; however, the structural degree of the raw food products,

TABLE 11.1
Summary of Particle Size Analysis of Model Food Chewed during 4, 9, and 14 Cycles

Number of Chewing Cycles	Number of Particles	Range	Median Diameter (mm)
4	104	0.04–2.84	1.66[a]
9	130	0.04–2.83	1.56[a]
14	180	0.04–2.80	1.31[b]

Source: Ramirez, C. et al., *Food Hydrocolloid*, 44, 328–332, 2015.

[a, b] Different letters in the median diameters indicate significant use of Kolmogorov–Smirnov test ($P < .05$).

such as cereals, vegetables, and meat, often reduce the nutrient release rate because many of the nutrients are entrapped in the cell structure, which functions as a natural controlled release system.

We have performed some studies on the effect of the number of chewing cycles on a particle size and nutrient release, using food models based on calcium alginate gels and gelatinized starch along with a chewing machine and *in vitro* digestion whose results will be discussed in this chapter. Table 11.1 indicates results obtained for the particle size in terms of median particle diameter (d_{50}) for each one of the three chewing cycles (4, 9, and 14) used for nonstructured and structured gels. The results showed that while chewing cycles increase, the particle size begins to become smaller. An analysis of the number of chewing cycles is depicted in Figure 11.1, which shows that an

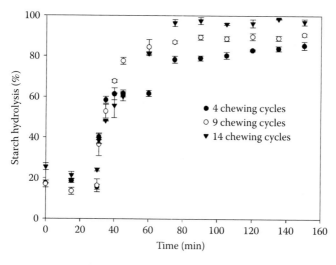

FIGURE 11.1 Starch hydrolysis based on *in vitro* digestion for model food applying 4, 9 and 14 chewing cycles.

increase in the number of chewing cycles increases the starch hydrolysis which varies from 81.80% to 96.92% when the number of chewing cycles increases from 4 to 14.

11.2.2 FOOD STRUCTURE AND ITS PROTECTIVE ROLE DURING DIGESTION

The structure of food plays an important role as a protective agent for food nutrients and as an agent that can influence the digestion rate, for example, delaying digestion (Parada and Aguilera, 2007; Mishra and Monro, 2012). In general, the study of food digestion has been performed with a focus on macronutrients isolated to study their behavior during digestion; however, foods belong to a complex system in which many nutrients can interact. For example, food matrices with physical structures that induce slow digestion can be found in pasta (such as spaghetti), in which a protein network limits the accessibility of amylose to the starch components (Zhang and Hamaker, 2009). Additionally, most legume foods are cooked in the whole grain form, which has substantial slowly digestible carbohydrates because of the entrapment of starch granules in the cell wall (Zhang and Hamaker, 2009). In this sense, the cooking time plays an important role by weakening the cell structure and allowing the nutrient release. Also, by weakening the cell structure, the chewing process could be affected positively through the nutrient release. We have performed some studies to evaluate the effect of cooking time of navy beans on particle size obtained after chewing and glucose release during an *in vitro* digestion. As we can see in Figure 11.2a and b, the glucose release during *in vitro* digestion of beans cooked at 100°C during 30 and 60 min was significantly affected by cooking time. When the beans were cooked for 30 min, the glucose release reached around 7 g/L; however, when they were cooked for 1 h, the release attained values close to 10 g/L. Similar results were obtained by Singh et al. (2014). Also, the chewing process was affected, showing that in general upper nine cycles were able to generate higher glucose releases. When four chewing cycles were applied to beans cooked for 30 and 60 min, the mean diameter of the

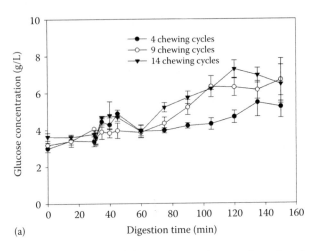

FIGURE 11.2 (a) Effect of chewing cicles number on glucose release from beans cooked at 100°C during 30 min. *(Continued)*

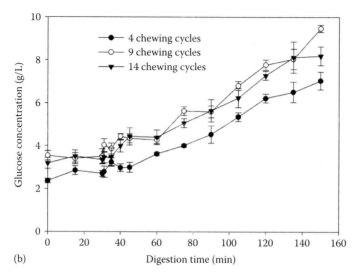

FIGURE 11.2 (Continued) (b) Effect of chewing cycle number on glucose release from beans cooked at 100°C during 60 min.

particles were of 0.255 and 0.180 cm, respectively. If the chewing cycles are increased to 14 cycles, the mean diameter for 30 and 60 min of cooking comes up to 0.182 and 0.144 cm, respectively. These results clearly support the idea that the cooking time not only affects the bean structure, but also affects the particle size in the chewing process as well as the amount of glucose released, which was higher in instances when upper nine chewing cycles were applied (smallest particle size).

According to Fardet (2015), if we compare two foods with equal chemical compositions but with different structures, these foods may provide substantially different health effects. For example, highly processed foods are commonly characterized by a loss of raw structure and excessive refinement of nutritional density, transforming them into highly energy-dense and poorly satiating foods (Fardet, 2015).

Mishra and Monro (2012) found that starchy foods, such as cereal grains that have lost their native structure due to processing could have a nutritionally deleterious effect on people and be harmful to health due to higher glycemic impact. In cases where the structure is not present, the development of a secondary structure by food processing technologies provides another good opportunity for creating slowly digestible carbohydrates by controlling the extent of starch gelatinization and the interactions between starch and other food components (Zhang and Hamaker, 2009). For instance, interactions between starch and nonstarch food components that produce new chemical or physical structures are another method for decreasing starch digestibility (Zhang and Hamaker, 2009; Srichamroen, 2014). For instance, guar gum plays a protective role in starch granules during swelling, enabling the leakage of amylose, which delays and decreases starch hydrolysis (Dartois et al., 2010; Bordoloi et al., 2012).

Castenmiller et al. (1999) conducted a study on β-carotene bioavailability from the food matrix of spinach and found that processing spinach had an effect on the matrix (cell disruption) and, therefore, on β-carotene bioavailability.

For example, Roman et al. (2012) studied the release and bioaccessibility of β-carotene from fortified almond butter using *in vitro* digestion studies. Their results showed that β-carotene that was added as oil into the almond butter was not completely released from the food matrix at any stage of digestion, which could be caused by protein and carbohydrate compounds in the almond butter that may emulsify or encapsulate the β-carotene.

For lipids, digestion is an interfacial process that is performed by the adsorption of lipases to the interface of emulsified fat drops (Wooster et al., 2014). Therefore, a method to control fat drop digestion is through the incorporation of structural agents that form a matrix around the drop. For example, Wooster et al. (2014) studied how the food structure can affect lipid digestion by incorporating different biopolymer networks during emulsion structuring. These authors found that a thermoreversible polymer network does not have a relevant effect on delaying fat digestion due to an increase in viscosity, which affects lipase movement and drop coalescence. However, although the polymer network based on starch delayed fat digestion, dilution during gastric digestion affected the viscosity, leading to normal lipid digestion. The authors reported that a simple biopolymer network could affect the emulsion structure and therefore emulsion digestion.

11.2.3 Study Case: Effect of Glucose Release from Sodium Alginate Gel–Starch Mixtures Based on *In Vitro* Digestion Studies

Foods are complex matrices formed by many compounds whose nutritional properties vary when they are mixed. Therefore, the study of ingredient interactions that occur during food formulation is necessary to understand the final behavior of food during digestion.

The structures of food at different levels (nano, micro, meso, and macro scale) have always presented a challenge for food engineers. Structure is a fundamental variable that influences the transport and physical properties of food (Aguilera, 2005). The perceived quality of food is impacted by its microstructure that contributes to attributes such as mouthfeel and creaminess. Food structure engineering has, therefore, become important to food scientists and engineers as a controlled effort to preserve, modify, or create structures through processing techniques and matrix composition (Kulozik et al., 2003). These food processes determine the structural, physical (color, size, and shape), mechanical (texture, stability, and flavor release), and nutritional (satiation and digestion) properties of food products (Lundin et al., 2008). There is evidence that the effect in the post-prandial elevation of blood glucose during carbohydrates consumption can cause physiological complications related to obesity and diabetes, and these complications could be managed by structure control during food design (Mishra and Monro, 2009).

Furthermore, the glycemic response to different starchy products, such as pasta or dough, varies according to the integrity of the polymeric network in the food material. Riccardi et al. (2003) found that the differences in the glycemic response to different types of pasta and dough were due to differences in the matrix microstructure, which enclosed the starch and prevented enzymatic access during digestion (Parada and Aguilera, 2011).

These studies have suggested that designing microstructures to control and tune the physiological responses that are triggered during the digestive process could be useful (Kaufmann and Palzer, 2011). For example, emulsions can trigger different responses during digestion. Emulsions with a droplet size of approximately 160 nm present faster lipolysis *in vitro* than emulsions with a droplet size of 200 μm; therefore, the droplet size affects the digestion rate and bioavailability (Lundin et al., 2008). Similar results were obtained by Troncoso et al. (2012), in which fatty acid content increased from 61% to 71% when the oil droplet radius was decreased from 86 to 30 nm.

Food design examines ingredient interactions that can improve or delay food digestion. The different biopolymer interactions that can occur on the nanoscale level are determinant in food development. Biopolymer properties can be modified when interactions occur between biopolymers. For example, Weber at al. (2009) studied the interactions of guar and xanthan gums with starch, and this study found that the pasting properties of starch were affected by the gum type and concentration. Guar gum significantly increased the viscosity of the starch gel, and xanthan gum significantly decreased the starch viscosity. These interactions can be attributed to hydrogen bonding because covalent bonding was not observed by infrared spectra analysis. Galactomannan-based gums can affect the water molecule availability, reducing the starch swelling and gelatinization and affecting the paste properties (Kaur and Singh, 2009).

If the rheological and textural properties are modified by the presence of other biopolymers, it is possible that glucose can be released from the starch molecules during digestion. Dartois et al. (2010) studied the influence of guar gum on the *in vitro* starch digestibility, and the results of this study showed that the starch digestibility was affected by the presence of guar gum. This change in the starch availability was attributed to the increase of viscosity due to the presence of guar gum, which inhibited the enzymatic action on the amylose chains.

The main objective of this study case was to show the effect of sodium alginate on starch solutions and to evaluate the effect of sodium alginate on glucose release during the *in vitro* digestibility of starch gel.

The results obtained from this work are presented below. Figure 11.3 shows the starch hydrolysis (%) obtained during the digestion time (150 min). The hydrolysis was initiated when the intestinal fluid was added, mainly by the presence of amylase, which can digest the amylose chains into small glucose molecules. No hydrolysis was observed during *in vitro* gastric digestion which is consistent because of the absence of amylase. In our study, the simulated digestion showed that approximately 72% of the starch was hydrolyzed into glucose by the enzymatic cocktail present in intestinal fluids during the first 10 min of intestinal digestion. Similar results were obtained by Dartois et al. (2010), who studied the effect of guar gum on the digestibility of waxy maize starch based on *in vitro* digestion studies, considering simulate gastric and intestinal condition. The results showed that guar gum delayed and decreased the starch hydrolysis, due to guar gum could form a barrier around the starch granules.

When the starch was blended with sodium alginate, we observed a reduction in the hydrolysis level, and the starch hydrolysis was dependent on the sodium alginate concentration. For example, we observed no significant difference ($p > .05$) from the

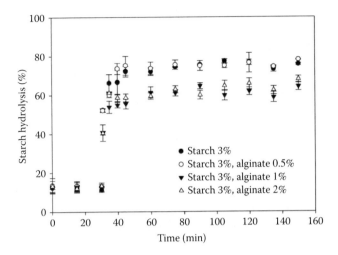

FIGURE 11.3 Effect of sodium alginate concentration on starch 3% hydrolysis during *in vitro* digestion. (From Ramirez, C. et al., *Food Hydrocolloid*, 44, 328–332, 2015. With permission.)

control when 0.5 g/100 g of alginate was blended with starch; however, increasing the alginate concentration to 1.0 and 2.0 g/100 g significantly decreased the starch hydrolysis with respect to the control. After 10 min of intestinal digestion, only 55% of the starch had been converted into glucose. Additionally, the results show that there is a critical concentration of alginate that affects the starch hydrolysis because there is no significant difference between 1.0 and 2.0 g/100 g alginate.

Similar results have been reported by other authors when starch is blended with other biopolymers, such as guar gum (Dartois et al., 2010), and a significant reduction in starch hydrolysis was observed when 1.0 g/100 g guar gum was added to the mixture, decreasing the starch hydrolysis to 15% by the end of the intestinal digestion. This reduction in starch hydrolysis is due to the changes in the viscosity induced by the presence of the hydrated galactomannan chain of the guar gum, which affects the ability of enzymes to hydrolyze the amylose chain.

In addition to changes in the viscosity of the starch solution, sodium alginate could affect the enzymatic hydrolysis by changing molecular interactions. Shi and BeMiller (2002) reported that the viscosity change induced by gums such as alginate was attributed to the retarding of starch granules gelatinization and of amylose leaching from the granule due to protective effect of the anionic gum. Microstructural analysis revealed that the starch granules remained intact when the starches were cooked at 95°C–100°C with anionic gum. Krüger et al. (2003) attributed this lower swelling of starch granules to a lower heating rate and the reduced mobility of water molecules.

According to *in vitro* digestion results, the starch solution reduced the hydrolysis percentage when mixed with sodium alginate. In order to understand the behavior of this mixture, the complex index (CI) was measured. The results showed that the CI increased with the sodium alginate concentration. In Table 11.2, it is observed that CI varied from 34.34% to 57.11% when the content of sodium alginate changed

TABLE 11.2
Complex Index (CI %) Measured in Starch–Sodium Alginate Mixtures

Starch (% w/w)	Alginate (% w/w)	Absorbance			CI (%)
3.0	0.0	0.977	±	0.011	
3.0	0.5	0.419	±	0.013	34.34
3.0	1.0	0.483	±	0.001	50.56
3.0	2.0	0.642	±	0.054	57.11

Source: Ramirez, C. et al., *Food Hydrocolloid*, 44, 328–332, 2015.

from 0.5 g/100 g to 2.0 g/100 g. These values suggest that the amount of free starch diminished when sodium alginate was incorporated to the mixture. This effect has been previously reported for starch mixed with fatty acids and emulsifiers. In this case, starch interacts with leaked amylose from the granule forming a stable complex which unable the unfolding of amylose chain (Guraya et al., 1997). However, the literature has also reported that guar gum reduces starch hydrolysis, because it acts as a barrier layer around the starch granule (Dartois et al., 2010). Similar results were reported by Mandala and Bayas (2004), where starch was mixed with xanthan gum, a film composed of xanthan gum and amylose was formed around the granule which enabled an increase of consistency at continuous phase.

On the other hand, the protector effect of sodium alginate on starch granules could be observed by the viscosity changes due to the incorporation of sodium alginate. Table 11.3 presents the values of the consistency index (K), flow behavior index (n), and R-square. The results show that the K value is lower when alginate is added to the starch solution. The consistency index varied from 0.80 to 4.59 when 0.5 g/100 g and 2.0 g/100 g alginate were added to the starch solution, respectively. The flow behavior index values suggest that the fluid has a pseudoplastic behavior.

Figure 11.4 shows the apparent viscosity versus the shear rate for the starch and starch-alginate solutions. The apparent viscosity for the starch solution is diminished with the presence of alginate. However, increasing the alginate concentration

TABLE 11.3
Ostwald–de Waele Parameters for Starch and Starch–Alginate Mixture Measured at 37°C

Samples (%)		K (Pa s^n)	n	R^2
Starch 3		12.64	0.42	0.995
Starch 3	Alginate 0.5	0.81	0.66	0.996
Starch 3	Alginate 1.0	1.93	0.58	0.996
Starch 3	Alginate 2.0	4.62	0.62	0.998

Source: Ramirez, C. et al., *Food Hydrocolloid*, 44, 328–332, 2015.

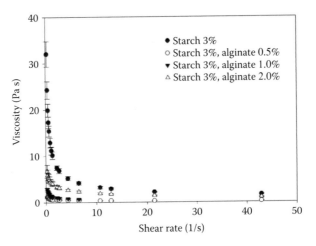

FIGURE 11.4 Viscosity-shear rate curves for starch and starch–alginate mixtures at 37°C. (From Ramirez, C. et al., *Food Hydrocolloid*, 44, 328–332, 2015. With permission.)

increases the apparent viscosity, but the reported viscosity values were lower than the control, as shown in Figure 11.4. The results suggest a protector role of alginate against starch granule swelling, which reduces the viscosity.

From a digestibility point of view, 0.5 g/100 g alginate did not significantly reduce starch hydrolysis. However, 1.0 g/100 g and 2.0 g/100 g alginate significantly lowered the starch hydrolysis compared to the control and 0.5% alginate, which suggests that the viscosity could play an important role during digestion and diminish the enzymatic action on the amylose chains. However, molecular interaction between the alginate and the amylose chains could determine the hydrolysis percentage.

Figure 11.5 shows the cryoscopic point depression produced by adding alginate to the starch solution. The cryoscopic temperature and the heat of fusion have a

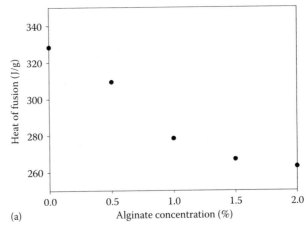

FIGURE 11.5 Changes in the cryoscopic point by alginate addition in starch solutions (3%). (a) Heat of fusion. *(Continued)*

Bioactive Compound Encapsulation

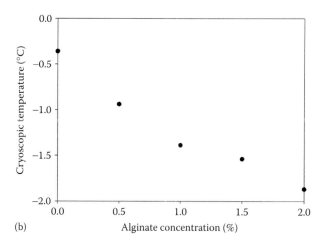

(b)

FIGURE 11.5 (Continued) Changes in the cryoscopic point by alginate addition in starch solutions (3%). (b) Cryoscopic temperature. (From Ramirez, C. et al., *Food Hydrocolloid*, 44, 328–332, 2015. With permission.)

strong relationship with the alginate concentration. However, the fitted curves in Figure 11.3 are nonlinear, suggesting a putative interaction between both polymers. The anionic charge of alginate can affect the behavior of other polymers (Yang and Wright, 2002). For example, the diffusion of substrates into alginate hydrogels increases with pH because of the increase of negative charges.

A factorial experiment design (two levels with three center points) was used to examine the statistical interactions. Factors were starch concentration (0%, 3%, and centre point of 1.5%), alginate concentration (0%, 2%, and centre point of 1%), and interaction between them. Responses studied were heat of fusion and cryoscopic temperature. The results show that the interaction between factors (statistical interaction) affects significantly the heat of fusion ($p = 0.037$), but it is not clear that cryoscopic temperature was affected by the interactions ($p = 0.730$). The latter strongly indicates the presence of a putative interaction between both polymers, affecting at least one thermal property.

These results indicate that molecular interactions between biopolymers must be considered when anionic molecules such as sodium alginate are included in the food formulation because these interactions can reduce the enzymatic activity and the release of glucose. Chemically modified starches (oxidation, acetylation, hydroxypropylation, and cross-linking) exhibit differences during hydrolysis, resulting in varied glucose concentrations. Chung et al. (2008) reported that these changes produce resistant starches that cannot be completely hydrolyzed by amylase. Considering these results, sodium alginate could prevent the normal swelling of starch granules by keeping the crystalline structure of the starch granules intact. Therefore, structural analysis based on electron scanning microscopy studies could be necessary to a better understanding of the effect of alginate on starch granules.

In general, the results show that alginate reduces starch hydrolysis from 72% to 55%. Alginate exerts a protector effect on starch granules which prevents the normal swelling and diminishes the action of enzymes on starch granules. This was observed through the CI values which were higher when the amount of sodium alginate was increased. These results suggest that the increased viscosity and molecular interactions that occur when alginate is added to a starch solution could affect the glucose release during *in vitro* digestion.

11.2.4 Microencapsulation of Food Nutrients

In general, the encapsulation of food nutrients such as β-carotene was developed to protect this type of nutrient from environmental conditions, to prevent oxidation through the formation of a coating around the core, and to increase the bioavailability of these compounds from degradation as part of the digestive process (Gallardo et al., 2013; Giroux et al., 2013).

However, because of the worldwide increase in highly processed foods, the design of micro or nanostructures was necessary to encapsulate nutrients and protect them during storage, processing, consumption, and digestion so that they can then be delivered at a specific place in the gastrointestinal tract (McClements et al., 2008).

Microcapsules produced by complex coacervation acquire beneficial controlled release characteristics, heat-resistance properties and high encapsulation efficiency (Jain et al., 2015). Complex coacervation involves phase separation of a single polyelectrolyte or mixture of polyelectrolytes from a solution and subsequent deposition of the newly formed coacervate phase around the active ingredient. (Anandharamakrishnan, 2014). According to Silva et al. (2012), this method has a high loading capacity (above 99%) and easily controlled liberation of the content by mechanical stress, temperature, or pH changes. This technique is appropriate for encapsulation of hydrophobic compounds such as lycopene, β-carotene, and oils that are rich in omega-3 fatty acids. The particle size depends on the type of raw material that is used (gelatin, maltodextrin, chitosan, and others) and can vary between 100 and 600 nm according to Anandharamakrishnan (2014). For example, Butstraen and Salaün (2014) used complex coacervation of Arabic gum and chitosan to microencapsulate a triglyceride blend. The microcapsule sizes ranged from 5 to 10 μm, with an encapsulation yield of 97%.

A typical method for producing microcapsules containing oil is based on spray-drying technology, which consists of converting liquid oils and flavor in the form of emulsions into dry and stable powders (Jafari et al., 2008). For example, another liposoluble and labile nutrient is β-carotene, which has been microencapsulated by spray drying for preservation and shelf-life stability with good results in terms of its bioavailability according to Donhowe et al. (2014).

Other techniques such as extrusion are based on the gelification properties of biopolymers such as alginate or pectin. For instance, alginate has been widely applied as an encapsulating agent because it is biocompatible and nontoxic; alginate also protects active compounds from heat and moisture, improving the stability of the compounds during processing and digestion (Lupo et al., 2014). Pectin is another biopolymer that is used for active compound encapsulation and for controlled release at a specific target in the digestive tract (colon) because pectin can protect the active compound from gastric and intestinal enzyme degradation by colon pectinases, allowing for the release of inert compounds (Yu et al., 2009).

Emulsification is another method of enclosing labile compounds such as vitamin B12, which was encapsulated into a double w/o/w emulsion by Giroux et al. (2013). This study demonstrated that double emulsion was very effective at preventing vitamin B12 release during *in vitro* gastric digestion, increasing vitamin retention compared with nonencapsulated vitamin.

11.2.5 Control Release of Nutrient during Digestion

The formulation of micro or nanocapsules has the potential to alter the rate at which microencapsulated compound are delivered and absorbed in the human body (Shen et al., 2011). For example, in a binary biopolymer system containing a protein and polysaccharide, the net interaction between the biopolymers can be associative (complexing) or segregative (repulsive), depending on the precise structures of the biopolymers present and the prevailing conditions (pH, ionic strength and mixing conditions) (Kurukji et al., 2016). According to Kurukji et al. (2016), one way to encapsulate functional compounds in protein-polysaccharide complexes involves *bottom-up* self-assembly of the constituent biopolymers. If an active compound is included during or after complex assembly it can become entrapped—physically or chemically—within the biopolymer matrix. The biopolymer contain ionizing groups, such as those found on polypeptide side chain and polyssacharides, and electrostatic force predominates that holds the newly formed complex together; complexes produced in this way have potential application as pH-responsive material that can be assembled and dissociated under pH control.

In this sense, the adequate design of capsule in terms of biopolymer mixture can allow to control the release of nutrient at the intestinal level and not during gastric digestion. In general, protein hydrogels can stabilize active payload compounds and modulate their release kinetics to avoid overdoses and to maximize health benefits. Also, food protein can be used and added to a wide variety of product without altering the sensory properties and are degraded by digestive enzymes in the human gastrointestinal tract (Chen et al., 2010). For example, Teng et al. (2013) observed that soy protein isolated nanoparticles exhibited highest encapsulation efficiency of 97.2% and a moderate to good stability (seta potential around—36 mV). However, soy protein isolate has exhibited a high digestibility in the gastrointestinal tract. Chen and Subirade (2009) studied the release properties of the soy protein isolate microspheres in simulated gastric (pH 1.2) and intestinal (pH 7.4) fluids using riboflavin as a nutrient model. They could observe two-phase pattern of release for SPI microspheres: a rapid burst release in both gastric and intestinal medium for the first 15 min, and then a steady release from 15 min until 4 h with near to zero-order kinetics ($r^2 > 0.991$). Also, faster release in simulated gastric fluids than in simulated intestinal fluids was noted, which was attributed to greater swelling of microspheres of SPI at pH 1.2. Certainly, early burst release is undesirable in controlled delivery system, suggesting that soy protein should not be used alone as coating material for controlled release of hydrophilic nutrients like riboflavin because of their low barrier capacity on the micrometer scale (Chen and Subirade, 2009).

11.2.6 Inclusion of Microcapsules in Food Matrices

The incorporation of microencapsulated compounds into food matrices allows controlled release of nutrients at specific stages of digestion. For example, Roman et al. (2012) incorporated microencapsulated β-carotene in almond butter and evaluated the behavior of β-carotene during *in vitro* digestion. The results showed that microencapsulated β-carotene remained in the capsules during gastric digestion and that the majority of its release was produced at the intestinal stage.

Additionally, the development of submicron particles in the range of several μm to nm may provide interesting new properties; however, these particles may be quickly degraded in the gut and lose their ability to protect the nutraceutical that is incorporated inside (Chen, 2009). In contrast, the inclusion of these particles into a food matrix can reduce degradation due to the protective role of the matrix.

However, the inclusion of the capsule compound into a food matrix could affect the texture properties of food. Even more interesting is how the incorporation of these particles can influence the breakdown of food during chewing, affecting the particle size of the bolus. Hutchings et al. (2011) have developed some macro-scale studies. In that study, they concluded that when heterogeneous food is chewed, one food component could influence the breakdown of another food component because the food matrix influences the chewing process (in terms of the chewing duration, frequency, and chewing number).

Thus, the perception of the food in terms of its hardness or water content can influence the number of chews required to form a bolus that is ready to be swallowed (Jalabert-Malbos et al., 2007). The number of chews influences the particle size that forms the bolus, which can directly affect how protected a nutrient is in the structure that is present in the particle. This aspect is commonly overlooked in bioavailability studies (Lemmens et al., 2010).

Such as in nature, where the nutrients are usually tramped into food matrix controlling its release, the microencapsulation has the potential to change the rate at which the compound or nutrient incorporated into the microcapsules is released, catalyzed, and absorbed in the human body. For example, Raatz et al. (2009) found that the ingestion of emulsified fish oil was compared with that of a fish oil gelatin capsule in humans, and the results showed a higher absorption of long chain n-3 polyunsaturated fatty acid (LC n-3 PUFAs) as evidenced by EPA and DHA levels in plasma phospholipids when the ingestion of the emulsified form was carried out. The results obtained were attributed to improved digestion and absorption because of the enhancement of the action of pancreatic lipase on long-chain fatty acid, considering that lipid emulsification in the stomach is a critical step in fat digestion mainly by the generation of a lipid–water interface which is essential for the interaction between water-soluble lipases and unsoluble lipids promoting in this way the bioavailability of dietary fat. Therefore, emulsification of fish oil bypasses this normal physiological step and improves its bioavailability (Raatz et al., 2009). In the same work, the author also attributed the differences in lipid absorption to the transport system (emulsified and gelatin capsules), which was lower for gelatin encapsulated liquid oil than emulsified oil, due to the possibility that the gelatin capsule breakdown affected the initial rate of absorption of the fatty acid from the capsular triglyceride fish oil.

Shen et al. (2011) studied the effect on *in vitro* digestion of different food matrices (cereal bar, yogurt and orange juice) fortified with microencapsulated tuna oil powder, considering that the component of a food and its microstructure affect the bioavailability of nutrients. The results show that microencapsulated tuna oil powder and fortified orange juice and yogurt samples had a similar and smaller droplet size after exposure to simulated gastric fluids and simulated intestinal fluids than cereal bar sample. This was caused due to the lipase action, which acts at the interface; therefore, it is expected that lipolysis will be related to the surface area available for lipase action. This implies that small lipid droplets have a larger surface area and therefore will be digested more readily than bigger droplets. The lower amount of free long chain n-3 poly unsaturated fatty acids released during digestion of cereal bar sample compared with the other food structures could be caused due to the presence of larger lipid drop in the digest of cereal bar. Also, the cereal bar was elaborated with high amount dietary fiber and solid fat that can affect the lipid digestion.

REFERENCES

Aguilera J.M. Why food microstructure? *Journal of Food Engineering*, 67: (2005) 3–11.

Anandharamkrishnan C. *Techniques for Nanoencapsulation of Food Ingredients. Springer Brief in Food, Health and Nutrition*. Springer, New York, p. 38 (2014).

Bordoloi, A., Singh, J., and Kaur, L. In vitro digestibility of starch in cooked potatoes as affected by guar gum: Microstructural and rheological characteristics. *Food Chemistry*, 133, (2012) 1206–1213.

Butstraen, C. and Salaün, F. Preparation of microcapsules by complex coacervation of gum arabic and chitosan. *Carbohydrate Polymers*, 99, (2014) 608–616.

Castenmiller, J.J.M., West, C.E., Linssen, J.P.H., van het Hof, K.H., and Voragen, A.G.J. The food matrix of spinach is a limiting factor in determining the bioavailability of β-carotene and to a lesser extent of lutein in human. *The Journal of Nutrition*, 129, (1999) 349–355.

Chen, L. Protein micro/nanoparticles for controlled nutraceutical delivery in functional foods. In *Designing Functional Foods*, McClements, D.J. and Decker, E.A. (eds.) CRC Press, Boca Raton, FL (2009) pp. 572–600.

Chen L., Hébrard, G., Beyssac, E., Denis, S., and Subirade, M. In vitro study of the release properties of Soy-Zein protein microspheres with a dynamic artificial digestive system. *Journal of Agricultural and Food Chemistry*, 58, (2010) 9861–9867.

Chen, L. and Subirade, M. Elaboration and characterization of Soy/Zein protein microspheres for controlled nutraceutical delivery. *Biomacromolecules*, 10, (2009) 3327–3334.

Chung, H-J., Shin, D-H., and Lim S-T. In vitro starch digestibility and estimated glycemic index of chemically modified corn starches. *Food Research International*, 41, (2008) 579–585.

Dartois, A., Sing, J., Kaur, L., and Singh, H. Influence of guar gum on the in vitro starch digestibility-rheological and microstructural characteristics. *Food Biophysics*, 5, (2010) 149–160.

Donhowe, E., Flores, F.P., Kerr, W.L., Wicker, L., and Kong, F. Characterization and *in vitro* bioavailability of β-carotene: Effects of microencapsulation method and food matrix. *LWT-Food Science and Technology*, 57, (2014) 42–48.

Fardet, A. Nutrient bioavailability and kinetics of release is a neglected key issue when comparing complex food versus supplement health potential. *Journal of Nutrition Health & Food Engineering*, 2, (2015) 45–46.

Gallardo, G., Guida, L., Martinez, V., López, MC., Bernhardt, D., Blasco, R., Pedroza-Islas, R., and Hermida, L.G. Microencapsulation of linseed oil by spray drying for functional foods application. *Food Research International*, 52, (2013) 473–482.

Giroux, H.J., Constantineau, S., Fustier, P., Champagne, C.P., St-Gelais, D., Lacroix, M., and Britten, M. Cheese fortification using water-in-oil-in water double emulsions as carrier for water soluble nutrients. *International Dairy Journal*, 29, (2013) 107–114.

Guraya, H., Kadan, R.S., and Champagne, E.T. Effect of rice starch-lipid complexes on in vitro digestibility, complexing index, and viscosity. *Cereal Chemistry*, 74, (1997) 561–565.

Hutchings, S., Foster, K.D., Bronlund, J.E., Lentle, R., Jones, J.R., and Morgensten, M.P. Mastication of heterogeneous foods: Peanuts inside two different food matrices. *Food Quality and Preference*, 22, (2011) 332–339.

Jafari, SM., Assadpoor, E., Bhandari, B., and He, Y. Nano-particle encapsulation of fish oil by spray drying. *Food Research International*, 41, (2008) 172–183.

Jain, A., Thakur, D., Ghoshal, G., Katare, O.P., and Shivhare, U.S. Microencapsulation by complex coacervation using whey protein isolated and gum acacia: An approach to preserve the functionality and controlled release of β-carotene. *Food Bioprocess and Technology*, 8, (2015) 1635–1644.

Jalabert-Malbos, M-L., Mishellany-Dutour, A., Woda, A., and Peyron, M.-E. Particle size distribution in the food bolus after mastication of natural foods. *Food Quality and Preference*, 18, (2007) 803–812.

Kaufmann, S.F. and Palzer, S. Food structure engineering for nutrition, health and wellness. *Procedia Food Science*, 1, (2011) 1479–1486.

Kaur, L. and Singh, J. The role of galactomannan seed gums in diet and health—A review. In *Recent Progress in Medicinal Plants: Standardization of Herbal/Ayurvedic Formulations*, vol. 24, Govil, J.N. and Singh, V.K. (eds.) Stadium Press LLC, Houston, TX (2009) pp. 429–467.

Krüger, A., Ferrero, C., and Zaritzky, N.E. Modelling corn starch swelling in bath system: Effect of sucrose and hydrocolloids. *Journal of Food Engineering*, 58, (2003) 125–133.

Kulozik, U., Tolkach, A., Bulca, S., and Hinrichs, J. The role of processing and matrix design in development and control od microstructures dairy food production—A survey. *International Dairy Journal*, 13, (2003) 621–630.

Kurukji, D., Norton, I., and Spyropoulus, F. Fabrication of sub-micron protein-chitosan electrostatic complexes for encapsulation and pH-modulated delivery of model hydrophilic active copounds. *Food Hydrocolloids*, 53, (2016) 249–260.

Lemmens, L., Van Buggenhout, S., Van Loey, A., and Hendricks, M. Particle size reduction leading to cell rupture is more important for the β-carotene bioaccessibility of raw compared to thermally processed carrots. *Journal of Agricultural and Food Chemistry*, 58(24), (2010) 12769–12776.

Lundin, L., Golding, M., and Wooster, TJ. Understanding food structure and function in developing food for appetite control. *Nutrition & Dietetics*, 65, (2008) S79–S85.

Lupo, B., Maestro, A., Porras, M., and Gutierrez, J.M. Preparation of alginate microspheres by emulsification/internal gelation to encapsulte cocoa polyphenols. *Food Hydrocolloids*, 38, (2014) 56–65.

Mandala, I.G. and Bayas, E. Xanthan effect on swelling, solubility and viscosity of wheat starch dispersions. *Food Hydrocolloids*, 18, (2004) 191–201.

McClements, D.J., Decker, E.A., Park, Y., and Weiss, J. Designing food structures to control stability, digestion, release and absorption of lipophilic food components. *Food Biophysics*, 3, (2008) 219–228.

Mishra, S. and Monro, J.A. Digestibility of starch fractions in wholegrain rolled oats. *Journal of Cereal Science*, 50, (2009) 61–66.

Mishra, S. and Monro, J. Wholeness and primary and secondary food structure effects on in vitro digestion patterns determine nutritionally distinct carbohydrate fractions in cereal foods. *Food Chemistry*, 135, (2012) 1968–1974.

Parada J. and Aguilera J.M. Food microstructure affects the bioavailability of several nutrients. *Journal of Food Science*, 72, (2007) R21–R32.

Parada, J. and Aguilera, J.M. Microstructure, mechanical propeorties and starch digestibility of a cooked dough made with potato starch and wheat gluten. *LWT-Food Science and Technology*, 44, (2011) 1739–1744.

Raatz, S.K., Redmon, B., Wimmergren, N., Donadio J.V., and Bibus, D. Enhanced absorption of n-3 fatty acids from emulsified compared with encapsulated fish oil. *Journal of the American Dietetic Association*, 21, (2009) 1076–1081.

Ramirez, C., Millón, C., Nuñez, H., Pinto, M., Valencia, P., Acevedo, C., and Simpson, R. Study of effect of sodium alginate on potato starch digestibility during in vitro digestion. *Food Hydrocolloids*, 44, (2015) 328–332.

Ranawana, V., Leow, M., and Henry, C. Mastication effects on the glycaemic index: Impact on variability and practical implications. *European Journal of Clinical Nutrition*, 68, (2014) 137–139.

Riccardi, G., Clemente, G., and Giacco, R. Glycemic index of local food and diets: The Mediterrean experience. *Nutrition Reviews*, 61, (2003) S56–S60.

Roman, M.J., Burri, B.J., and Singh, R.P. Release and bioaccessibility of β-carotene from fortified almond butter during in vitro digestion. *Journal of Agricultural and Food Chemistry*, 60, (2012) 9659–9666.

Shen, A., Apriani, C., Weerakkody, R., Sanguansri, L., and Augustin, M.A. Food matrix effects on in vitro digestion of microencapsulated Tuna oil powder. *Journal of Agricultural and Food Chemistry*, 59, (2011) 8442–8449.

Shi, X. and BeMillier, J.N. Effect on food gum on viscosity of starches suspension during pasting. *Carbohydrate Polymers*, 50, (2002) 7–18.

Silva, D.F., Favaro-Trindade, C.S., Rocha, G.A., and Thomazini, M. Microencapsulation of lycopene by gelatin-pectin complex coacervation. *Journal of Food Processing and Preservation*, 36, (2012) 185–190.

Singh, J., Berg, T., Hardacre, A., and Boland, M. Cotyledon cell structure and in vitro starch digestion in navy beans. In *Food Structures, Digestion and Heath*, Boland, M., Golding, M., and Singh, H. (eds.) Academic Press, Elsevier, US, (2014) pp. 223–242.

Srichamroen, A. Physical quality and *in vitro* starch digestibility of bread as affected by addition of extracted malva nut gum. *LWT-Food Science and Technology*, 59, (2014) 486–494.

Teng, Z., Luo, Y., and Wang, Q. Carboxymethyl chitosan-soy protein complex nanoparticles for the encapsulation and controlled release of vitamin D_3. *Food Chemistry*, (2013) 524–532.

Troncoso, E., Aguilera, J.M., and McClements, D.J. Fabrication, characterization and lipase digestibility of food-grade nanoemulsions. *Food Hydrocolloids*, 27, (2012) 355–363.

Turgeon, S.L. and Rioux, L.-E. Food matrix impact on macronutrients nutritional properties. *Food Hydrocollois*, 25, (2011) 1915–1924.

Van der Bilt, A., Engelen, L., Pereira, L.J., van der Glas, H.W., and Abbink, J.H. Oral physiology and mastication. *Physiology and Bahavior*, 89, (2006) 22–27.

Weber, F.H., Clerici, M.T.P.S., Collares-Queiroz, F.P., and Chang, Y.K. Interaction of guar and xanthan gums with starch in the gels obtained from normal, waxy and high-amylose corn starches. *Starch/Stärke*, 61, (2009) 28–34.

Wooster, T.J., Day, L., Xu, M., Golding, M., Oiseth, S., Keog, J., and Clifton. P. Impact of different biopolymer network on the digestion of gastric structured emulsion. *Food Hydrocolloids*, 36, (2014) 102–114.

Yang H. and Wright J.R. Microencapsulation methods: Alginate (Ca^{2+} induced gelation) In: *Methods of Tissue Engineering*. Atala, A and Lanza, R. (eds.) Academic Press, San Diego, CA (2002).

Yu, C.-Y., Yin, B.-C., Zhang, W., Cheng, S.-X., Zhang, X.-Z., and Zhuo, R.-X. Composite microparticle drug delivery system based on chitosan, alginate and pectin with improved pH-sensitive drug release property. *Colloids and Surfaces B: Biointerfaces*, 68, (2009) 245–249.

Zhang, G. and Hamaker, B.R. Slowly digestible starch: Concept, mechanism a proposed extended glycemic index. *Critical Reviews in Food Science and Nutrition*, 49, (2009) 852–867.

12 Electrospinning as a Novel Delivery Vehicle for Bioactive Compounds in Food Nanotechnology

Behrouz Ghorani, Ali Alehosseini, and Nick Tucker

CONTENTS

12.1	Introduction	259
12.2	Fundamentals of Electrospinning and Electrospraying	261
12.3	The Electrospinning Process	264
12.4	Polymer Solution Properties	266
	12.4.1 Polymer Concentration, Molecular Weight, and Fiber Morphology	267
	12.4.2 Effect of Solvents	270
12.5	Processing Conditions	270
	12.5.1 Voltage	270
	12.5.2 Flow Rate	272
	12.5.3 Diameter of Spinneret Orifice	273
	12.5.4 Tip to Collector Distance	273
12.6	Ambient Conditions	274
12.7	Electrospinning versus Electrospray	275
12.8	How to Provide a Safe Delivery System?	276
	12.8.1 Protein-Based Encapsulating Materials Used in Electrospinning	277
	12.8.2 Carbohydrate-Based Encapsulating Materials Used in Electrospinning	279
	12.8.3 Recent Advances in Encapsulating of Probiotics by Electrospinning	280
12.9	Conclusions and Future Trends	281
References		283

12.1 INTRODUCTION

Demand for healthy foods has increased substantially over the past decade due to growth in the world population and an increased perception of unhealthy lifestyles. Globally, by 2050, the number of people aged over 65 is expected to reach a total of about 1.5 billion, which is equivalent to 16% of the world population: in 1950, this proportion was only 5% (Haub, 2011). As the population ages, there is expected to be an increasing demand for foodstuffs that have a potentially positive effect on health beyond the basic requirements of nutrition—functional foods. There is now a broad range of cholesterol-lowering functional foods available in the market (e.g., BENECOL® margarine spreads and cream cheese), which contain added esterised fat soluble forms of phytosterols or stanols (plant extracts) (Chen et al., 2011). Omega-3 fatty acids, which occur naturally in foods such as oily fish and some plant and seed oils, are another substance that is added to a variety of food products including margarine, milk, fruit juice, and egg-based recipes to make foods functional to reduce the risk of cardiovascular illnesses (Kaushik et al., 2010; Torres-Giner et al., 2010). Vitamin D (sometimes with calcium salts) is added to fruit juice to raise the dietary level of the vitamin in specific target populations, such as postmenopausal women to mitigate the risk of developing osteoporosis (Heaney, 2007). It is also worth pointing out that intestinal complaints such as constipation, flatulence, and bloating are common in older people and can have a considerable impact on their quality of life (Donini et al., 2009). Increasing dietary fiber along with the use of probiotic (or prebiotic) supplements or functional foods, have been suggested to improve digestive and immune health in older people (Donini et al., 2009). Thus, there is a strong motivation to progressively improve the performance of food products to provide this type of consumer benefit. One approach that could help in the development of a wide range of genuinely functional foods is the process of encapsulation of food ingredients that are rich in functional, but biochemically fragile vitamins and antioxidants. Encapsulated ingredients can be formulated to survive travel through the gastrointestinal (GI) system to deliver their payload at a particular point, thus maximizing the beneficial effect. In the case of nonsolid and semisolid foods, it is also essential to decrease the matrix size to allow their incorporation without affecting food sensory qualities (López-Rubio and Lagaron, 2012). More importantly, by decreasing the matrix size from micrometers to nanometers, biochemical vehicles with highly controllable delivery rates can be developed (López-Rubio and Lagaron, 2012). Clearly, the delivery rate of any bioactive compound to various sites within the body is directly affected by the particle size. In some cell types, only submicron nanoparticles can be absorbed efficiently (Hughes, 2005; Ezhilarasi et al., 2013). Larger particles generally release encapsulated compounds more slowly and over longer time periods. In addition, particle size reduction introduces several bioadhesive improvement factors, principally increased adhesive effect due to the increase in surface to volume ratio which thence leads to prolonged GI transit time, and higher bioavailability of the encapsulated compound (Chen et al., 2006). Controlled and targeted release of micronutrients improves their effectiveness, broadens the application range as food ingredients, and ensures optimal dosage, thereby improving the cost-effectiveness of the product (Mozafari et al., 2006). A widely used method for the production of materials on this scale is electrospinning, a process producing fiber diameters that are commonly less than 1 µm. These nanofiber materials have attracted particular attention because of their high specific surface area and the ability to modify the bulk properties of a material when introduced as multilayer fibrous assemblies (Ghorani et al., 2013).

Such fibrous materials have been studied as potential vehicles for encapsulation of bioactive compounds, drug delivery, as bimolecular sensors and as ultrafiltration media (Anu Bhushani and Anandharamakrishnan, 2014).

This chapter presents a comprehensive review of the fundamentals of electrospinning and spraying as methods to produce nanoscale fibers or particulates suitable for application in food technology by encapsulation to form nanoscale delivery systems.

12.2 FUNDAMENTALS OF ELECTROSPINNING AND ELECTROSPRAYING

Electrospinning is the process of using electrostatic forces to draw a pendant droplet of polymer solution into a fine fiber that will then deposit onto the nearest grounded collector (Frenot and Chronakis, 2003). The polymer in liquid phase is extruded from the needle tip at a constant rate by a syringe pump or at a constant pressure from a header tank, forming a droplet at the tip (Figure 12.1).

When a small volume of polymer liquid is exposed to an electric field, the droplet stretches toward the nearest lower potential point, forming into a structure known as the *Taylor cone* (Reneker et al., 2000; Yarin et al., 2001). When the electric field reaches a critical value at which the electrical forces overcome the surface tension of the droplet from the tip of the Taylor cone, a charged jet of liquid polymer is ejected toward the nearest lower potential (often electrically grounded) surface (Doshi and Reneker, 1995). The jet is then elongated partly by means of the *whipping instability* (more correctly described as an *expanding helix*) during its transit from the tip to the collector. The jet is also subject to drag forces, which can also be

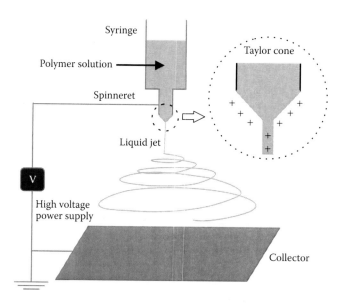

FIGURE 12.1 Schematic diagram of a basic electrospinning process.

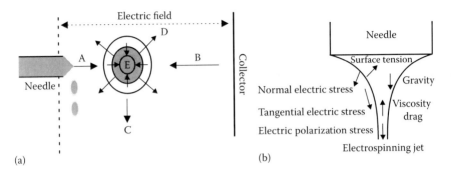

FIGURE 12.2 (a,b) Forces acting on the charged droplet during the electrospinning. (A) electrostatic force, (B) drag force, (C) gravity, (D) coulombic repulsion force, and (E) surface tension and viscoelastic force.

expected to contribute to its attenuation by drawing. Electrostatic force, drag force, gravity, columbic repulsion force, surface tension and viscoelastic forces all act on the charged jet (Ding et al., 2006) (Figure 12.2).

The drawing and thinning of the polymer jet continues until the solvent is evaporated and the continuous fiber is deposited on the grounded collector (Doshi and Reneker, 1995; Shin et al., 2001; Dabirian et al., 2007). When the charged strand is ejected by electrostatic force from the spinning tip toward the collector, a rising viscous friction-related drag force is imposed on the accelerating jet as a result of this interaction with the surrounding air. The expansive forces acting on the droplet are attributed to repulsive columbic force related to electrostatic charge, which is acting in opposition to the contractive forces attributed to surface tension and viscoelastic properties. Clearly, the presence of the electric field directly influences the electrostatic and columbic repulsion forces that are observed in the process. The electric field leads to the observed distortion and expansion of the charged droplet from spherical-like to a spindle-like structure and also influences the transition from an electrospraying (liquid atomization) to an electrospinning process (Grimm and Beauchamp, 2005; Almería et al., 2010).

The dynamics of electrically charged fluids (electrohydrodynamic) is the foundation for understanding electrospinning and electrospraying (Chakraborty et al., 2009). The electrospinning process is based on the same physical principles of the ejection of a continuous fiber forming stream, but molecular cohesion below a critical level leads to the formation of droplets from the Taylor cone rather than a fiber (Bock et al., 2011). There are three modes commonly found during electrospraying of viscous polymer solutions that include a dripping, intermittent jetting mode, and single cone jet mode accompanied by the formation of a Taylor cone (Jain et al., 2015). Electrospraying is characterized by the formation of fine charged droplets of a size close to one-half of the Rayleigh limit. This size limit is related to the magnitude of charge upon the drop that overcomes the opposing contractile force of surface tension, leading to splitting of the droplet into sizes smaller than would be expected from the effects of surface tension alone (Jaworek, 2007). Barrero and Loscertales determined the minimum flow rate at which the cone-jet mode was able to operate at a steady state (Barrero and Loscertales, 2007). The minimum flow rate (Q_{min}) can be obtained from

$$Q_{min} \approx (\sigma_l \varepsilon_0 \varepsilon_r)/(\rho_l \gamma_l) \qquad (12.1)$$

where:
Q_{min} is minimum liquid volume flow rate
σ_l is surface tension of the liquid
ε_0 is permittivity of the free space
ρ_l is mass density of the liquid, and γ_l is liquid bulk conductivity

Both theoretical considerations and experimental observations indicate that there is a consensus that the droplet diameter scales with the liquid flow rate, and is inversely proportional to the liquid conductivity to a certain power (Ogata et al., 1978; Tomita et al., 1986; La Mora and Fernandez, 1994; Gañán-Calvo, 1999; Hartman et al., 2000). When the solution concentration or the molecular chain entanglement in the liquid is high, the jet from Taylor cone is stabilized, and a staged process of elongation takes place as the fiber elongates in flight through an initial straight flight path and then after the onset of the *whipping instability* (Bock et al., 2012). In other words, the jet formed in electrospinning does not break into droplets but produces a micro or nanofiber (Jaworek and Sobczyk, 2008) (Figure 12.3).

If the solution concentration is low, the jet is destabilized due to varicose instability and hence fine spherical microparticles are formed (Jaworek and Sobczyk, 2008). These highly charged droplets self-disperse in space due to electrostatic repulsion, thereby preventing droplet agglomeration and coalescence (Jaworek, 2007). The size distribution of the droplets is usually narrow, with low standard

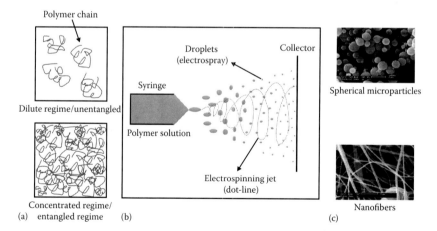

FIGURE 12.3 (a) Schematic of physical representation at the molecular level of entanglement regimes for dilute and concentrated polymer concentration. (b) Schematic diagram of a basic electrospinning (jet formation) and electrospraying (liquid-droplet atomization) process. (c) Examples of SEM image of microspheres (electrospraying) and nanofibers (electrospinning).

deviation and the droplets are smaller (often less than 1 µm) than those available from conventional mechanical atomizers (Jaworek, 2007). The disruption of droplets by electrostatic force was observed as long ago as 1882 (Rayleigh, 1882). The evaporation of the solvent leads to contraction and solidification of droplets resulting in solid polymeric particles that are finally deposited on the grounded collector. The potential of electrospraying for drug delivery systems (Chakraborty et al., 2009; Bock et al., 2012), film coating on foods (Khan et al., 2012), chocolate processing (Luo et al., 2012), preparation of solid lipid nanoparticles containing active compound (Eltayeb et al., 2013), stabilization of nutraceuticals (Torres-Giner et al., 2010), and specifically encapsulation of bioactives and probiotics (see Section 8.3) (López-Rubio and Lagaron, 2012; López-Rubio et al., 2012) have all been recently reported. The bioactive compounds encapsulated in electrospun fibers or electrosprayed particles are shown to possess enhanced stability and functionality and may be used as ingredients in functional foods (Pérez-Masiá et al., 2014a). It is to be expected that, as with electrospinning, benign process conditions will make electrospraying an attractive route for processing temperature sensitive materials, especially if the difficulties of manufacturing in bulk can be overcome.

12.3 THE ELECTROSPINNING PROCESS

Electrospinning is an effective route to produce polymer fibers with diameters of submicron or nanometre scale. A variety of fiber cross-sectional shapes and sizes can be obtained from a selection of different polymer solutions and control of processing conditions. Fibers prepared by electrospinning are usually monolithic with cylindrical cross sections; however, under some specific conditions other forms are observed (Spasova et al., 2006). Branched fibers, flat ribbons, and bent ribbon morphologies have all been reported (Koombhongse et al., 2001; Frenot and Chronakis, 2003), as well as hollow cylinder configurations (Koombhongse et al., 2001; Koski et al., 2004). Fiber branching is associated with high levels of charge in the spun solution or melt (Filatov et al., 2007), being a process related to the shattering of individual droplets as mentioned previously. Circular cross sections indicate that the fiber was in the solid state at the time of impact with the target. Ribbon geometries are due to the collapse of a liquid or gel state strand upon impact. Other noncircular section geometries may be due to the shrinkage upon solidification after impact, when the fiber is no longer under tension. Microscale helical coils (Kessick and Tepper, 2004) have also been produced (Koski et al., 2004). Helical structures can be produced on a conductive substrate by electrospinning from a two-component solution. The two-component solution consists of one conductive component such as poly(aniline sulfonic acid) (PASA) and one nonconductive polymer poly(ethylene oxide) (PEO) (Kessick and Tepper, 2004). Depending on the process conditions, the average diameter of electrospun fibers varies from about 5 nm to 10 µm and is always characterized by a distribution of fiber diameters that is usually wider than that achieved using conventional fiber spinning processes. Melt spinning tends to produce larger fiber sizes than solution spinning, because heat loss occurs faster than solvent loss. Structural

FIGURE 12.4 Defects in electrospun fibers: (a) beaded fibers and (b) spindle-like defects, controlled morphology and mechanical characterization of electrospun cellulose acetate (CA) fiber webs produced from single and binary solvent systems based on solubility parameters was discussed.

defects in the webs such as colloid beads, *beads-on-string*, or spindle morphologies can be present among the resultant fibers or even as the major product, depending on the operating conditions and the material properties (Spasova et al., 2006) (Figure 12.4).

Note that the formation of beads and spindles is due to the uneven thinning of the fiber during the drawing process, and not by stress relaxation after deposition of the fibers. Bead formation is observed due to the fibers that are not drawn by electrostatic force (e.g., glass fibers) (Mohr and Rowe, 1978).

Viscosity, surface tension of the polymer solution, and the net charge density carried by the jet are the main factors influencing the formation of beads (Fong et al., 1999; Spasova et al., 2006). The applied electrostatic field and conductivity of the solution influence the net charge density carried by the moving jet in the process. As the net charge density increases, the beads become smaller and more spindle-like, while the diameter of fibers becomes smaller; however, it should again be noted that bead or spindle formation is a phenomenon observed in nonelectrostatic fiber drawing (Walczak, 2002). Decreasing the surface tension gradually makes the beads disappear (Fong et al., 1999). Additives that modify the conductivity may also be included in the spinning solution (Stanger et al., 2009). Previously, NaCl has been added to a PEO/water solution to increase the net charge density carried by the jet (Angammana and Jayaram, 2008). Zong et al. investigated the effects of salt addition on the resulting morphologies of poly(D,L-lactic acid) (PDLA) electrospun fibers (Zong et al., 2002). It was demonstrated that no bead-on-string structures were formed when spinning from solutions containing 1% salt because the addition of salts results in a higher charge density on the surface of the jet during spinning—and consequently more electric charges are carried by the jet. Doshi and Reneker remarked that by reducing surface tension of a polymer solution, fibers without any beads could be produced (Doshi and Reneker, 1993). A possible way to reduce the surface tension and to eliminate beaded fibers is to add surfactant to the solution (Lin et al., 2004; Manee-in et al., 2006). As the charge carried by a unit length of the jet (charge density) increases, so do the elongation forces experienced by the jet. There is a limiting surface charge density, after which charge will be lost to the surroundings (Cross, 1987). The effect on fiber morphology of charge density increase is that any beads present will become smaller and more spindle-like. The fiber diameter also substantially decreases (Zong et al., 2002). Similarly, Uyar and Besenbacher, demonstrated that even slight differences in the conductivity of dimethylformamide (DMF) solutions can significantly affect the morphology of resulting electrospun polystyrene (PS) fibers, increasing the conductivity of solution results in the elimination of beads producing uniform, thin fibers (Uyar and Besenbacher, 2008). Recently, low molecular weight carbohydrate (e.g., maltodextrin, resistant starch) fibers were obtained by use of the commercial surfactants Tween 20, Span 20, and lecithin (Pérez-Masiá et al., 2014b). Changes in polymer concentration and solvent system can affect the solution viscosity and the surface tension coefficient. As the viscosity of the solution is increased, the beads become bigger, the average distance between beads becomes longer, the fiber diameter becomes larger, and the shape of the beads changes from spherical to spindle-like. The main parameters that affect the electrospinning process include solution properties, processing conditions, and ambient conditions (Table 12.1).

It is possible to produce fibers with different morphologies by varying these parameters (Deitzel et al., 2001; Heikkila and Harlin, 2008; Tehrani et al., 2010). Understanding how each of these processing parameters influences fiber morphology and dimensions is essential for proper process control.

12.4 POLYMER SOLUTION PROPERTIES

One of the main parameters influencing fiber diameter and morphology is the nature of the polymer solution (Teo and Ramakrishna, 2006). Fiber diameter and the propensity for bead formation along individual fibers are governed by surface tension and the viscosity of the solution (Fong et al., 1999; Spasova et al., 2006). Solution

TABLE 12.1
Processing Parameters in Electrospinning

Electrospinning Parameters	
Solution properties	Viscosity
	Polymer concentration
	Molecular weight of polymer
	Electrical conductivity
	Elasticity
	Surface tension
Processing conditions	Applied voltage
	Distance from needle to collector
	Volume feed rate
	Needle diameter
Ambient conditions	Temperature
	Humidity
	Atmospheric pressure

properties are controlled by polymer concentration, molecular weight of polymer, surface tension, and solution conductivity.

12.4.1 POLYMER CONCENTRATION, MOLECULAR WEIGHT, AND FIBER MORPHOLOGY

As a general trend, the solubility of synthetic polymers increases as the molecular weight decreases. However, other important physical properties such as viscosity, strength, flexibility, and the degree of molecular chain entanglement (Shenoy et al., 2005a, b) increase with molecular weight (Treloar, 1970). So long as a polymer can be electrospun into fiber, the ideal situation is for the polymer solution to consist of the highest possible percentage of polymer while still having a low enough viscosity to permit jet formation. In other words, a certain minimum polymer concentration is required to facilitate fiber formation during electrospinning (Huang et al., 2003; Shenoy et al., 2005a, b). Below this critical value, application of voltage results in electrospraying and bead formation primarily due to Rayleigh instability, also known as *capillary wave break up*. At such low polymer concentrations, the degree of entanglement of the network of polymer chains is insufficient for successful fiber spinning. As the polymer concentration is increased, a mixture of beads and fibers is obtained. Any further increase in polymer concentration results in the formation of a continuous fiber. However, the polymer concentration cannot be increased beyond a certain upper limit because although it is not commonly reported, the resulting high solution viscosity impedes continuous electrospinning (Shenoy et al., 2005a, b). Shenoy et al. (2005b) have developed a semiempirical theory in which a critical parameter, the entanglement number for a solution, $(\eta_e)_{soln}$, is defined to indicate whether electrospraying (beads) or electrospinning (fiber) will take place. The value of $(\eta_e)_{soln}$ can be obtained from (Shenoy et al., 2005b):

$$(\eta_e)_{soln} = \frac{\varnothing M_w}{M_e} \quad (12.2)$$

where:
M_w is the molecular weight of the polymer in solution
M_e is the polymer entanglement molecular weight

The latter is a function of polymer chain topology and morphology (Shenoy et al., 2005b; Munir et al., 2009). An entanglement number of two for a polymer solution corresponds to approximately one entanglement per chain, and for pure fiber formation by electrospinning it is typically necessary to use a polymer solution with an entanglement number greater than three and a half. At an entanglement number between two and three and a half, there will be a transition from beaded fiber to pure fiber. At an entanglement number less than two, electrospraying occurs and only particles are formed. These are ultimately visible as solid beads on the material collector (Shenoy et al., 2005b; Munir et al., 2009). As an example of how the entanglement number influences fiber morphology, the work of Munir et al. predicts the formation of beads, beaded fibers, and pure fibers that are formed when spinning

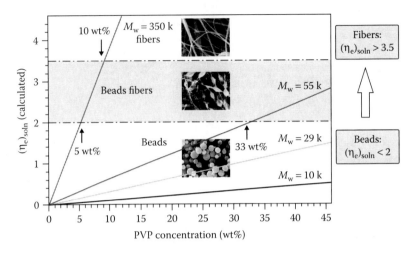

FIGURE 12.5 Plot of the calculated entanglement number $(\eta_e)_{soln}$ as a function of polymer concentration. In this example, the spinning solution was PVP/(water/ethanol) with various molecular weights.

polymer at various molecular weights and concentrations using the entanglement number theory (Munir et al., 2009) (Figure 12.5).

Below 5 wt% of polyvinyl pyrrolidone (PVP) (in water/ethanol) and with $(\eta_e)_{soln} \sim 2$, only beaded morphologies were obtained. Increasing the concentration above $(\eta_e)_{soln} \sim 2$, led to beaded fibers and a mixture of fibers and beads. When the polymer concentration was increased to 10 wt%, corresponding to $(\eta_e)_{soln} \sim 3.5$, pure fibers were obtained (Munir et al., 2009). Aqueous polysaccharide solutions that are suitable for fiber formation by electrospinning were found to meet at least two conditions: weak shear thinning at shear rates below 1000 s^{-1}, and an overlap concentration in the range of 10 and 20 (g. 100 mL^{-1}) (Stijnman et al., 2011).

Gupta et al. suggested that the electrospinning of dilute solutions resulted in the formation of polymer droplets due to insufficient overlap between molecular chains (Gupta et al., 2005). As the concentration was increased, polymer droplets and some beaded fibers were observed. Upon further increases in concentration (giving a semidilute entangled regime), beaded fibers were obtained. Uniform fiber formation was observed at higher polymer concentrations and the fiber diameter was also found to vary according to the viscosity of the solutions (Gupta et al., 2005). Tan et al. systematically studied factors affecting the production of ultrafine fibers by the electrospinning process (Tan et al., 2005). The main parameters were categorized into two groups. The first included parameters affecting the mass flow rate of polymer fed from the needle tip (e.g., polymer concentration, applied voltage, and volume feed rate) and the second, parameters affecting the electrical force during electrospinning (e.g., the electrical conductivity of the solvents). They indicate that polymer concentration, molecular weight, and solution electrical conductivity play a primal role in determining the morphology of electrospun fibers. Polymer fibers with smaller diameter can be electrospun with higher electrical conductivity of solution and lower polymer concentration which can be further decreased by use of a higher molecular weight of polymer. The electrical conductivity of a solution is related to the charge density on the jet, and thence the degree of elongation of the jet by electrical force. Therefore, under the same applied voltage and spinning distance, a solution with a higher electrical conductivity may experience a higher rate of elongation of the jet along its axis, resulting in smaller diameter fibers (Tan et al., 2005). Natural polymers are generally polyelectrolytic in nature, as each repeating unit bears an electrolyte group, forming ions in solution, increasing the charge carrying ability of the polymer jet, and subjecting the fiber strand to higher tension under the electric field. This results in a poor degree of fiber formation compared to synthetic polymers (Zong et al., 2002). As previously mentioned, to improve the conductivity of the solution, salt can be added, which will function as a charge carrier in the spinning solution and consequently may play an important role in increasing the production rate of electrospinning (Chaobo et al., 2006). High solution conductivity can be also achieved by using organic acid as the solvent: (Chaobo et al., 2006) used formic acid as the solvent along with small amount of pyridine to dissolve the nylon, and obtained ultrathin (3 nm) electrospun nylon fibers. Demir et al. showed that viscosity, solution concentration, and temperature are the dominant solution properties, affecting both the electrospinning process and the properties of the resulting fibers (Demir et al., 2002). They concluded that the morphology of electrospun fibers is strongly correlated with viscosity, concentration, and temperature. Low concentration solutions typically result in the formation of fibers with beads, whereas increased concentration favors the formation of curly, wavy, and straight fiber structures. Fiber diameter was found to be proportional to the cube of the polymer concentration (Demir et al., 2002). Lin et al. demonstrated that changes in polymer concentration and solution viscosity are directly dependent upon each other. It was observed that electrospinning of dilute polymer solutions produced beads or beads-on-string structure and that no fibers could be produced from the most dilute polymer solutions (Lin et al., 2005). Lim et al. suggested that due to the reduced viscosity of dilute solutions, polymer chains have more mobility and therefore are more readily able to align in the direction of the fiber axis during electrospinning (Lim et al., 2008), indicating that this may promote increased molecular orientation. Fibers produced with smaller diameters from dilute solutions were found to have a higher degree of molecular orientation, crystallinity, stiffness, and strength but exhibited low ductility (Lim et al., 2008). Zong et al. also emphasized that the polymer concentration or the corresponding viscosity of the polymer solution was one of the most effective variables to control the fiber morphology (Zong et al., 2002). Bead diameters were found to increase and the average distance between beads increased as the viscosity of the solution increased. Furthermore, the shape of the beads changed from spherical to spindle-like formations. They concluded that uniform fibers could be obtained at higher polymer concentrations and lower electrical field strengths (Zong et al., 2002). More recently researchers have developed modeling techniques to relate the main process parameters such as polymer concentration to fiber morphology and/or diameter (Theron et al., 2004; Thompson et al., 2007; Yördem et al., 2008).

12.4.2 Effect of Solvents

The morphology of electrospun fibers is influenced by both the physical and electrical properties of the solution, which are to a large extent determined by type of solvent (Eda et al., 2007). A typical spinning solution will consist of 80–90 wt% solvent and 10–20 wt% polymer, so the selection of solvents for electrospinning and their influence on fiber morphology has been one of the main areas of research in recent years (Lu et al., 2006; Ghorani et al., 2013). Wannatong et al. showed that both the density of electrospun webs and the presence of beads can be influenced by the choice of solvent (Wannatong et al., 2004). Uyar and Besenbacher indicated that differences in the degree of dryness of the collected fibers could be attributed to the density and boiling point of the solvent (Uyar and Besenbacher, 2008). The dryness of the collected fibers increased as the density and boiling point of the solvent decreased (Wannatong et al., 2004). Similar work on solvent selection was reported by Jarusuwannapoom et al. (Jarusuwannapoom et al., 2005). Solvent type can also influence fiber morphology. Yang et al. investigated the effects of dichloromethane (MC), ethanol, N,N-dimethylformamide (DMF), and mixtures of these solvents on the formation of ultrathin uniform PVP fibers during electrospinning process and noted substantial differences in the collected fiber morphologies ranging from smooth to helical texture (Yang et al., 2004). Van der Schueren et al. showed that substantially smaller fibers can be electrospun by selecting a binary rather than a single solvent system (Van der Schueren et al., 2011). The mixed solvent systems have been used by various researchers to balance electrospinning process requirements, for example, achieving high productivity coupled with small fiber diameters. Recently, the variation of the morphology of electrospun cellulose acetate (CA) fibers produced from single and binary solvent systems was interpreted based on solubility parameters to identify processing conditions for the production of defect-free CA fibrous webs by electrospinning (Ghorani et al., 2013). The Hildebrand solubility parameter (δ) and the radius of the sphere in the Hansen space ($D_{(s-p)}$) of acetone, acetic acid, water, N, N-dimethylacetamide (DMAc), methanol, and chloroform were examined and discussed for the electrospinning of CA. The Hildebrand solubility parameter (δ) of acetone and DMAc were found to be within an appropriate range for the dissolution of CA. The suitability of the binary solvent system of acetone: DMAc (2:1) for the continuous electrospinning of defect-free CA fibers was confirmed (Ghorani et al., 2013).

12.5 PROCESSING CONDITIONS

Other important parameters that effect the electrospinning process are the voltage, distance from needle to collector (tip to collector distance), volume feed rate, needle diameter and ambient conditions such as temperature, humidity, and atmospheric pressure (Ramakrishna, 2005; Pham et al., 2006; Mazoochi and Jabbari, 2011).

12.5.1 Voltage

The most fundamental element in the electrospinning process is the application of high voltage to the polymer solution. Variation of the voltage controls the electrical field strength between the spinneret and the collection point, and thus the strength of the drawing force. The degree of charge in the solution controls the initiation of the electrospinning process as the repulsive charges overcome the surface tension of the solution (Mazoochi and Jabbari, 2011). Depending on the feed rate of the solution, a higher voltage may be required to replenish this charge so that a stable Taylor cone is formed (Zong et al., 2002). Generally, both negative and positive voltages above about 6 kV are able to distort a polymer droplet into the diagnostic shape of the Taylor cone during the process of jet initiation (Taylor, 1964). Note that the balance between surface tension and the electrical force is critical in determining the initial cone shape of the polymer solution at the needle tip (Zong et al., 2002). As the driving voltage increases, the length of the single jet tends to decrease slightly and the apex angle of the Taylor cone increases. There is strong acceleration and low variation in direction in the stable region close to the tip of the cone (Buer et al., 2001). If the applied voltage is higher, the increase in the amount of charge present will cause the jet to accelerate faster and a greater volume flow rate of solution will be drawn from the tip of the needle. This may result in a smaller and less stable Taylor cone (Mazoochi and Jabbari, 2011). As the viscosity increases, more force is required to overcome both the surface tension and the viscoelastic force to attenuate the jet and form fibers (Doshi and Reneker, 1993). The threshold voltage needed to eject a charged jet from the drop at the nozzle depends mainly on the solution concentration. As both the voltage supplied and the resultant electrical field have an influence on the stretching and acceleration of the jet, they will also have an influence on the morphology and diameter of the obtained fibers (Buer et al., 2001; Megelski et al., 2002; Zhang et al., 2005; Zhuo et al., 2008). It is generally accepted that greater splittability of the droplet or bifurcation of the fiber strand are a consequence of a higher applied voltage leading to a higher internal electrostatic force (Filatov et al., 2007), resulting in a decrease in the average diameters of the electrospun fibers (Lee et al., 2003; Wannatong et al., 2004; Tan et al., 2005; Lin et al., 2008; Mazoochi and Jabbari, 2011). Therefore, the average fiber diameter will decrease following an increase in the applied voltage. Lin et al. have explained the influence of applied voltage and the tip-collector distance (TCD) on resultant fiber diameter (Lin et al., 2008). They indicate that the average fiber diameter reaches a minimum value after an initial increase in the applied voltage and then becomes larger as the applied voltage increases (Lin et al., 2008) (Figure 12.6).

This trend has been attributed to the shorter flight duration that is available to form fibers as the voltage increases. This is frequently referred to as the *flight time* during electrospinning (Lin et al., 2008). As a result, at higher voltage, some studies report the production of electrospun fibers with relatively large diameters and a higher tendency for bead formation (Zong et al., 2002; Deitzel et al., 2001; Lin et al., 2008). The fiber arrangement and crystallinity are also thought to be governed by the applied voltage (Zhao et al., 2004; Linnemann et al., 2005). At higher voltage, the fiber arrangement becomes more irregular because of jet instability in the high strength electrical field (Yuan et al., 2004; Linnemann et al., 2005). Similarly, the crystallinity of collected fibers has been found to increase with increasing voltage until a critical point after which crystallinity decreases with further increases in the voltage (Zhao et al., 2004).

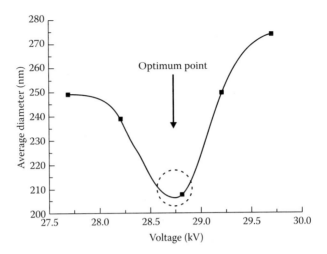

FIGURE 12.6 Example of the relationship between the applied voltage and the average fiber diameter during electrospinning.

12.5.2 Flow Rate

The polymer solution flow rate determines the amount of solution available at the needle tip during electrospinning. For a given voltage, there is a corresponding flow rate if a stable Taylor cone is to be maintained (Rutledge et al., 2001). A power law dependence exists between flow rate and applied voltage, that is, the flow rate is proportional to the cube of the voltage supplied (Demir et al., 2002). Consequently, any change in flow rate in a given electrical field may be expected to substantially effect fiber morphology (Jalili et al., 2005). However, the experimental evidence does not consistently support this. Zhang et al. showed that the morphological structure was only slightly changed by varying flow rate and that this was the least important effect of the process parameters studied (Zhang et al., 2005). Fibers with relatively high diameters are commonly electrospun at high flow rates, and a low flow rate frequently yields uniform electrospun fibers (Zong et al., 2002; Megelski et al., 2002; Yuan et al., 2004; Dotti et al., 2007; Ballengee and Pintauro, 2011). The reasons for this have been discussed by Zong et al. in terms of differences in the polymer jet velocity and polymer drying time (Zong et al., 2002). When the droplet suspended at the end of the needle tip is large and there is a high feeding rate, there is a larger volume of solvent to remove prior to fiber collection and the velocity of the jet in transit means the time for solvent evaporation is very limited. This can result in large beads and fused fiber–fiber intersections between fibers in the collected web (Zong et al., 2002). Low flow rates provide more time for solvent evaporation, and so bead formation can be reduced (Yuan et al., 2004). It has also been observed that the mean pore size in electrospun webs increases as the polymer flow rate increases (Megelski et al., 2002), which is likely to be related to variation of the fiber diameters in each web.

Electrospinning as a Novel Delivery Vehicle

12.5.3 Diameter of Spinneret Orifice

In the case of electrospinning systems that use a needle as spinneret, the internal diameter of the needle or the pipette orifice can influence the process. As the needle diameter increases, clogging may occur at the needle tip (Mo et al., 2004). This is the result of surface tension effects. Decreasing the radius of the droplet by reducing the needle diameter increases the surface tension of the droplet. Assuming the electrostatic field strength is constant, the initial acceleration of the jet decreases with increasing polymer surface tension and the average velocity is therefore decreased. Effectively, this means the jet flight time from the needle tip to the collector increases. This increases the time for solvent evaporation, jet attenuation, and splitting. As a result, fibers with smaller diameter and a narrower size distribution can be obtained by using a small diameter needle. However, if the diameter of the orifice is too small, it may be difficult for a satisfactory polymer droplet to form, but this also depends on the surface tension of the solution (Zhao et al., 2004). Although, it may seem logical to expect thinner electrospun fibers to be formed from a smaller diameter needle, previous studies have concluded that there is no correlation between the nozzle diameter and the fiber diameter (Yamashita et al., 2006).

12.5.4 Tip to Collector Distance

The elongation of the polymer jet takes place between the needle tip, where the Taylor cone is formed, and the collector. As may be expected, the tip to collector distance (TCD) has a direct influence on jet flight time and the electrostatic field strength. Inadequate drying of the fiber is attributed to insufficient distance between the needle tip and the collector. In this instance, the drying time is not long enough to evaporate the solvent before the fibers are deposited on the collector and consequently partially dried fibers with fused fiber–fiber intersections are observed as well as a densely packed structure (Barhate et al., 2006). Increasing the TCD results in longer flight times and solvent evaporation time, which tends to both decrease the rate of bead formation and mean fiber diameter (Doshi and Reneker, 1993; Yuan et al., 2004; Jarusuwannapoom et al., 2005; Ahn et al., 2006; Lin et al., 2008). However, at a fixed voltage, an increase in the TCD contributes to decreased electrostatic field strength and thus both jet splitting and attenuation are affected. Therefore, as the TCD is progressively increased from zero, the average fiber diameter reaches a minimum value before increasing as the TCD is increased further (Lin et al., 2008) (Figure 12.7).

It may be argued that the effect of decreasing TCD has a similar effect on fiber morphology and fiber diameter as increasing the voltage (Frenot and Chronakis, 2003; Lee et al., 2003; Zhao et al., 2004; Chowdhury and Stylios, 2010). Ying et al. reported that if the ambient humidity is low, increasing the TCD can lead to a reduction in fiber diameter as more solvent evaporates over the longer distance (Ying et al., 2006). Conversely, when the ambient humidity is high, the increased distance does not increase the evaporation rate but does reduce the field strength. The type of material selected for the grounded collector effects the degree of surface charge built-up

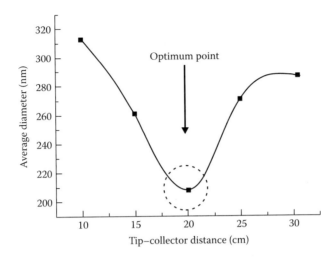

FIGURE 12.7 Example of the relationship between the tip-collector distance (cm) and the average fiber diameter during electrospinning.

during the electrospinning process (Chowdhury and Stylios, 2010). To increase the density of the deposited electrospun fibers, the use of a more highly conductive collector has been suggested. This arises because the deposition of electrospun fibers on the grounded collector is inversely proportional to the surface charge accumulation on the collector (Chowdhury and Stylios, 2010).

The term *electrospinning number* is defined as the ratio of electrical energy (Vq) to the surface energy (γR^2), $Vq/\gamma R^2$, where V is the voltage, q is the charge, γ is the surface tension and R is the radius of a Taylor cone droplet (Shenoy et al., 2005a). When $Vq/\gamma R^2 > 1$, a jet is ejected from the Taylor cone. As the applied voltage increases, the electrostatic forces acting on the polymer droplet also increase, which provides an additional force to overcome viscoelastic and surface tension forces exerted by the polymer solution. Therefore, an increase in the electrospinning number is normally expected to increase elongation of the polymer jet and lead to finer fibers (Subramanian et al., 2010).

12.6 AMBIENT CONDITIONS

Ambient parameters, particularly humidity and temperature, are thought to influence fiber morphology and the productivity of the electrospinning process (Chen and Yu, 2010). This is because a direct relationship exists between solvent evaporation and temperature as well as between the conductivity of the solvent and temperature. Both can influence the electrospinnability of a polymer solution (Chen and Yu, 2010). Furthermore, polymer viscosity and surface tension of the polymer solution are influenced by temperature (De Vrieze et al., 2009; Bhardwaj and Kundu, 2010). De Vrieze et al. showed that the average fiber diameter of electrospun fibers is influenced by humidity but these trends cannot be generalized (De Vrieze et al., 2009). In the case of electrospun CA fibers, an increase in diameter was observed with increasing humidity, while for PVP, there was a decrease in diameter. It has also been suggested that an increase in humidity can effect pore diameter and pore size distribution (Casper et al., 2003). Demir et al. reported differences in the uniformity of fiber diameters with temperature. In his work concerned with electrospinning polyurethane, high temperatures led to greater uniformity in diameter regardless of the applied electrostatic field (Demir et al., 2002).

12.7 ELECTROSPINNING VERSUS ELECTROSPRAY

The difference between electrospinning and electrospraying (electrohydrodynamic spraying), which are related technologies, is based on the degree of molecular cohesion in the raw material—a property which is most readily controlled by variation in concentration of the polymer solution (Chakraborty et al., 2009; Enayati et al., 2011). The electrospraying process is being used as a novel nanoencapsulation method where instead of production of nanofibers, nanoparticles are generated. In this technique, the electrostatic force induced by high voltage atomizes liquid into finer droplets than would result from surface tension alone, and evaporation of the solvent is performed during the flight of the droplets toward the ground electrode (Zhang and Kawakami, 2010; Hao et al., 2013). The role of chain entanglements of polymer solutions on fiber or bead formation during electrospinning/spraying process is very important (see Figure 12.3). The principle of electrospraying is essentially the same as that of electrospinning. The most important variable that allows us to distinguish between electrospraying and electrospinning is the polymer concentration used in the process (Shenoy et al., 2005a; Chakraborty et al., 2009). Changes in polymer concentration and molecular weight affect the viscosity and surface tension of the solution, and so greatly affect the electrosprayed product. The size of particles can be controlled by changing the concentration of the dissolved or suspended fraction (Jaworek, 2007). It has been acknowledged that the percentage of chain entanglement in the polymer in solution has a significant effect upon fiber or bead formation (Shenoy et al., 2005a; Zhang et al., 2012). The lack or limited development of such chain entanglement leads to the generation of particles in lieu of fibers at a given crucial limit with some overlap among the two resulting types of products (Almería et al., 2010). In other words, at low chain entanglement density, electrospraying of droplets instead of electrospinning of fibers will occur (Shenoy et al., 2005a; Bock et al., 2011). Also, as with electrospinning, for the electrospraying process, the polymer concentration should also be optimized. Otherwise, with increasing solution concentration, spindle-like and finally uniform fibers with increased diameters are formed (Fathi et al., 2014). With biopolymers, the formation of a nanofiber or nanodroplets (beads) depends on parameters such as biopolymer molecular mass, biopolymer chains entanglement, and solvent evaporation rate. A nanofiber is only formed when the product of biopolymer concentration (c) and solution intrinsic viscosity (η) known as the *Berry number*, $Be = [\eta]c$, is more than a certain critical value (Be_{cr}) which are specific characteristics of each biopolymer (Jaworek et al., 2009). If the Berry number of a solution is less than the critical value, the surface instabilities compelled by electrospun jet motion and charge interaction might lead to the production of a nanoparticle (bead) instead of nanofiber.

Unlike electrospinning, electrospraying results from the interaction of bulk and surface electrohydrodynamic forces that break the jet into droplets (Chakraborty et al., 2009). If the viscosity (polymer concentration) is not high enough to resist the forces of Rayleigh shattering, which assist in partial particle formation, the jet produced will not result in fiber formation, but instead particles interconnected by thin threads (Shin et al., 2001). In simple terms, polymer solutions with high viscosity generates continuous fibers, but the jet of a dilute polymer solution breaks into droplets. Particle engineering is based on an understanding of the underlying particle generation mechanisms. Electrospraying and electrospinning lead to the production of particles and fibers, respectively. These are two similar techniques based on the same physical rules with solutions of higher viscosity being used for electrospinning (Bohr et al., 2015). The dripping mode changes to a cone-jet, as the electric field is increased. However, with the increase of the electric field to a certain level, the liquid ejection becomes unstable and produces a multijet spray that reduces the uniformity of the particles. Cone-jet stability is required for the generation of uniform sized droplets. The voltage domain of the cone-jet mode becomes broader as the viscosity of the solution increases. Several researchers have reported that the electric field strength has a significant effect on the mode of spray production (Jaworek and Krupa, 1999; Grigoriev and Edirisinghe, 2002; Smith et al., 2006). Jaworek et al. reported that if the solvent evaporates very slowly or the polymer concentration was very low, the surface volatility induced by electrospun jet motion and charge interaction could led to the generation of beads on the fiber or to disruption of the jet into beads (Jaworek et al., 2009). The electrical permittivity and conductivity values are constant for a given polymer solution, so consequently, the flow rate is the main parameter available to control the size of electrosprayed droplets.

12.8 HOW TO PROVIDE A SAFE DELIVERY SYSTEM?

Recently reported research has shown that both electrospraying and electrospinning are excellent methods for the manufacture of encapsulated delivery systems. As a result, there is clear potential to design and develop electrosprayed particles as novel improved functionality products and delivery systems for complementary food compounds (Ghorani and Tucker, 2015). Encapsulation allows the safekeeping of environmentally sensitive core bioactive components, such as drugs, probiotic bacteria, vitamins, flavors, dyes, enzymes, sweeteners, acidulates, nutrients, preservatives, and the like. Also, the outer shell can be made from proteins, polysaccharides, fats, gums, or waxes. For example, by encapsulation, a drug applied as a core material can be targeted and released in a controlled method (Gouin, 2004; Bocanegra et al., 2005; Arya et al., 2009).

For development of a delivery system, a particular application must be selected from the most appropriate combination of compounds and procedures. Some of the most important factors to consider are (Ghorani and Tucker, 2015; Gómez-Mascaraque et al., 2015):

1. *Characteristics* such as transparency or opacity and color
2. *Rheological properties* such as viscosity and the rheological behavior of the final product

3. *Sensory properties*: the mouthfeel characteristics of the nanoparticles through the oral processing
4. *Stability*: the physical and chemical stability of the nanoparticles throughout the shelf life of products
5. *Biological functionality*: the release of bioactive components from the delivery system after ingestion at an appropriate site in the GI tract

Also some of the particle characteristics that can be designed into a delivery system include (McClements and Li, 2010; Donsì et al., 2011; Khayata et al., 2012; Sáiz-Abajo et al., 2013)

1. Digestibility
2. The chemical composition of particles
3. Particle size distribution
4. Particle charge
5. Particle permeability
6. Resistance to adverse environmental conditions such as ability to resist changes to pH, ionic strength, temperature, enzymes

Another factor to consider is the change in particle diameter distribution that occurs as a consequence of transformations in, say, concentration, or molecular weight, which can be a disrupting factor that can affect the payload (e.g., drugs or probiotic bacteria) release kinetics. The degree resistance of the polymer–biocomponents to the environment is another variable that significantly influences the extent and rate of the release (Chakraborty et al., 2009). Even though a number of different types of delivery systems have been developed and reported, there is still no complete understanding of the main factors and thus an effective design for systems for specific applications is still somewhat of an art (Fathi et al., 2014). There are still concerns about the extent to which some very sensitive compounds (proteins, DNA etc.) may be damaged by exposure to high voltages. However, some research demonstrates that electrosprayed cells do remain viable, and that any effects of high voltage on protein/DNA structure and function may only be temporary (Clarke and Jayasinghe, 2008).

12.8.1 Protein-Based Encapsulating Materials Used in Electrospinning

A growing interest in the use of electrospun fibers in the food industries has seen the demonstration of electrospinning of biopolymers and the encapsulation of food ingredients, enzymes, and other active compounds of particular interest to the food industry (Rezaei et al., 2015). The unique characteristics of biopolymers such as biodegradability, biocompatibility, the particular physical, and chemical behavior of biopolymers, and antibacterial activity have created an enormous demand for these products, mainly as encapsulation and delivery systems (Fathi et al., 2014). Proposed specific applications of such composites are active packaging or preservation of nutrient activity for consumption (Anu Bhushani and Anandharamakrishnan, 2014). Another benefit of electrospinning food ingredients is for the introduction of different textures and mouthfeel of the food (Nieuwland et al., 2013). Preservation of active compounds through encapsulation in electrospun fibers is probably the most widely investigated field in the application of food technology (Alborzi et al., 2013). Encapsulation of active compounds into fibers is often achieved by mixing the active compounds into the polymer solution and electrospinning. To avoid the use of toxic solvents for the generation of food-related products, most polymers used for electrospinning of food-based application are dissolved in water or ethanol (Kayaci and Uyar, 2012). As a carrier for proteins or active compounds, the selected polymer for electrospinning should be natural, edible, should not require the use of toxic solvents, and should be electrospinnable to give reliable and consistent fibers without the need to introduce a man-made polymer as a spinning aid into the mixture (López-Rubio and Lagaron, 2012). Polymer-based delivery systems are widely recognized in the biomedical and pharmaceutical sectors (Liechty et al., 2010). Likewise, the use of food-grade polymers and biopolymers as bioactive molecule delivery devices in food systems is widely being investigated (López-Rubio and Lagaron, 2012; Nieuwland et al., 2013). In this regard, the natural biopolymers, proteins, and carbohydrates are commonly used for encapsulation because of the controlled and sustained release properties that can be achieved to deliver the incorporated compound in the right place and at the right time (Fathi et al., 2014). Proteins are preferred over synthetic polymers (more importantly carbohydrate polymers) because they form significant component parts of the human body, and are often in themselves, valuable dietary supplements and functional food enhancers (Nieuwland et al., 2013). Proteins are infamously difficult to electrospin, mainly because of their complex secondary and tertiary structures. To be spinnable, the proteins should be well-dissolved in a random coil conformation. Globular proteins have too little internal interaction to entangle during the spinning process (Nieuwland et al., 2013). For example, the spinning of pure caseinate was reported as not possible in water due to clustering of the molecules (Pitkowski et al., 2008). Common protein-based encapsulating materials used in electrospinning are whey protein isolate (WPI) (López-Rubio et al., 2012), WPC (López-Rubio and Lagaron, 2012), soy protein isolate (Xu et al., 2012), egg albumen (EA) (Wongsasulak et al., 2010), collagen (Bürck et al., 2012), gelatin (Songchotikunpan, Tattiyakul, and Supaphol, 2008), zein (Brahatheeswaran et al., 2012), and casein (Nieuwland et al., 2013). López-Rubio et al. showed the feasibility of using electrospinning for functional food applications by encapsulating *Bifidobacterium* strains in food hydrocolloids, specifically WPC and carbohydrate (pullulan) (López-Rubio et al., 2012). WPC demonstrated greater protective ability as an encapsulation material than pullulan, as it effectively prolonged the survival of the cells even at high relative humidities (López-Rubio et al., 2012). The usefulness of the WPC and WPI capsules was also demonstrated through the encapsulation of the antioxidants β-carotene and β-lactoglobulin (López-Rubio and Lagaron, 2012; Sullivan et al., 2014). Recently, protein isolated from the natural substance amaranth (amaranth protein isolate or API) was electrospun into ultrathin fibers and the importance of morphologies and textures for the protein as bioactive encapsulation matrices for functional food applications was revealed (Aceituno-Medina et al., 2013). The properties of zein, a hydrophobic protein extracted from corn by solvation in ethanol, in enhancing the stability and bioavailability of various gallic acids (Neo et al., 2013), β-carotenes (Fernandez et al., 2009), curcumins

(Brahatheeswaran et al., 2012), (−)-epi-gallocatechingallates (EGCGs) (Li et al., 2009), α-tocopherols (Wongsasulak et al., 2014), ferulic acids (Yang et al., 2013), and tannin (de Oliveira Mori et al., 2014) have also been demonstrated.

12.8.2 Carbohydrate-Based Encapsulating Materials Used in Electrospinning

Natural and modified polysaccharides are also promising vehicles for nano and microencapsulation of active food ingredients since they are biocompatible, biodegradable, and possess a high potential to be modified to achieve the required properties (Schiffman and Schauer, 2008). Moreover, carbohydrate-based delivery systems can interact with a wide range of bioactive compounds via their functional groups, which make them versatile carriers to bind and entrap a variety of hydrophilic and hydrophobic bioactive food ingredients. In addition, they are considered as suitable shells to impart high temperature stability compared to lipid- or protein-based delivery systems that might require the additional stage of being malted or denatured (Fathi et al., 2014). Carbohydrates consist of monosaccharide, oligosaccharide, and polysaccharides, and can form delivery systems that are conventionally categorized according to their biological origins: higher plant origin (e.g., starch, cellulose, pectin, and guar gum); animal origin (e.g., chitosan); algal origin (e.g., alginate and carrageenan), and microbial origin (e.g., xanthan, dextran, and cyclodextrins) (Fathi et al., 2014). Chitosan (Desai et al., 2008), alginate (Alborzi et al., 2010), cellulose and its derivatives (Ghorani et al., 2013), starch (Kong and Ziegler, 2012), pullulan (Karim et al., 2009; Fabra et al., 2014), inulin (Jain et al., 2014), guar-gum (Lubambo et al., 2013), and dextrans (Sun et al., 2013) are reported as satisfactory for the electrospinning of outer wall materials. Edible nanofibrous thin films were fabricated by electrospinning from blended solutions of CA in 85% acetic acid and EA in 50% formic acid (Wongsasulak et al., 2010). Similarly, biocompatible composite fibers were electrospun from EA and PEO at identified optimum electrospinning conditions (Wongsasulak et al., 2007). Vitamins A and E were immobilized onto electrospun CA nanofibers with a smooth and round cross-sectional morphology (Taepaiboon et al., 2007). The immobilization of these vitamins onto the nanofibers resulted in the controlled release of vitamins over the test period compared to the burst release that was observed in control cast films of CA (Taepaiboon et al., 2007). Stijnman et al. investigated the ability to form fibers using electrospinning from aqueous solutions of polysaccharides derived from plant sources (galactomannansand glucomannans), seaweeds (carrageenans and alginate), microbial origin materials with nonbranched (pullulan) and brush-like (dextran) structures and other vegetable sources such as starch, and methylcellulose (Stijnman et al., 2011). Encapsulation improves the stability, preservation, bioavailability, and controlled release properties of the subject biomolecule, and also can mask unwanted odors or the taste of the delivered compound (Mascheroni et al., 2013). Folic acid (vitamin B) without any coating is susceptible to degradation when exposed to light and acidic condition. However, when it is encapsulated within sodium alginate-pectin- PEO nanofibers, almost 100% of the folic acid is retained after 41 days of storage in the dark at pH 3. This contrasts against 8% recovery after one day of storage at pH 3 (Alborzi et al., 2013). Perillaldehyde aroma compound immobilized directly in an edible polysaccharide nanofibrous matrix of pullulan-cyclodextrin (Mascheroni et al., 2013). Inclusion complexes of vanilla, menthol, and eugenol flavor compounds, and cyclodextrins (α-CD, β-CD, γ-CD) makes the preparation more stable at high temperatures and increases shelf life if they are electrospun into polyvinyl alcohol (PVA) nanofibers (Kayaci and Uyar, 2012). Similarly, the successful application of emulsion electrospinning for the encapsulation of highly volatile fragrances, such as limonene in a fibrous matrix, has recently been presented. The use of composite blends of carbohydrates with polymers that are compatible for the formation of electrospun fibers with enhanced material properties such as rheological properties and higher tensile strength have recently been taken into account (Anu Bhushani and Anandharamakrishnan, 2014). The addition of PVA was reported as a highly crucial factor to improve the solution spinnability and the rheological profiles of the agar/PVA blends (Sousa et al., 2015). Synthetic polymers such as PEO, PVA, when combined with biopolymers, improve the fiber-forming ability of the blended solution. The soluble fiber fraction obtained by alkaline treatment of cereal wastes (corncob and wheat straw) has been electrospun along with 6%(w/v) PVA to obtain ultrafine fibers (200–800 nm) (Kuan et al., 2011).

12.8.3 Recent Advances in Encapsulating of Probiotics by Electrospinning

Apart from these food bioactive compounds, the interest for biological preservation of foods brings about the need for encapsulation techniques that maintain the viability and stability of probiotic bacteria and bacteriocins during food processing and storage particularly using proteins and carbohydrates as the encapsulating materials (Burgain et al., 2011). Encapsulation of bioactive compounds and probiotic bacteria within prebiotic substances to protect or even enhance their survival while passing upper GI tract and during food processing and storage is an area of great interest for both academia and the food industries (López-Rubio et al., 2012). Different methods for micro and nanoencapsulation of food bioactives (e.g., spray drying, freeze drying, emulsification, coacervation, nanoprecipitation, liposome preparation) have been suggested, examined, and applied to the encapsulation and drying of probiotics and bioactive compounds; clearly each technique has its own advantages and disadvantages (Anu Bhushani and Anandharamakrishnan, 2014). The main drawbacks of these techniques are that the use of high temperature or organic agents in at least one of the production steps leads to some destruction of sensitive encapsulated nutrients as well as toxicity problems associated with residual organic agents (López-Rubio and Lagaron, 2012; López-Rubio et al., 2012). For instance, encapsulation by spray drying significantly reduces the viability of bacteria or damages the structure of the target molecules (López-Rubio et al., 2012). Electrospinning is thus attractive as it is a conceptually simple technology that does not involve any severe conditions of temperature pressure and chemistry, and is therefore suitable for encapsulation of sensitive compounds. Electrospinning allows the use of a wide range of food grade, biodegradable, biocompatible, or conducting polymeric substances as wall materials for the encapsulation of bioactives (Anu Bhushani and Anandharamakrishnan, 2014). In one attempt, the feasibility and potential of soluble dietary fiber (SDF) from certain agricultural waste streams-okara (soybean solid waste), oil palm trunk (OPT), and oil palm frond (OPF) obtained via alkali treatment, in the nanoencapsulation of *Lactobacillus acidophilus* was reported (Fung et al., 2011). Viability studies showed good bacterial survivability (78.6%–90%) after electrospinning and retained viability at refrigeration temperatures during the 21-day storage study (Fung et al., 2011). In another work, the feasibility of using the electrospinning technique to encapsulate *Bifidobacterium* strains in food hydrocolloids for functional food applications was highlighted (López-Rubio et al., 2012). Specifically, a protein (WPC) and a carbohydrate (pullulan) were used as encapsulation materials due to their ability to form micro, submicron, and nanocapsules through an electrospraying process. Encapsulation through electrospraying substantially increased the viability of the bifidobacterial strain, especially at 20°C. WPC demonstrated greater protection ability as encapsulation material than pullulan as it effectively prolonged the survival of the cells even at high relative humidity (López-Rubio et al., 2012). Salalha et al. increased the viability of encapsulated bacteria (*Escherichia coli* and *Staphylococcus albus*) in an electrospun structure made with PVA (Salalha et al., 2006). The bacteria remained viable for three months stored at −20°C and −55°C without deterioration in viability (Salalha et al., 2006). Heunis et al. showed that electrospun nanofibersspun from poly(D, L-lactide) (PDLLA) and PEO have the potential to serve as a carrier matrix for bacteriocins (proteinaceous toxins produced by bacteria to inhibit the growth of similar or closely related bacterial strain, in this case *Plantaricin 423*, produced by *Lactobacillus plantarum*, and the bacteriocin ST4SA produced by *Enterococcus mundtii*) (Heunis et al., 2011). They believe that this technology opens a new field in developing controlled antimicrobial delivery systems for various applications (Heunis et al., 2011). Core/shell-structured electrospun fibers were also produced by coaxial electrospinning using a PEO and cellulose diacetate (CDA) mixture for the shell and a T4 phage/buffer suspension for the core (Korehei and Kadla, 2014). *In vitro* release studies revealed that the water-soluble PEO fibers exhibited a burst release over a period of 30 min that resulted in almost 100% release of the encapsulated T4 phage (Korehei and Kadla, 2014). Consequently, there is significant scope to develop electrospun fibrous assemblies that have the potential to advance the design and performance of novel delivery systems for functional products (Anu Bhushani and Anandharamakrishnan, 2014). The incorporation of biomolecules as supplementary components within a polymeric electrospun fiber by first solution blending prior to electrospinning is an emerging technique to increase the performance of functional materials in food by the application of nanotechnology (Ghorani, 2012). To optimize production conditions and maximize throughput, a clear understanding of the mechanisms of electrospinning is essential.

12.9 CONCLUSIONS AND FUTURE TRENDS

Encapsulation of bioactive compounds and probiotic bacteria within prebiotic substances to protect or even enhance their survival while passing upper GI tract, is an area of great interest for both academia and the food industries. Encapsulation technology is based on packaging of bioactive compounds in micro or nanoscaled particles that isolate them and control their release upon applying specific conditions.

Electrospun nanofibers can also be used as the delivery system in foods for nutrients to protect them during the processing and storage or in delivery systems for transferring the components to the target site in the body. The main advantages of electrospun nanofibers for the encapsulation of food bioactives are sustained and controlled release, room temperature processes, reduced denaturation, efficient encapsulation, enhanced stability of bioactives achieved by the use of food grade polymers and biopolymers. Moreover, the size of the final product can be controlled by manipulating processing parameters.

This chapter provides a comprehensive discussion presented on the fundamentals of electrospinning/spraying polymer nanofibers/particles including processing, structure, property characterization and applications. Increased chain entanglement and longer relaxation times as a consequence of increased polymer concentration are thought to be responsible for fiber formation. A bead dominant morphology is generally observed when a minimum degree of chain entanglement is not present. When the viscosity of the solution is too low, electrospraying (liquid atomization) may occur and beads are formed instead of fibers. Information of the processing conditions for electrospinning of ultrafine fibers is also discussed. The numbers of applications of electrospinning and its potentials in the food systems (laboratory scale) have been recognized in the recent years; however, the industrial scale application of this technology in the food industry as a delivery vehicle is still limited and needs to be extended. The main reason why the potentialities of nanofibers have not been widely adopted in the food industry can be attributed to the prevalence of studies on synthetic polymers rather than on biopolymers which has limited the appeal of this material. Clearly, the proposed application of electrospun fibers must be available as practical industrial scale processes. Production of new materials based on electrospun nanofibers by using polymer for food use can be highly significant for all food sectors, particularly: new food formulas; new food ingredients; novel packaging, sensors, biosensors, or delivery systems. Secondly, the low throughput of the electrospinning system restricts its commercial exploitation at a large scale and therefore increasing the production efficiency indicates research and development priority for wider application. However, it should be noted that recent work in the area of nanocomposites indicates that the high surface to volume ratio of nanomaterials means that surface energy effects are much more significant than at the macro scale. The formation of a zone of influence around the nanofibers (the interphase zone) means that only small fractions (say <1 wt%) of nanomaterials) are needed in the product to produce highly significant effects. The effect of the interphase on food properties has not been significantly investigated: it is possible that this property may open up new areas of flavor intensification. In addition, the solidification of polymers in a strong electrical field results in the formation of an electret in electrostatic terms, the equivalent of a permanent magnet. This effect has been noted in electrospun fiber formation, and means that there is a long lasting (in the order of years—Filatov et al., 2007) residual charge intrinsic in the fiber: this is in addition to the transient surface charge which is more commonly discussed (Filatov et al., 2007). The effect of this charge on the adhesion to the fiber of food materials such as proteins, has not been widely investigated, but again, is likely to be of significant use in constructing, say nutraceutical delivery systems. Finally, many food-grade polymer systems (e.g., natural food hydrocolloids) are difficult to optimize with electrospinning, due to their poor viscoelastic behavior, lack of sufficient molecular entanglement, limited solubility, and, more generally, because only a few processing parameters can be controlled directly as some of the involved parameters are either highly interdependent or derive from the properties of the used polymer solution. Industrially produced electrospun nanofibers reach a dimensional spread of the order of 10% points or worse among nominally identical samples. Therefore, a main challenge related with the mass production of nanofiber fabrics is the implementation of methods allowing increase of the process and product reproducibility and to extend the classes of utilizable materials. Currently, fabrication of electrospun fibers at industrial scale is feasible, and has been done on a commercial scale since the late 1930s. Many companies supplying either laboratory and large-scale electrospinning equipment or nanofibrous products (mainly for filtration and medical purpose) already exist and possess well-defined market segments; however, the authors believe that the next steps in progress toward a new generation of industrial applications will be based on controlled morphology and alignment of electrospun structures. A tighter interaction between the academic and the industrial communities and a critical assessment of the weaknesses and strength of electrospinning technologies for industrial processes can certainly be useful to stimulate future developments. Moreover, the electrospun fibers, typically produced as nonwoven mats with intertwined fibrous structures, will need parallel development of both their physical form and methods of incorporation into food stuffs. This will take a considerable development effort if the advantages of these materials are to be incorporated in the functional foods of the near future.

REFERENCES

Aceituno-Medina, M., Lopez-Rubio, A., Mendoza, S., and Lagaron, J. M. (2013). Development of novel ultrathin structures based in amaranth (*Amaranthus hypochondriacus*) protein isolate through electrospinning. *Food Hydrocolloids, 31*(2), 289–298. doi: 10.1016/j.foodhyd.2012.11.009.

Ahn, Y. C., Park, S. K., Kim, G. T., Hwang, Y. J., Lee, C. G., Shin, H. S., and Lee, J. K. (2006). Development of high efficiency nanofilters made of nanofibers. *Current Applied Physics, 6*(6), 1030–1035. doi: 10.1016/j.cap.2005.07.013.

Alborzi, S., Lim, L.-T., and Kakuda, Y. (2010). Electrospinning of sodium alginate-pectin ultrafine fibers. *Journal of Food Science, 75*(1), C100–C107. doi: 10.1111/j.1750-3841.2009.01437.x.

Alborzi, S., Lim, L. T., and Kakuda, Y. (2013). Encapsulation of folic acid and its stability in sodium alginate-pectin-poly(ethylene oxide) electrospun fibres. Journal of Microencapsulation, *30*(1), 64–71. doi: 10.3109/02652048.2012.696153.

Almería, B., Deng, W., Fahmy, T. M., and Gomez, A. (2010). Controlling the morphology of electrospray-generated PLGA microparticles for drug delivery. *Journal of Colloid and Interface Science, 343*(1), 125–133. doi: 10.1016/j.jcis.2009.10.002.

Angammana, C. J. and Jayaram, S. H. (2008). Analysis of the effects of solution conductivity on electro-spinning process and fiber morphology. In *Industry Applications Society Annual Meeting, IEEE,* Edmonton, Canada, pp. 1–4.

Anu Bhushani, J. and Anandharamakrishnan, C. (2014). Electrospinning and electrospraying techniques: Potential food based applications. *Trends in Food Science and Technology, 38*(1), 21–33. doi: 10.1016/j.tifs.2014.03.004.

Arya, N., Chakraborty, S., Dube, N., and Katti, D. S. (2009). Electrospraying: A facile technique for synthesis of chitosan-based micro/nanospheres for drug delivery applications. *Journal of Biomedical Materials Research Part B: Applied Biomaterials,* 88(1), 17–31.

Ballengee, J. B. and Pintauro, P. N. (2011). Morphological control of electrospun nafion nanofiber mats. *Journal of the Electrochemical Society,* 158(5), B568–B572. doi: 10.1149/1.3561645.

Barhate, R. S., Loong, C. K., and Ramakrishna, S. (2006). Preparation and characterization of nanofibrous filtering media. *Journal of Membrane Science,* 283(1–2), 209–218. doi: 10.1016/j.memsci.2006.06.030.

Barrero, A. and Loscertales, I. G. (2007). Micro-and nanoparticles via capillary flows. *Annual Review of Fluid Mechanics,* 39, 89–106.

Bhardwaj, N. and Kundu, S. C. (2010). Electrospinning: A fascinating fiber fabrication technique. *Biotechnology Advances,* 28(3), 325–347. doi: 10.1016/j.biotechadv.2010.01.004.

Bocanegra, R., Gaonkar, A. G., Barrero, A., Loscertales, I. G., Pechack, D., and Marquez, M. (2005). Production of cocoa butter microcapsules using an electrospray process. *Journal of Food Science-Chicago,* 70(8), E492.

Bock, N., Dargaville, T. R., and Woodruff, M. A. (2012). Electrospraying of polymers with therapeutic molecules: State of the art. *Progress in Polymer Science,* 37(11), 1510–1551. doi: 10.1016/j.progpolymsci.2012.03.002.

Bock, N., Woodruff, M. A., Hutmacher, D. W., and Dargaville, T. R. (2011). Electrospraying, a reproducible method for production of polymeric microspheres for biomedical applications. *Polymers,* 3(1), 131–149.

Bohr, A., Wan, F., Kristensen, J., Dyas, M., Stride, E., Baldursdottír, S., Edirisinghe, M., and Yang, M. (2015). Pharmaceutical microparticle engineering with electrospraying: The role of mixed solvent systems in particle formation and characteristics. *Journal of Materials Science: Materials in Medicine,* 26(2), 1–13.

Brahatheeswaran, D., Mathew, A., Aswathy, R. G., Nagaoka, Y., Venugopal, K., Yoshida, Y., Maekawa, T., and Sakthikumar, D. (2012). Hybrid fluorescent curcumin loaded zein electrospun nanofibrous scaffold for biomedical applications. *Biomed Mater,* 7(4), 045001. doi: 10.1088/1748-6041/7/4/045001.

Buer, A., Ugbolue, S. C., and Warner, S. B. (2001). Electrospinning and properties of some nanofibers. *Textile Research Journal,* 71(4), 323–328. doi: 10.1177/004051750107100408.

Bürck, J., Heissler, S., Geckle, U., Ardakani, M. F., Schneider, R., Ulrich, A. S., and Kazanci, M. (2012). Resemblance of electrospun collagen nanofibers to their native structure. *Langmuir,* 29(5), 1562–1572. doi: 10.1021/la3033258.

Burgain, J., Gaiani, C., Linder, M., and Scher, J. (2011). Encapsulation of probiotic living cells: From laboratory scale to industrial applications. *Journal of Food Engineering,* 104(4), 467–483. doi: 10.1016/j.jfoodeng.2010.12.031.

Casper, C. L., Stephens, J. S., Tassi, N. G., Chase, D. B., and Rabolt, J. F. (2003). Controlling surface morphology of electrospun polystyrene fibers: Effect of humidity and molecular weight in the electrospinning process. *Macromolecules,* 37(2), 573–578. doi: 10.1021/ma0351975.

Chakraborty, S., Liao, I. C., Adler, A., and Leong, K. W. (2009). Electrohydrodynamics: A facile technique to fabricate drug delivery systems. *Advanced Drug Delivery Reviews,* 61(12), 1043–1054. doi: 10.1016/j.addr.2009.07.013.

Chaobo, H., Shuiliang, C., Chuilin, L., Darrell, H. R., Haiyan, Q., Ying, Y., and Haoqing, H. (2006). Electrospun polymer nanofibres with small diameters. *Nanotechnology,* 17(6), 1558.

Chen, Z.-Y., Ma, K. Y., Liang, Y., Peng, C., and Zuo, Y. (2011). Role and classification of cholesterol-lowering functional foods. *Journal of Functional Foods,* 3(2), 61–69. doi: 10.1016/j.jff.2011.02.003.

Chen, L., Remondetto, G. E., and Subirade, M. (2006). Food protein-based materials as nutraceutical delivery systems. *Trends in Food Science and Technology, 17*(5), 272–283. doi: 10.1016/j.tifs.2005.12.011.

Chen, H.-M. and Yu, D.-G. (2010). An elevated temperature electrospinning process for preparing acyclovir-loaded PAN ultrafine fibers. *Journal of Materials Processing Technology, 210*(12), 1551–1555. doi: 10.1016/j.jmatprotec.2010.05.001.

Chowdhury, M. and Stylios, G. (2010). Effect of experimental parameters on the morphology of electrospun Nylon 6 fibres. *International Journal of Basic and Applied Sciences, 10*(6), 116–131.

Clarke, J. D. and Jayasinghe, S. N. (2008). Bio-electrosprayed multicellular zebrafish embryos are viable and develop normally. *Biomedical Materials, 3*(1), 011001.

Cross, J. (1987). *Electrostatics, Principles, Problems and Applications*: Taylor & Francis, Boca Raton, FL.

Dabirian, F., Hosseini, Y., and Ravandi, S. A. H. (2007). Manipulation of the electric field of electrospinning system to produce polyacrylonitrile nanofiber yarn. *Journal of the Textile Institute, 98*(3), 237–241. doi: 10.1080/00405000701463979.

De Oliveira Mori, C. L. S., dos Passos, N. A., Oliveira, J. E., Mattoso, L. H. C., Mori, F. A., Carvalho, A. G., de Souza Fonseca, A., and Tonoli, G. H. D. (2014). Electrospinning of zein/tannin bio-nanofibers. *Industrial Crops and Products, 52*(0), 298–304. doi: 10.1016/j.indcrop.2013.10.047.

De Vrieze, S., Van Camp, T., Nelvig, A., Hagström, B., Westbroek, P., and De Clerck, K. (2009). The effect of temperature and humidity on electrospinning. *Journal of Materials Science, 44*(5), 1357–1362. doi: 10.1007/s10853-008-3010-6.

Deitzel, J. M., Kleinmeyer, J., Harris, D., and Tan, B. N. C. (2001). The effect of processing variables on the morphology of electrospun nanofibers and textiles. *Polymer, 42*(1), 261–272. doi: 10.1016/s0032-3861(00)00250-0.

Demir, M. M., Yilgor, I., Yilgor, E., and Erman, B. (2002). Electrospinning of polyurethane fibers. *Polymer, 43*(11), 3303–3309. doi: 10.1016/s0032-3861(02)00136-2.

Desai, K., Kit, K., Li, J., and Zivanovic, S. (2008). Morphological and surface properties of electrospun chitosan nanofibers. *Biomacromolecules, 9*(3), 1000–1006. doi: 10.1021/bm701017z.

Ding, B., Li, C., Miyauchi1, Y., Kuwaki, O., and Shiratori, S. (2006). Formation of novel 2D polymer nanowebs via electrospinning. *Nanotechnology, 17*(15), 3685–3691.

Donini, L. M., Savina, C., and Cannella, C. (2009). Nutrition in the elderly: Role of fiber. *Arch Gerontol Geriatr, 49(Suppl 1)*, 61–69. doi: 10.1016/j.archger.2009.09.013.

Donsì, F., Annunziata, M., Sessa, M., and Ferrari, G. (2011). Nanoencapsulation of essential oils to enhance their antimicrobial activity in foods. *LWT-Food Science and Technology, 44*(9), 1908–1914.

Doshi, J. and Reneker, D. H. (1993, 2–8 Oct 1993). *Electrospinning process and applications of electrospun fibers*. In Paper presented at the *Industry Applications Society Annual Meeting, Conference Record of the 1993 IEEE*, Toronto, ON, pp. 1698–1703.

Doshi, J. and Reneker, D. H. (1995). Electrospinning process and applications of electrospun fibers. *Journal of Electrostatics, 35*(2–3), 151–160. doi: 10.1016/0304-3886(95)00041-8.

Dotti, F., Varesano, A., Montarsolo, A., Aluigi, A., Tonin, C., and Mazzuchetti, G. (2007). Electrospun porous mats for high efficiency filtration. *Journal of Industrial Textiles, 37*(2), 151–162. doi: 10.1177/1528083707078133.

Eda, G., Liu, J., and Shivkumar, S. (2007). Solvent effects on jet evolution during electrospinning of semi-dilute polystyrene solutions. *European Polymer Journal, 43*(4), 1154–1167. doi: 10.1016/j.eurpolymj.2007.01.003.

Eltayeb, M., Bakhshi, P. K., Stride, E., and Edirisinghe, M. (2013). Preparation of solid lipid nanoparticles containing active compound by electrohydrodynamic spraying. *Food Research International, 53*(1), 88–95. doi: 10.1016/j.foodres.2013.03.047.

Enayati, M., Chang, M.-W., Bragman, F., Edirisinghe, M., and Stride, E. (2011). Electrohydrodynamic preparation of particles, capsules and bubbles for biomedical engineering applications. *Colloids and Surfaces A: Physicochemical and Engineering Aspects, 382*(1–3), 154–164. doi: 10.1016/j.colsurfa.2010.11.038.

Ezhilarasi, P. N., Karthik, P., Chhanwal, N., and Anandharamakrishnan, C. (2013). Nanoencapsulation techniques for food bioactive components: A review. *Food and Bioprocess Technology, 6*(3), 628–647. doi: 10.1007/s11947-012-0944-0.

Fabra, M. J., López-Rubio, A., and Lagaron, J. M. (2014). On the use of different hydrocolloids as electrospun adhesive interlayers to enhance the barrier properties of polyhydroxyalkanoates of interest in fully renewable food packaging concepts. *Food Hydrocolloids, 39*(0), 77–84. doi: 10.1016/j.foodhyd.2013.12.023.

Fathi, M., Martin, A., and McClements, D. J. (2014). Nanoencapsulation of food ingredients using carbohydrate based delivery systems. *Trends in Food Science and Technology, 39*(1), 18–39.

Fernandez, A., Torres-Giner, S., and Lagaron, J. M. (2009). Novel route to stabilization of bioactive antioxidants by encapsulation in electrospun fibers of zein prolamine. *Food Hydrocolloids, 23*(5), 1427–1432. doi: 10.1016/j.foodhyd.2008.10.011.

Filatov, Y., Budyka, A., and Kirichenko, V. (2007). *Electrospinning of Micro- and Nanofibers: Fundamentals and Applications in Separation and Filtration Processes.* New York: Begell House.

Fong, H., Chun, I., and Reneker, D. H. (1999). Beaded nanofibers formed during electrospinning. *Polymer, 40*(16), 4585–4592. doi: 10.1016/s0032-3861(99)00068-3.

Frenot, A. and Chronakis, I. S. (2003). Polymer nanofibers assembled by electrospinning. *Current Opinion in Colloid and Interface Science, 8*(1), 64–75. doi: 10.1016/s1359-0294(03)00004-9.

Fung, W.-Y., Yuen, K.-H., and Liong, M.-T. (2011). Agrowaste-based nanofibers as a probiotic encapsulant: Fabrication and characterization. *Journal of Agricultural and Food Chemistry, 59*(15), 8140–8147. doi: 10.1021/jf2009342.

Gañán-Calvo, A. M. (1999). The surface charge in electrospraying: Its nature and its universal scaling laws. *Journal of Aerosol Science, 30*(7), 863–872.

Ghorani, B. (2012). *Production and Properties of Electrospun Webs for Therapeutic Applications.* University of Leeds (School of Design), Leeds, UK.

Ghorani, B., Russell, S. J., and Goswami, P. (2013). Controlled morphology and mechanical characterisation of electrospun cellulose acetate fibre webs. *International Journal of Polymer Science,* 12. doi: 10.1155/2013/256161.

Ghorani, B. and Tucker, N. (2015). Fundamentals of electrospinning as a novel delivery vehicle for bioactive compounds in food nanotechnology. *Food Hydrocolloids, 51*, 227–240.

Gómez-Mascaraque, L. G., Lagarón, J. M., and López-Rubio, A. (2015). Electrosprayed gelatin submicroparticles as edible carriers for the encapsulation of polyphenols of interest in functional foods. *Food Hydrocolloids, 49*, 42–52.

Gouin, S. (2004). Microencapsulation: Industrial appraisal of existing technologies and trends. *Trends in Food Science and Technology, 15*(7), 330–347.

Grigoriev, D. and Edirisinghe, M. (2002). Evaporation of liquid during cone-jet mode electrospraying. *Journal of Applied Physics, 91*(1), 437–439.

Grimm, R. L. and Beauchamp, J. L. (2005). Dynamics of field-induced droplet ionization: Time-resolved studies of distortion, jetting, and progeny formation from charged and neutral methanol droplets exposed to strong electric fields. *The Journal of Physical Chemistry B, 109*(16), 8244–8250. doi: 10.1021/jp0450540.

Gupta, P., Elkins, C., Long, T. E., and Wilkes, G. L. (2005). Electrospinning of linear homopolymers of poly(methyl methacrylate): Exploring relationships between fiber formation, viscosity, molecular weight and concentration in a good solvent. *Polymer, 46*(13), 4799–4810. doi: 10.1016/j.polymer.2005.04.021.

Hao, S., Wang, Y., Wang, B., Deng, J., Liu, X., and Liu, J. (2013). Rapid preparation of pH-sensitive polymeric nanoparticle with high loading capacity using electrospray for oral drug delivery. *Materials Science and Engineering: C, 33*(8), 4562–4567.

Hartman, R., Brunner, D., Camelot, D., Marijnissen, J., and Scarlett, B. (2000). Jet break-up in electrohydrodynamic atomization in the cone-jet mode. *Journal of Aerosol Science, 31*(1), 65–95.

Haub, C. (2011). World population aging: Clocks illustrate growth in population under age 5 and over age 65, Population Reference Bureau (PRB): http://www.prb.org/Articles/2011/agingpopulationclocks.aspx.

Heaney, R. P. (2007). Bone health. *The American Journal of Clinical Nutrition, 85*(1), 300S–303S.

Heikkila, P. and Harlin, A. (2008). Parameter study of electrospinning of polyamide-6. *European Polymer Journal, 44*(10), 3067–3079. doi: 10.1016/j.eurpolymj.2008.06.032.

Heunis, T., Bshena, O., Klumperman, B., and Dicks, L. (2011). Release of bacteriocins from nanofibers prepared with combinations of poly(D, L-lactide) (PDLLA) and poly(ethylene oxide) (PEO). *International Journal of Molecular Sciences, 12*(4), 2158–2173. doi: 10.3390/ijms12042158.

Huang, Z.-M., Zhang, Y. Z., Kotaki, M., and Ramakrishna, S. (2003). A review on polymer nanofibers by electrospinning and their applications in nanocomposites. *Composites Science and Technology, 63*(15), 2223–2253. doi: 10.1016/s0266-3538(03)00178-7.

Hughes, G. A. (2005). Nanostructure-mediated drug delivery. *Nanomedicine, 1*(1), 22–30. doi: 10.1016/j.nano.2004.11.009.

Jain, E., Scott, K. M., Zustiak, S. P., and Sell, S. A. (2015). Fabrication of polyethylene glycol-based hydrogel microspheres through electrospraying. *Macromolecular Materials and Engineering, 300*, 823–835.

Jain, A. K., Sood, V., Bora, M., Vasita, R., and Katti, D. S. (2014). Electrosprayed inulin microparticles for microbiota triggered targeting of colon. *Carbohydrate Polymers, 112*(0), 225–234. doi: 10.1016/j.carbpol.2014.05.087.

Jalili, R., Hosseini, S. A., and Morshed, M. (2005). The effects of operating parameters on the morphology of electrospun polyacrilonitrile nanofibres. *Iranian Polymer Journal, 14*(12), 1074–1081.

Jarusuwannapoom, T., Hongrojjanawiwat, W., Jitjaicham, S., Wannatong, L., Nithitanakul, M., Pattamaprom, C., Koombhongsed, P., Rangkupan, R., and Supaphol, P. (2005). Effect of solvents on electro-spinnability of polystyrene solutions and morphological appearance of resulting electrospun polystyrene fibers. *European Polymer Journal, 41*(3), 409–421. doi: 10.1016/j.eurpolymj.2004.10.010.

Jaworek, A. (2007). Micro- and nanoparticle production by electrospraying. *Powder Technology, 176*(1), 18–35. doi: 10.1016/j.powtec.2007.01.035.

Jaworek, A. and Krupa, A. (1999). Classification of the modes of EHD spraying. *Journal of Aerosol Science, 30*(7), 873–893.

Jaworek, A., Krupa, A., Lackowski, M., Sobczyk, A., Czech, T., Ramakrishna, S., Sundarrajan, S., and Pliszka, D. (2009). Nanocomposite fabric formation by electrospinning and electrospraying technologies. *Journal of Electrostatics, 67*(2), 435–438.

Jaworek, A. and Sobczyk, A. T. (2008). Electrospraying route to nanotechnology: An overview. *Journal of Electrostatics, 66*(3–4), 197–219. doi: 10.1016/j.elstat.2007.10.001.

Karim, M. R., Lee, H. W., Kim, R., Ji, B. C., Cho, J. W., Son, T. W., Oh, W., and Yeum, J. H. (2009). Preparation and characterization of electrospun pullulan/montmorillonite nanofiber mats in aqueous solution. *Carbohydrate Polymers, 78*(2), 336–342. doi: 10.1016/j.carbpol.2009.04.024.

Kaushik, P., Dowling, K., Barrow, C. J., and Adhikari, B. (2015). Microencapsulation of omega-3 fatty acids: A review of microencapsulation and characterization methods. *Journal of Functional Foods, 19*(Part B), 868–881. doi: 10.1016/j.jff.2014.06.029.

Kayaci, F. and Uyar, T. (2012). Encapsulation of vanillin/cyclodextrin inclusion complex in electrospun polyvinyl alcohol (PVA) nanowebs: Prolonged shelf-life and high temperature stability of vanillin. *Food Chemistry, 133*(3), 641–649. doi: 10.1016/j.foodchem.2012.01.040.

Kessick, R. and Tepper, G. (2004). Microscale polymeric helical structures produced by electrospinning. *Applied Physics Letters, 84*(23), 4807–4809.

Khan, M. K. I., Schutyser, M. A. I., Schroën, K., and Boom, R. (2012). The potential of electrospraying for hydrophobic film coating on foods. *Journal of Food Engineering, 108*(3), 410–416. doi: 10.1016/j.jfoodeng.2011.09.005.

Khayata, N., Abdelwahed, W., Chehna, M., Charcosset, C., and Fessi, H. (2012). Preparation of vitamin E loaded nanocapsules by the nanoprecipitation method: From laboratory scale to large scale using a membrane contactor. *International Journal of Pharmaceutics, 423*(2), 419–427.

Kong, L. and Ziegler, G. R. (2012). Role of molecular entanglements in starch fiber formation by electrospinning. *Biomacromolecules, 13*(8), 2247–2253. doi: 10.1021/bm300396j.

Koombhongse, S., Liu, W., and Reneker, D. H. (2001). Flat polymer ribbons and other shapes by electrospinning. *Journal of Polymer Science Part B: Polymer Physics, 39*(21), 2598–2606. doi: 10.1002/polb.10015.

Korehei, R. and Kadla, J. F. (2014). Encapsulation of T4 bacteriophage in electrospun poly(ethylene oxide)/cellulose diacetate fibers. *Carbohydrate Polymers, 100*(0), 150–157. doi: 10.1016/j.carbpol.2013.03.079.

Koski, A., Yim, K., and Shivkumar, S. (2004). Effect of molecular weight on fibrous PVA produced by electrospinning. *Materials Letters, 58*(3–4), 493–497. doi: 10.1016/s0167-577x(03)00532-9.

Kuan, C.-Y., Yuen, K.-H., Bhat, R., and Liong, M.-T. (2011). Physicochemical characterization of alkali treated fractions from corncob and wheat straw and the production of nanofibres. *Food Research International, 44*(9), 2822–2829. doi: 10.1016/j.foodres.2011.06.023.

La Mora, D. and Fernandez, J. (1994). The current emitted by highly conducting Taylor cones. *Journal of Fluid Mechanics, 260*, 155–184.

Lee, K. H., Kim, H. Y., Bang, H. J., Jung, Y. H., and Lee, S. G. (2003). The change of bead morphology formed on electrospun polystyrene fibers. *Polymer, 44*(14), 4029–4034. doi: 10.1016/s0032-3861(03)00345-8.

Li, Y., Lim, L. T., and Kakuda, Y. (2009). Electrospun zein fibers as carriers to stabilize (−)-epigallocatechin gallate. *Journal of Food Science, 74*(3), C233–C240. doi: 10.1111/j.1750-3841.2009.01093.x.

Liechty, W. B., Kryscio, D. R., Slaughter, B. V., and Peppas, N. A. (2010). Polymers for drug delivery systems. *Annual Review of Chemical and Biomolecular Engineering, 1*, 149–173. doi: 10.1146/annurev-chembioeng-073009-100847.

Lim, C. T., Tan, E. P. S., and Ng, S. Y. (2008). Effects of crystalline morphology on the tensile properties of electrospun polymer nanofibers. *Applied Physics Letters, 92*(14), 141908.

Lin, T., Wang, H., Wang, H., and Wang, X. (2004). The charge effect of cationic surfactants on the elimination of fibre beads in the electrospinning of polystyrene. *Nanotechnology, 15*(9), 1375–1381.

Lin, T., Wang, H., Wang, H., and Wang, X. (2005). Effects of polymer concentration and cationic surfactant on the morphology of electrospun polyacrylonitrile nanofibres. *Journal of Materials Science and Technology, 21*, 9–12.

Lin, Y., Yao, Y., Yang, X., Wei, N., Li, X., Gong, P., Li, R., and Wu, D. (2008). Preparation of poly(ether sulfone) nanofibers by gas-jet/electrospinning. *Journal of Applied Polymer Science, 107*(2), 909–917. doi: 10.1002/app.26445.

Linnemann, B., Rana, A., and Gries, T. (2005). Electrospinning: Nanofiber from polycaprolactone (PCL) *Chemical Fibres International, 55*(6), 370–372.

López-Rubio, A. and Lagaron, J. M. (2012). Whey protein capsules obtained through electrospraying for the encapsulation of bioactives. *Innovative Food Science and Emerging Technologies, 13*(0), 200–206. doi: 10.1016/j.ifset.2011.10.012.

López-Rubio, A., Sanchez, E., Wilkanowicz, S., Sanz, Y., and Lagaron, J. M. (2012). Electrospinning as a useful technique for the encapsulation of living bifidobacteria in food hydrocolloids. *Food Hydrocolloids, 28*(1), 159–167. doi: 10.1016/j.foodhyd.2011.12.008.

Lu, C., Chen, P., Li, J., and Zhang, Y. (2006). Computer simulation of electrospinning. Part I. Effect of solvent in electrospinning. *Polymer, 47*(3), 915–921. doi: 10.1016/j.polymer.2005.11.090.

Lubambo, A. F., de Freitas, R. A., Sierakowski, M.-R., Lucyszyn, N., Sassaki, G. L., Serafim, B. M., and Saul, C. K. (2013). Electrospinning of commercial guar-gum: Effects of purification and filtration. *Carbohydrate Polymers, 93*(2), 484–491. doi: 10.1016/j.carbpol.2013.01.031.

Luo, C. J., Loh, S., Stride, E., and Edirisinghe, M. (2012). Electrospraying and electrospinning of chocolate suspensions. *Food and Bioprocess Technology, 5*(6), 2285–2300. doi: 10.1007/s11947-011-0534-6.

Manee-in, J., Nithitanakul, M., and Supaphol, P. (2006). Effects of solvent properties, solvent system, electrostatic field strength, and inorganic salt addition on electrospun polystyrene fibres. *Iranian Polymer Journal, 15*(4), 341–354.

Mascheroni, E., Fuenmayor, C. A., Cosio, M. S., Di Silvestro, G., Piergiovanni, L., Mannino, S., and Schiraldi, A. (2013). Encapsulation of volatiles in nanofibrous polysaccharide membranes for humidity-triggered release. *Carbohydrate Polymers, 98*(1), 17–25. doi: 10.1016/j.carbpol.2013.04.068.

Mazoochi, T. and Jabbari, V. (2011). Chitosan nanofibrous scaffold fabricated via electrospinning: The effect of processing parameters on the nanofiber morphology. *International Journal of Polymer Analysis and Characterization, 16*(5), 277–289. doi: 10.1080/1023666x.2011.587943.

McClements, D. J. and Li, Y. (2010). Structured emulsion-based delivery systems: Controlling the digestion and release of lipophilic food components. *Advances in Colloid and Interface Science, 159*(2), 213–228.

Megelski, S., Stephens, J. S., Chase, D. B., and Rabolt, J. F. (2002). Micro- and nanostructured surface morphology on electrospun polymer fibers. *Macromolecules, 35*(22), 8456–8466. doi: 10.1021/ma020444a.

Mo, X. M., Xu, C. Y., Kotaki, M., and Ramakrishna, S. (2004). Electrospun P(LLA-CL) nanofiber: A biomimetic extracellular matrix for smooth muscle cell and endothelial cell proliferation. *Biomaterials, 25*(10), 1883–1890. doi: 10.1016/j.biomaterials.2003.08.042.

Mohr, J. G. and Rowe, W. P. (1978). *Fiber Glass*. New York: Van Nostrand Reinhold.

Mozafari, M. R., Flanagan, J., Matia-Merino, L., Awati, A., Omri, A., Suntres, Z. E., and Singh, H. (2006). Recent trends in the lipid-based nanoencapsulation of antioxidants and their role in foods. *Journal of the Science of Food and Agriculture, 86*(13), 2038–2045. doi: 10.1002/jsfa.2576.

Munir, M. M., Suryamas, A. B., Iskandar, F., and Okuyama, K. (2009). Scaling law on particle-to-fiber formation during electrospinning. *Polymer, 50*(20), 4935–4943. doi: 10.1016/j.polymer.2009.08.011.

Neo, Y. P., Ray, S., Jin, J., Gizdavic-Nikolaidis, M., Nieuwoudt, M. K., Liu, D., and Quek, S. Y. (2013). Encapsulation of food grade antioxidant in natural biopolymer by electrospinning technique: A physicochemical study based on zein–gallic acid system. *Food Chemistry, 136*(2), 1013–1021. doi: 10.1016/j.foodchem.2012.09.010.

Nieuwland, M., Geerdink, P., Brier, P., van den Eijnden, P., Henket, J. T. M. M., Langelaan, M. L. P., Stroeksb, N., van Deventer, H. C., and Martin, A. H. (2013). Food-grade electrospinning of proteins. *Innovative Food Science and Emerging Technologies, 20*(0), 269–275. doi: 10.1016/j.ifset.2013.09.004.

Ogata, S., Hara, Y., and Shinohara, H. (1978). Break-up mechanism of a charged liquid jet. *International Chemical Engineering, 18*(3), 482–488.

Pérez-Masiá, R., Lagaron, J., and López-Rubio, A. (2014a). Development and optimization of novel encapsulation structures of interest in functional foods through electrospraying. *Food and Bioprocess Technology, 7*(11), 3236–3245. doi: 10.1007/s11947-014-1304-z.

Pérez-Masiá, R., Lagaron, J. M., and López-Rubio, A. (2014b). Surfactant-aided electrospraying of low molecular weight carbohydrate polymers from aqueous solutions. *Carbohydrate Polymers, 101*(0), 249–255. doi: 10.1016/j.carbpol.2013.09.032.

Pham, Q. P., Sharma, U., and Mikos, A. G. (2006). Electrospinning of polymeric nanofibers for tissue engineering applications: A review. *Tissue Engineering, 12*(5), 1197–1211. doi: 10.1089/ten.2006.12.1197.

Pitkowski, A., Durand, D., and Nicolai, T. (2008). Structure and dynamical mechanical properties of suspensions of sodium caseinate. *Journal of Colloid and Interface Science, 326*(1), 96–102. doi: 10.1016/j.jcis.2008.07.003.

Ramakrishna, S. (2005). *An Introduction to Electrospinning and Nanofibers*. Hackensack, NJ: World Scientific.

Rayleigh, L. (1882). XX. On the equilibrium of liquid conducting masses charged with electricity. *Philosophical Magazine Series 5, 14*(87), 184–186. doi: 10.1080/14786448208628425.

Reneker, D. H., Yarin, A. L., Fong, H., and Koombhongse, S. (2000). Bending instability of electrically charged liquid jets of polymer solutions in electrospinning. *Journal of Applied Physics, 87*(9), 4531–4547.

Rezaei, A., Nasirpour, A., and Fathi, M. (2015). Application of cellulosic nanofibers in food science using electrospinning and its potential risk. *Comprehensive Reviews in Food Science and Food Safety*, doi: 10.1111/1541-4337.12128.

Rutledge, G. C., Li, Y., Fridrikh, S., Warner, S. B., Kalayci, V. E., and Patra, P. (2000). Electrostatic spinning and properties of ultrafine fibers. *National Textile Center Annual Report (M98-D01)*, National Textile Center, pp. 1–10.

Sáiz-Abajo, M.-J., González-Ferrero, C., Moreno-Ruiz, A., Romo-Hualde, A., and González-Navarro, C. J. (2013). Thermal protection of β-carotene in re-assembled casein micelles during different processing technologies applied in food industry. *Food Chemistry, 138*(2), 1581–1587.

Salalha, W., Kuhn, J., Dror, Y., and Zussman, E. (2006). Encapsulation of bacteria and viruses in electrospun nanofibres. *Nanotechnology, 17*(18), 4675.

Schiffman, J. D. and Schauer, C. L. (2008). A review: Electrospinning of biopolymer nanofibers and their applications. *Polymer Reviews, 48*(2), 317–352. doi: 10.1080/15583720802022182.

Shenoy, S. L., Bates, W. D., Frisch, H. L., and Wnek, G. E. (2005). Role of chain entanglements on fiber formation during electrospinning of polymer solutions: Good solvent, non-specific polymer-polymer interaction limit. *Polymer, 46*(10), 3372–3384. doi: 10.1016/j.polymer.2005.03.011.

Shenoy, S. L., Bates, W. D., and Wnek, G. (2005). Correlations between electrospinnability and physical gelation. *Polymer, 46*(21), 8990–9004. doi: 10.1016/j.polymer.2005.06.053.

Shin, Y. M., Hohman, M. M., Brenner, M. P., and Rutledge, G. C. (2001). Electrospinning: A whipping fluid jet generates submicron polymer fibers. *Applied Physics Letters, 78*(8), 1149–1151.

Smith, K., Alexander, M., and Stark, J. (2006). Voltage effects on the volumetric flow rate in cone-jet mode electrospraying. *Journal of Applied Physics, 99*(6), 064909.

Songchotikunpan, P., Tattiyakul, J., and Supaphol, P. (2008). Extraction and electrospinning of gelatin from fish skin. *International Journal of Biological Macromolecules, 42*(3), 247–255. doi: 10.1016/j.ijbiomac.2007.11.005.

Sousa, A. M. M., Souza, H. K. S., Uknalis, J., Liu, S.-C., Gonçalves, M. P., and Liu, L. (2015). Electrospinning of agar/PVA aqueous solutions and its relation with rheological properties. *Carbohydrate Polymers, 115*(0), 348–355. doi: 10.1016/j.carbpol.2014.08.074.

Spasova, M., Mincheva, R., Paneva, D., Manolova, N., and Rashkov, I. (2006). Perspectives on: Criteria for complex evaluation of the morphology and alignment of electrospun polymer nanofibers. *Journal of Bioactive and Compatible Polymers, 21*(5), 465–479. doi: 10.1177/0883911506068495.

Stanger, J., Tucker, N., Kirwan, K., and Staiger, M. P. (2009). Effect of charge density on the Taylor cone in electrospinning. *International Journal of Modern Physics B, 23*(06), 1956–1961. doi: 10.1142/S0217979209061895.

Stijnman, A. C., Bodnar, I., and Hans Tromp, R. (2011). Electrospinning of food-grade polysaccharides. *Food Hydrocolloids, 25*(5), 1393–1398. doi: 10.1016/j.foodhyd.2011.01.005.

Subramanian, C., Weiss, R. A., and Shaw, M. T. (2010). Electrospinning and characterization of highly sulfonated polystyrene fibers. *Polymer, 51*(9), 1983–1989. doi: 10.1016/j.polymer.2010.02.052.

Sullivan, S. T., Tang, C., Kennedy, A., Talwar, S., and Khan, S. A. (2014). Electrospinning and heat treatment of whey protein nanofibers. *Food Hydrocolloids, 35*(0), 36–50. doi: 10.1016/j.foodhyd.2013.07.023.

Sun, X.-Z., Williams, G. R., Hou, X.-X., and Zhu, L.-M. (2013). Electrospun curcumin-loaded fibers with potential biomedical applications. *Carbohydrate Polymers, 94*(1), 147–153. doi: 10.1016/j.carbpol.2012.12.064.

Taepaiboon, P., Rungsardthong, U., and Supaphol, P. (2007). Vitamin-loaded electrospun cellulose acetate nanofiber mats as transdermal and dermal therapeutic agents of vitamin A acid and vitamin E. *European Journal of Pharmaceutics and Biopharmaceutics, 67*(2), 387–397. doi: 10.1016/j.ejpb.2007.03.018.

Tan, S. H., Inai, R., Kotaki, M., and Ramakrishna, S. (2005). Systematic parameter study for ultra-fine fiber fabrication via electrospinning process. *Polymer, 46*(16), 6128–6134. doi: 10.1016/j.polymer.2005.05.068.

Taylor, G. (1964). Disintegration of water drops in an electric field. *Proceedings of the Royal Society of London: Series A, Mathematical and Physical Sciences, 280*(1382), 383–397.

Tehrani, A. H., Zadhoush, A., Karbasi, S., and Khorasani, S. N. (2010). Experimental investigation of the governing parameters in the electrospinning of poly(3-hydroxybutyrate) scaffolds: Structural characteristics of the pores. *Journal of Applied Polymer Science, 118*(5), 2682–2689. doi: 10.1002/app.32620.

Teo, W. E. and Ramakrishna, S. (2006). A review on electrospinning design and nanofibre assemblies. *Nanotechnology, 17*(14), 89–106.

Theron, S. A., Zussman, E., and Yarin, A. L. (2004). Experimental investigation of the governing parameters in the electrospinning of polymer solutions. *Polymer, 45*(6), 2017–2030. doi: 10.1016/j.polymer.2004.01.024.

Thompson, C. J., Chase, G. G., Yarin, A. L., and Reneker, D. H. (2007). Effects of parameters on nanofiber diameter determined from electrospinning model. *Polymer, 48*(23), 6913–6922. doi: 10.1016/j.polymer.2007.09.017.

Tomita, Y., Ishibashi, Y., and Yokoyama, T. (1986). Fundamental studies on an electrostatic ink jet printer: 1st report, electrostatic drop formation. *Bulletin of JSME, 29*(257), 3737–3743.

Torres-Giner, S., Martinez-Abad, A., Ocio, M. J., and Lagaron, J. M. (2010). Stabilization of a nutraceutical omega-3 fatty acid by encapsulation in ultrathin electrosprayed zein prolamine. *Journal of Food Science, 75*(6), N69–N79. doi: 10.1111/j.1750-3841.2010.01678.x.

Treloar, L. R. G. (1970). *Introduction to Polymer Science*. London, UK: Wykeham Publications.

Uyar, T. and Besenbacher, F. (2008). Electrospinning of uniform polystyrene fibers: The effect of solvent conductivity. *Polymer, 49*(24), 5336–5343. doi: 10.1016/j.polymer.2008.09.025.

Van der Schueren, L., De Schoenmaker, B., Kalaoglu, Ö. I., and De Clerck, K. (2011). An alternative solvent system for the steady state electrospinning of polycaprolactone. *European Polymer Journal, 47*(6), 1256–1263. doi: 10.1016/j.eurpolymj.2011.02.025.

Walczak, Z. K. (2002). *Processes of Fibre Formation*. Oxford, UK: Elsevier.

Wannatong, L., Sirivat, A., and Supaphol, P. (2004). Effects of solvents on electrospun polymeric fibers: Preliminary study on polystyrene. *Polymer International, 53*(11), 1851–1859. doi: 10.1002/pi.1599.

Wongsasulak, S., Kit, K. M., McClements, D. J., Yoovidhya, T., and Weiss, J. (2007). The effect of solution properties on the morphology of ultrafine electrospun egg albumen–PEO composite fibers. *Polymer, 48*(2), 448–457. doi: 10.1016/j.polymer.2006.11.025.

Wongsasulak, S., Patapeejumruswong, M., Weiss, J., Supaphol, P., and Yoovidhya, T. (2010). Electrospinning of food-grade nanofibers from cellulose acetate and egg albumen blends. *Journal of Food Engineering, 98*(3), 370–376. doi: 10.1016/j.jfoodeng.2010.01.014.

Wongsasulak, S., Pathumban, S., and Yoovidhya, T. (2014). Effect of entrapped α-tocopherol on mucoadhesivity and evaluation of the release, degradation, and swelling characteristics of zein–chitosan composite electrospun fibers. *Journal of Food Engineering, 120*(0), 110–117. doi: 10.1016/j.jfoodeng.2013.07.028.

Xu, X., Jiang, L., Zhou, Z., Wu, X., and Wang, Y. (2012). Preparation and properties of electrospun soy protein isolate/polyethylene oxide nanofiber membranes. *ACS Applied Materials and Interfaces, 4*(8), 4331–4337. doi: 10.1021/am300991e.

Yamashita, Y., Tanaka, A., and Ko, F. (2006). *Electrospinning for the industrial nanofiber technology.* Paper presented at the *Proceedings of International Fiber Conference*, Seoul, Korea.

Yang, Q., Li, Z., Hong, Y., Zhao, Y., Qiu, S., Wang, C., and Wei, Y. (2004). Influence of solvents on the formation of ultrathin uniform poly(vinyl pyrrolidone) nanofibers with electrospinning. *Journal of Polymer Science Part B: Polymer Physics, 42*(20), 3721–3726. doi: 10.1002/polb.20222.

Yang, J.-M., Zha, L.-S., Yu, D.-G., and Liu, J. (2013). Coaxial electrospinning with acetic acid for preparing ferulic acid/zein composite fibers with improved drug release profiles. *Colloids and Surfaces B: Biointerfaces, 102*(0), 737–743. doi: 10.1016/j.colsurfb.2012.09.039.

Yarin, A., Koombhongse, S., and Reneker, D. (2001). Bending instability in electrospinning of nanofibers. *Journal Applied Physics, 89*(5), 3018–3026.

Ying, Y., Zhidong, J., Qiang, L., and Zhicheng, G. (2006). Experimental investigation of the governing parameters in the electrospinning of polyethylene oxide solution. *Dielectrics and Electrical Insulation, IEEE Transactions on, 13*(3), 580–585.

Yördem, O. S., Papila, M., and Menceloglu, Y. Z. (2008). Effects of electrospinning parameters on polyacrylonitrile nanofiber diameter: An investigation by response surface methodology. *Materials and Design, 29*(1), 34–44. doi: 10.1016/j.matdes.2006.12.013.

Yuan, X., Zhang, Y., Dong, C., and Sheng, J. (2004). Morphology of ultrafine polysulfone fibers prepared by electrospinning. *Polymer International, 53*(11), 1704–1710. doi: 10.1002/pi.1538.

Zhang, W., Chen, M., Zha, B., and Diao, G. (2012). Correlation of polymer-like solution behaviors with electrospun fiber formation of hydroxypropyl-β-cyclodextrin and the adsorption study on the fiber. *Physical Chemistry Chemical Physics, 14*(27), 9729–9737.

Zhang, S. and Kawakami, K. (2010). One-step preparation of chitosan solid nanoparticles by electrospray deposition. *International Journal of Pharmaceutics, 397*(1), 211–217.

Zhang, C., Yuan, X., Wu, L., Han, Y., and Sheng, J. (2005). Study on morphology of electrospun poly(vinyl alcohol) mats. *European Polymer Journal, 41*(3), 423–432. doi: 10.1016/j.eurpolymj.2004.10.027.

Zhao, S., Wu, X., Wang, L., and Huang, Y. (2004). Electrospinning of ethyl–cyanoethyl cellulose/tetrahydrofuran solutions. *Journal of Applied Polymer Science, 91*(1), 242–246. doi: 10.1002/app.13196.

Zhuo, H., Hu, J., and Chen, S. (2008). Electrospun polyurethane nanofibres having shape memory effect. *Materials Letters, 62*(14), 2074–2076. doi: 10.1016/j.matlet.2007.11.018.

Zong, X., Kim, K., Fang, D., Ran, S., Hsiao, B. S., and Chu, B. (2002). Structure and process relationship of electrospun bioabsorbable nanofiber membranes. *Polymer, 43*(16), 4403–4412. doi: 10.1016/s0032-3861(02)00275-6.

Index

Note: Page numbers followed by f and t refer to figures and tables, respectively.

A

AA. *See* Ascorbic acid (AA)
Absorption enhancers, 168–169
ABTS·+ radical, 136–137
Activation volume (V_a), 125–126
Aerobic oxidation pathway, 113f
Agglomerates, 68–70, 68f
Alginate, sodium, 246–252
    protector effect of, 249
    on starch solutions, 247, 248f
Amadori compounds, 203
Amaranth protein isolate (API), 278
Ambient parameters, electrospinning process, 274–275
Anaerobic pathway, 87
Analyses of variance (ANOVA), 137, 139, 141
Andes berry (*Rubus glaucus* Benth) pulp, 190
Animal models, 171
Anomalous diffusion, 162–163
ANOVA. *See* Analyses of variance (ANOVA)
Anthocyanidins, 36, 36f, 224
Anthocyanin, 36
    HPP-treated fruit purée, 115, 116f
    hydrostatic HPP and HPT effects, 224–225, 226f
Antioxidant(s)
    activity, 231–234, 232t
    capacity, HPP-treated fruit purée, 119t, 121
    defined, 61
    encapsulation of
        advantages and disadvantages, 68–71
        application of, 71–72, 74–75
        considerations about source, 62–66
        in food system, 66–68
        future trends, 75
        level in food, 67–68
        mixtures of, 62
Anti-pernicious anemia factor, 85
Aqueous polysaccharide solution, 268
Aromas, volatile, 189
Arrhenius equation, 88–89, 125
Ascorbic acid (AA), 40, 85, 96, 134, 196–197, 205–207. *See also* Vitamin C (ascorbic acid)
    for calibration curve, 136
    degradation, 91
        under chemical oxidation, 87–89
        electric field frequency effect, 92–93, 93f
        kinetics, 89
        during sonication, 138–139
    HPP fruit purée, 111–113
        change in, 112t–113t
        content during storage, 122, 124f
        degradation, 126
    impact of USNs on, 137–139, 138t
    interactions between free radicals and, 139
Aspergillus niger, inactivation, 16
Autoxidation, polyunsaturated lipids, 63

B

Bacillus subtilis, inactivation, 16
Bacteriocin, 281
Baking, IR, 6, 7t
Batch-type infrared heater, 21, 21f
Batch-wise HPP system, 107–108, 108f. *See also* High pressure process (HPP)
β-carotene, 68, 117, 252, 254
    bioaccessibility of, 242, 246
    bioavailability, 245
    degrading effect on, 206–207
    molecule, 227, 227f
B-complex vitamins, 40–41, 82, 84
Bead fiber formation, 265–266, 265f, 269
Beriberi, 40. *See also* Vitamin B1 (thiamine)
Berry number, 275
BHT. *See* Butylated hydroxytoluene (BHT)
BI. *See* Browning index (BI)
Bifidobacterium strains, 278, 281
Binary biopolymer system, 253
Bioactive compound, 32–43
    application of microencapsulated, 66
    characteristics of delivery system for, 151
    chemical stability of, 156
    classes and fruit sources, 107f
    defined, 133
    effect of HPP and HPT treatments, 235
    electrospinning for, 259–281
    encapsulation in digestion, 242–255
        case study based on, 246–252
        chewing, 242–244

293

Index

Bioactive compound (*Continued*)
    food structure and its protective role, 244–246
        microcapsules in food matrices, 254–255
        microencapsulation of food nutrients, 252–253
        release of nutrient, 253–254
    encapsulation of, 150, 157t, 158f, 160t, 171
    enrichment, 51–53
    entrapped, 166
    factors affecting preservation in vegetable products, 234–235
    in fruits, 196–199
        carotenoids, 197–198
        phenolic compounds, 198–199
        purées, 106
        vitamin C, 196–197
    HPP-treated fruit purée, 107–108
        degradation after, 124–126
        effects, 110–121
        stability under, 108–110
        storage stability in, 121–124, 123t
    lipophilic, 167
    minerals, 41–43
        calcium, 42
        iron, 43
        magnesium, 42–43
        zinc, 43
    multilayer nanocapsules as carriers of, 156–160
        EE and loading capacity, 159
        encapsulation techniques, 157–159
    multilayers impact in release of, 161–165
    OH impact in protection of, 54–55
    oral bioavailability of, 171
    osmotic pretreatment effect, 199–202
    in plant, 106
    polyphenols, 32–38
        classification, 33f
        flavonoids, 33–34, 36–38
        nonflavonoids, 33–34
        structure, 33f
    relationship between TAC and, 143–144
    release systems, 161, 162t
    release-tailored, 161
    stability, 108–110, 121–124, 123t, 156
    uptake/absorption of, 168–169
    vegetable matrix as carriers of, 51–54
        enrichment with bioactive compounds, 51–53
        enzymatic inactivation, 53–54
    vitamins, 38–41
        B complex, 40–41
        biotin, 39, 41
        choline, 39, 41
        classification of, 39f
        liposoluble vitamins (A, D, E, and K), 39–40
        vitamin C, 39–40
        water-soluble, 157
Biochemical reactions, 125t
Biopolymer, 253, 275
    characteristics, 277
    food-grade, 150
    mixture, 253
    molecular interaction, 251
    properties of, 247, 252–253
Biotin, 39, 41, 84
Blackbody radiation
    laws for
        Planck's law, 4–5
        Stefan–Boltzmann law, 5
        Wien's displacement law, 5
    spectral characteristics of, 4, 4f
Blanching, IR, 5–6
Block freeze concentration, 187–188, 190
Bottom-up self-assembly, 253
Browning index (BI), 54, 54f
Butylated hydroxyanisol (BHA), 63, 63f, 74–75
Butylated hydroxytoluene (BHT), 63, 63f, 72, 74–75

C

Calciferol (vitamin D), 39, 83, 260
Calcium, 42, 202
Calcium alginate, 70, 72, 243
Capillary wave break up, 267
Carbohydrate-based encapsulation, 279–280
Carbohydrates, 84, 279
    hydrophilic, 69
Carotenes, 197, 227
Carotenoids, 197–198, 227–229
    in chromoplasts, 227
    effects on, 198
    factors affect, 206
    functional groups, 197
    HPP-treated fruit purée, 115–118, 116t, 122
    hydrostatic high pressure treatment on, 227–229
    structure, 227
    vitamin A, 197
    xanthophylls, 227
Carrot structure (cell wall), 242, 244
Carry-through property, antioxidants, 63
Case II transport, 162–163
Casein protein, 278
Cavitations, USN, 132
Cell disruption, 242, 245
Cellulose acetate (CA) fibers, 270, 274–275, 279
Centrifugation method, 152, 188
Cereals, dehydration of, 14t–15t

Index

Chewing in digestion, 242–244
Chitosan, 70–71, 164, 169
  water-soluble, 154
Choline, 39, 41
CI. *See* Complex index (CI)
Cobalamin (vitamin B12), 39, 41, 85
Co-encapsulation, 71
Coffee bean roaster, 8
Collagen protein, 85, 197
Colloidal templates, 153–156
  hollow nanocapsules, 153–154
  liposomes, 154
  nanoemulsions, 154–155
Columbic repulsion force, 262
Complete block cryoconcentration technology, 187–188
Complex coacervation, 74–75, 252
Complex index (CI), 248–249, 249t, 252
Concentrated juices, quality of, 190
Concentration of liquid foods, freeze, 184, 184f
Condensed tannins, 36
Cone-jet mode, 262–263, 276
Consistency index (K), starch, 249
Consumption of functional foods, 150
Continuous-flow OH process, 49–50, 50f
Continuous process, 132
Controlled release technologies, 161
Conventional drying, 3, 8
Conventional equipment, 132
Conventional heating, 3, 3f, 50–51, 86, 93f
Core–shell microcapsule, 68f
Corrin nucleus, 85
Cryoconcentration process, 186
  complete block, 187–188
  partial block, 187
Cyanidin, 36, 225, 225f
Cytotoxicity, 170–171

D

Deformation–relaxation phenomena (DRP), 47, 48f
Degradation after HPP, bioactives, 124–126
  AA, 126
  anthocyanin, 115
Dehydration, 6
  of cereals, grains, and seeds, 14t–15t
  of fruits, 9t–10t
  osmotic, 196
  partial, 47
  of vegetables, 11t–13t, 18
Dehydroascorbic acid (DHAA), 206–207
Deterioration processes, 44, 66
Dextran sulfate, 154
DHA, 254
Diffusion of bioactive compounds, 162
Diketogulonic acid hydrolysis, 87, 206

1,1-Diphenyl-2-picryl-hydrazyl (DPPH), 201, 205, 207
Discontinuous process, 132
Diverse functions, vitamins, 82
DPPH. *See* 1,1-Diphenyl-2-picryl-hydrazyl (DPPH)
Drag force, 261–262
Drying air properties, effect of, 202–208
  air velocity, 207
  drying temperature, 204–207
  humidity, 208
Drying, IR, 8

E

EE. *See* Encapsulation efficiency (EE)
Egg albumen (EA), 278–279
Electrical conductivity (EC), food, 46
Electric field, 46, 261–262
  in cone-jet mode, 276
  frequency, 92–93
  strength effect in
    OH, 90–92
    PET, 94–96
  treatments, vitamins
    degradation mechanisms during, 85–89
    stability in fruits pulp during, 89–98
Electrochemical reactions, 87–88, 96
Electrohydrodynamic fluid, 262
Electromagnetic wave, 4
  spectrum, 2f
Electronic generation unit, 132
Electropermeabilization, 45, 53
Electrophoretic method, 152
Electroporation, 45, 46f, 81–82
Electrospinning, 261–264, 280
  ambient conditions, 274–275
  for bioactive compounds, 259–281
  carbohydrate-based encapsulation material, 279–280
  defined, 261
  electrospray *vs.*, 275–276
  future trends, 281–283
  helical structures, 264
  number, 274
  polymer solution properties, 266–270
  principle, 262
  probiotics encapsulation by, 280–281
  processing conditions (parameters), 266t, 270–274
    flow rate, 272
    spinneret orifice diameter, 273
    TCD, 273–274
    voltage, 270–271
  process of, 261f, 263–266, 265f, 266t

Index

Electrospinning (Continued)
  protein-based encapsulation material, 277–279
  safe delivery system, 276–281
    carbohydrate-based encapsulation materials, 279–280
    factors to, 276–277
    particle characteristics, 277
    probiotics encapsulation, 280–281
    protein-based encapsulation materials, 277–279
Electrospraying, 261–264
  electrospinning and, 275–276
  encapsulation through, 281
  of polymer solutions, 262–263
  principle, 275
Electrospun fibers, 264–265, 269
  CA, 274–275, 279
  core/shell-structured, 281
  in flow rate, 272
  nanofibers application, 282
  polystyrene, 266
  production of, 271
  solvent effect on, 270
Electrostatic force, 253, 261, 262f, 275
Electrostatic interactions, 156, 159, 165
Elution, 136
Emerging technologies, food processing, 32, 44, 45f, 51, 81
Emulsification, 155, 253–254
Encapsulated polyphenols, 62
Encapsulation, 276, 278
  of antioxidants, 61–75
    advantages and disadvantages, 68–71
    application of, 71–72, 73t, 74–75
    considerations about source, 62–66
    as food additives, 66–67
    in food system, 66–68, 71–75
    future trends, 75
    as health supplement, 67–68, 67f
    natural, 64–66, 64f
    synthetic, 62–64, 63f
  of bioactive compounds, 150, 157t, 158f, 160t, 171, 280
    during digestion, 242–255
  carbohydrate-based, 279–280
  in multilayer nanocapsules, 157–159
  of probiotics, 280–281
  protein-based, 277–279
  technologies in food industry, 62
Encapsulation efficiency (EE), 70, 70f, 159, 160t
Enrichment with bioactive compounds, 51–53
Entanglement number, 267, 268f
Environmental triggers, 153
Enzymatic browning, 53, 197, 199, 203
Enzymatic digestion stability on PDS, 168
Enzymatic inactivation, 53–54
Enzyme inhibitors in nanocapsules, 169

EPA, 254
Epigallocatechin gallate (EGCG), encapsulated antioxidants, 74, 279
Epithelium, 167–168
Escherichia coli, 40, 281
  inactivation, 16
European Food Safety Authority (EFSA), 171
Evaporation, freeze concentration, 189–190
Eyring equation, 126

F

Far-IR
  dry-blanching process, 6
  dryer with vacuum extractor, 18–19, 20f
  heater, 22, 22f
  radiation, 3, 9t, 15–16
  thawing, 16
Fat-soluble vitamins, 83–84
Ferric reducing antioxidant power (FRAP), 205
Fibers, 264
  arrangement and crystallinity, 271
  biocompatible composite, 279
  continuous, 262
  diameter, 266–267, 269
  electrospun, 264–265, 265f
  morphology, 266–269
  polymer, 268–269
Fickian diffusion, 162–164
Film freeze concentration method, 186–187
Filtration method, 152
Fish and flaxseed oils, encapsulated antioxidants, 74
Flavanols (flavan-3-ols), 33, 36–37, 37f
Flavanones, 33, 37
Flavones, 37–38
Flavonoids, 33–34, 36–38, 64, 64f, 204, 224
  anthocyanidins and anthocyanins, 33, 36, 36f
  flavanols, 33, 36–37, 37f
  flavanones, 33, 37
  flavones, 37–38
  flavonols, 33, 37, 38f
  generic structures of, 35f
  HPP-treated fruit purée, 112t–113t, 114–115
  isoflavones, 33, 38
Flavonols, 33, 37, 38f
Flight time, 271
Flow behavior index (n), starch, 249
Flow rate, polymer solution, 272
Folate, 41, 85, 113
Folic acid (vitamin B), 39, 52, 279
Folin–Ciocalteau method, 136
Follicle-associated epithelium (FAE), 168
Food
  additives, antioxidants as, 66–67
  digestion of, 241–242
  formulation, 241

Index

fragmentation, 167
heterogeneous, 254
industry, encapsulation technologies in, 62
nanotechnology, 62
nutrients, microencapsulation of, 252–253
packaging materials, encapsulated antioxidants, 74
perception of, 242, 254
processing, 241
    IR. *See* IR radiation
protein, 253
structure and its role in digestion, 244–246
structure levels, 246
system, encapsulation of antioxidants in, 66–68
    as food additives, 66–67
    as health supplement, 67–68, 67f
Food-engineer processes, 52
Food-grade biopolymers, 150, 278
Free radicals, 65
    oxygen-derived, 65
    scavenging, 66
Freeze concentration technique, 183
    energy consumption of, 183, 189
    evaporation, 189–190
    fundamentals of, 184–186
    future trends, 192
    general claim of, 185
    heat treatment advantages to concentrate liquid foods, 189–190
    maple syrup, 186
    methods of, 186–188
        complete block cryoconcentration, 187–188
        film freeze concentration, 186–187
        partial block cryoconcentration, 187
        PFC, 188
        suspension crystallization, 186
    protection of heat-labile components, 190
    stages of, 185f
    volatile aromas, 189
    yerba mate infusion, 190
Fruit(s)
    dehydration of, 9t–10t
    juices, 133–134
        mixed with milk beverages, 134
    purées, HPP treated
        advantage in, 107, 109
        degradation of bioactive, 124–126
        industrial application, 108
        storage stability of bioactive, 121–124
Functional foods, consumption of, 150
Fusarium proliferatum, inactivation, 16
Fused fiber–fiber intersections, 272–273

G

Galactomannan-based gums, 247
Gastric juice, 167
Gastrointestinal (GI)
    digestion process, 168
    system, 260, 280
    tract, 66, 71, 150, 253
        benefits associated with colon delivery, 166
        features and challenges, 166–168
        microbiota, 167
        multilayer nanocapsules behavior in, 165–171
        parts, 166
Gelatin–gum arabic system, 74–75
Gelatinized starch, 243, 245
Generic structures of flavonoids, 35f
GI. *See* Gastrointestinal (GI)
Goldenberry, antioxidant of, 205
Grains, dehydration of, 14t–15t
Granny Smith apple purée, AA in, 111
    degradation, 126
    storage of, 122
Green tea catechin, encapsulated antioxidants, 74
Guar gum, 245, 247–248, 279
Guava purée, AA in, 111, 113f

H

HDM. *See* Hydrodynamic mechanisms (HDM)
Health supplement, antioxidants as, 67–68, 67f
Heating unit, dryer, 18
Heat-labile components, 183–184
    freeze concentration protection of, 190, 191t
Hielscher UP200S ultrasonic processor, 135
High hydrostatic pressure. *See* High pressure process (HPP)
High pressure process (HPP), 107–108, 220
    application, 220
    batch-wise system, 107–108, 108f
    benefits, 221
    bioactive compound under, 108–126
    comparison between HPP/HPT treatments, 220t
    defined, 107
    effect on bioactive compounds, 110–121, 222–224
        anthocyanin, 115, 116f
        antioxidant capacity, 121
        carotenoids, 115–118
        flavonoids, 114–115
        phenolics, 118–120
        vitamins, 111–114
    industrial application in fruit purée, 108
    pressure–temperature profile, 109f, 114f
    steps, 108, 124
High pressure thermal (HPT) treatment, 220
    benefits, 221–222
    comparison between HPP/HPT treatments, 220t
    effect on bioactive compounds, 222–224
    effect on pumpkin, 235

High-temperature short-time (HTST) process, 43, 80, 86
Hildebrand solubility parameter (δ), 270
Hollow nanocapsules, 153–154
Hot air drying system, 18, 19f
HPP. *See* High pressure process (HPP)
HPT. *See* High pressure thermal (HPT) treatment
Human microbiota, 167
Hydrodynamic mechanisms (HDM), 47–48, 48f
Hydrophilic–hydrophobic interactions, 154
Hydrophilic polymers, 163–164
Hydrostatic high-pressure treatment effects
　on anthocyanin compounds, 224–226
　on carotenoid compounds, 227–229
　on compounds with antioxidant activity, 231–234
　future trends, 235–236
　in liquid and solid foods, 221
　on phenolic compounds, 229–231
Hydroxybenzoic acid, 33–34, 35f
Hydroxycinnamic acid, 33–34, 34f, 204
Hydroxylated aromatic ring, 134

I

Ice crystals, 185–186, 188
Ice nucleus, 184
Inaudible sound waves, 132
Infrared (IR) heating, 1, 3. *See also* IR radiation
　in baking industry, 6
　conventional and, 3, 3f
Innovative functional foods, 62
Insulin, 168–169
Integrated mid-IR system, 18, 19f
Intrinsic viscosity (η), 275
In vitro cytotoxicity, 170, 172t
In vitro digestion
　bioactive compound encapsulation, 242–255
　　case study, 246–252
　　chewing, 242–244
　　food structure and its protective role, 244–246
　　microcapsules in food matrices, 254–255
　　microencapsulation of food nutrients, 252–253
　　overview, 241–242
　　release of nutrient, 253–254
　　sodium alginate gel–starch mixtures in, 246–252
Iron, 43
IR radiation, 1
　basic principles governing, 4–5
　　Planck's law, 4–5
　　Stefan–Boltzmann law, 5
　　Wien's displacement law, 5
　effect on dehydration of
　　cereals, grains, and seeds, 14t–15t
　　fruits, 9t–10t
　　vegetables, 11t–13t
　in food processing, 5–16
　　advantages of, 3
　　equipment for, 18–22
　　IR baking, 6, 7t
　　IR blanching, 5–6
　　IR drying, 8, 9t–10t, 11t–13t, 14t–15t
　　IR microbial inactivation, 16
　　IR roasting, 8
　　IR thawing, 15–16
　　limitations of, 18
　　liquid foods, 16, 17t
Isoflavones, 33, 38

L

Laboratory-scale IR heating system, 20, 21f
Laboratory-scale OH, 49, 49f
Lab-scale infrared heating system, 22, 22f
Lactobacillus acidophilus, 281
L-ascorbic acid, guava purée, 113f
Layer-by-layer (LbL) electrostatic deposition technique, 150–152, 154, 156, 158, 168
Le Chatelier's principle, 125–126
Lethal effect, 220
Lignans, 33
Linear superposition model, 163–165
Lipases, 246, 254–255
Lipid, 61
　autoxidation of polyunsaturated, 63
　emulsification, 254
　matrix, 158
　oxidation, 61, 189–190
　preoxidation, 65
Lipophilic bioactive compounds, 151, 167
Lipophilic vitamins, deficiency syndromes, 84
Liposoluble (lipid-soluble) vitamins (A, D, E, and K), 39–40, 83
Liposomes, 154
　preparation of, 74
Liquid foods
　concentration of, 184, 184f
　IR processing of, 16, 17t
Loading capacity (LC), 159, 160t
Long chain n-3 polyunsaturated fatty acid (LC n-3 PUFA), 254–255
LSD test, 137
Lutein, 197, 227
Lycopene, 117, 197–198, 252

M

Magnesium, 42–43
Magnetostrictive transducers, 132

Index

Maillard reaction, 139, 203, 205
Mathematical models, 162–163
Matrix systems, 69
Median particle diameter (d_{50}), chewing cycles, 243
MEF. *See* Moderate electric field (MEF) process
Microbial inactivation, IR, 16
Microbiota, human, 167
Microbubbles, 132
Microcapsules, 68, 68f, 252
    in food matrices, 254–255
    method for producing, 252
Microencapsulated bioactive compounds, application, 66
Microencapsulation, 62, 65, 254
    of food nutrients, 252–253
Microfluidization, 74
Microparticles, morphologies of, 68, 68f
Microsphere, 68f, 253
Microwave heating, 6, 50–51
Minerals, 41–43
    calcium, 42
    iron, 43
    magnesium, 42–43
    zinc, 43
Minimal processing and OH, 43–46, 44f
Minimum flow rate (Q_{min}), 263
Moderate electric field (MEF) process, 32, 45, 50–51, 81, 92
Molecular chain entanglement, 263, 263f
    degree of, 267
    entanglement number for solution, 267, 268f
Mononuclear core and multishell microcapsule, 68f
Mouth, GI tract, 166
Mucoadhesive polymer, 169
Multilayer liposomes, 158
Multilayer nanocapsules
    after LbL deposition, 155f
    behavior in GI tract, 165–171
        changes in delivery properties, 169–171
        main features and challenges, 166–168
        uptake/absorption of bioactive compounds, 168–169
    as carriers of bioactive compounds, 156–160
        EE and loading capacity, 159, 160t
        encapsulation techniques, 157–159
    characteristics on toxicity, 169, 170f
    colloidal templates, 153–156
        hollow nanocapsules, 153–154
        liposomes, 154
        nanoemulsions, 154–155
    construction, 156
    development, 151–156
        advantages and limitations, 155–156
        colloidal templates, 153–155
    factors affecting layers' properties, 152–153
    LbL electrostatic deposition technique, 151–152
    future perspectives, 171, 173
    impact in release of bioactive compounds, 161–165
    insertion of enzyme inhibitors in, 169
    insulin, 168–169
    mucoadhesive, 169
    size of, 169–170
    step-wise formation of, 150
    transport routes in FAE, 168
    in vitro and *in vivo* toxicity studies of, 172t
Multiphase food system, 47

N

Nano-based delivery systems, 173
Nanocapsules. *See* Multilayer nanocapsules
Nanoemulsions, 154–155
    squalene, 155
    templates, 155, 158
Nanoencapsulation, 150, 275, 280–281
Nanofiber(s), 260, 275
    electrospun, 279, 282–283
    production, 263, 275
    SEM image of, 263f
Nanotechnology, food, 62
Natural antioxidants, 64–66
    benefits, 65
    free radicals, 65
    phenolic and polyphenolic compounds, 64, 64f
Natural controlled release system, 242–243
Natural foods, 32
Natural polymers, 269
Niacin (vitamin B3), 39–40, 84
Nonenzymatic browning reactions, 141, 203
Nonflavonoids, 33–34
    lignans, 33
    phenolic acids, 33–34
    stilbene, 33–34, 35f
Non-nutritional substances, 133
Nonthermal cell disintegration. *See* Pulsed electric field (PEF)
Nonthermal technologies, 44, 96
Novel nanoencapsulation method, 275
Nutrients
    bioavailability of, 255
    control release of, 253–254
    delivery rate of, 242
    hydrophilic, 253–254
    microencapsulation of food, 252–253
Nutrition-related diseases, 150

O

OD. *See* Osmotic dehydration (OD)
OH. *See* Ohmic heating (OH)

Index

Ohmic heating (OH), 43–50, 80–81
　advantages and limitations, 50–51
　ascorbic acid degradation, 93f
　continuous-flow, 49–50, 50f
　defined, 32, 43, 80
　effect of electric field
　　frequency, 92–93
　　strength, 90–92
　electrochemical reactions in, 87
　in food industry, 43
　future perspectives, 55
　impact in protection of bioactive compounds, 54–55
　laboratory-scale, 49, 49f
　minimal processing and, 43–46, 44f
　OD/VI and, 46–50
　　advantages, 47
　　applications, 48–49
　　disadvantages, 47
　　DRP, 47, 48f
　　HDM, 47–48, 48f
　　pretreatments in osmodehydrated apple, 49
　schematic diagram of, 43f
　shelf life effect on, 93–94
Omega-3 fatty acids, 260
One-way ANOVA, 137
o-quinones, 53, 203
Oral bioavailability of bioactive compounds, 171
Orange juice skimmed milk beverage, 134–135
　ascorbic acid values for, 137, 138t
　TAC and vitamin C of, 144f
　TAC values in, 140t, 142
　total phenolics content in unprocessed, 139
　TPC values in, 143t
Oregano and sage extracts, encapsulated antioxidants, 74
Osmotic
　agent, 199–200
　dehydration, 196, 199–201
　pretreatment effect on bioactive compounds, 199–202
　　additional mechanisms, 201–202
　　agitation and material geometry, 200–201
　　concentration and type of agent, 199–200
　　temperature, 201
　solution, 199–201
　treatment, 199, 201–202
Osmotic dehydration (OD), 32
　OH and, 46–50
　　advantages, 47
　　applications, 48–49
　　disadvantages, 47
　　DRP, 47, 48f
　　HDM, 47–48, 48f
　　pretreatments in osmodehydrated apple, 49
　principle, 196
Ostwald–de Waele parameter, 249, 249t

Oxidation in vitamins, 86
Oxidative degradation of lipids, 61

P

Pantothenic acid (vitamin B5), 40, 84
Parenchyma tissue, fresh apples, 52, 53f
Pasteurization effect, 220, 234
Peanut kernel, encapsulated antioxidants, 74
Pectin, 252–253
PEF. See Pulsed electric field (PEF)
Permeabilization process, 46, 81–82
PFC. See Progressive freeze concentration (PFC)
Phenolic acids, 33–34, 64, 64f, 204
Phenolic compounds, 66, 134, 198–199, 207
　drying temperature effect on, 204–205
　in high-pressure-treated fruit purées, 118–120, 119t, 120f
　hydrostatic high pressure treatments on, 229–231, 230t
　natural antioxidants, 64, 64f
　synthetic antioxidants, 63, 63f
pH, multilayer nanocapsules, 152
Phylloquinone (vitamin K), 39–40, 83–84
Phytochemical, 222
Piezoelectric USN generator, 132
Planck's law, 4–5
Polarization, 46, 92
Poly(D, L-lactide) (PDLLA), 265, 281
Polyelectrolytes, 151–152
　characteristics, 153
　nanocapsule shell, 158
　selection of, 153
Polyelectrolytes delivery system (PDS), 168
Poly(ethylene oxide) (PEO), 264, 279, 281
Poly-L-lysine (PLL) release, 164, 165f
Polymer
　liquid, 261
　solution properties, 266–270
　　chain entanglements, 275
　　concentration, molecular weight, and fiber morphology, 267–269
　　effect of solvents, 270
　synthetic, 280
Polymeric matrix, properties of, 163
Polynuclear core and homogeneous shell microcapsules, 68f
Polyphenol, 32–38
　application, 62
　classification, 33f
　defined, 32
　flavonoids, 33–34, 36–38
　　anthocyanidins and anthocyanins, 33, 36, 36f
　　flavanols, 33, 36–37, 37f
　　flavanones, 33, 37
　　flavones, 37–38

Index

  flavonols, 33, 37, 38f
  generic structure, 35f
  isoflavones, 33, 38
  nonflavonoids, 33–34
    lignans, 33
    phenolic acids, 33–34
    stilbene, 33–34, 35f
  structures of, 33f
  tea, 71
  from yerba mate, 72
Polyphenolic compounds, 67, 224
  natural antioxidants, 64, 64f
Polyphenoloxidase (PPO), 204, 226
  enzymatic browning by, 53–54, 203, 229
  enzyme inactivation, 226, 231, 233, 235
Polyunsaturated lipids, autoxidation of, 63
Polyvinyl alcohol (PVA) nanofiber, 280–281
Potential toxicity effects, 169–171
Power law model, 163, 272
PPO. *See* Polyphenoloxidase (PPO)
Pressure-assisted thermal processing (PATP), 220
Probiotics, encapsulation of, 280–281
Progressive freeze concentration (PFC), 188, 190
Propyl gallate (PG), 63, 63f
Proteins. *See specific proteins*
Protein-based encapsulation material, 277–279
Protein–polysaccharide complex, 253
Provitamin A, 83, 197, 206
Pseudo-first order, 124
Pulsed electric field (PEF), 81–82
  effects of, 94–98
    electric field strength, 94–96
    shell life, 96–98
  electrochemical reactions in, 87–88
  process parameters, 94
Pumpkin
  carotenoids in, 227
  effect of HPT treatment, 235
  model matrices, 222–224, 223f–224f
Pyridoxal (vitamin B6), 39–41, 84
Pyridoxal phosphate (PLP), 84
Pyridoxamine/pyridoxine (vitamin B6), 39–41, 84

R

Radiation laws, 4
Reactive oxygen species (ROS), 43, 65, 83, 85
Red plum
  anthocyanins in, 225
  cyanidin, 225
  model matrices, 222–224, 223f–224f
Release-tailored bioactive compounds, 161
Reservoirs, 68–69
Response surface plot, AA loss, 114f
Retinoid, 83
Retinol (vitamin A), 39, 68, 83, 96, 197

Riboflavin (vitamin B2), 39–40, 84–85, 97, 253–254
Roasting, IR, 8
ROS. *See* Reactive oxygen species (ROS)
R-square, starch, 249

S

Safe delivery system, electrospinning, 276–281
  carbohydrate-based encapsulation materials, 279–280
  factors to, 276–277
  particle characteristics, 277
  probiotics encapsulation, 280–281
  protein-based encapsulation materials, 277–279
Sage and oregano extracts, encapsulated antioxidants, 74
Saliva, 166–167
Salt, multilayer nanocapsules, 152–153
Saturation method, 152, 155
Secondary antioxidants, 66–67
Secondary metabolites, 106
Seeds, dehydration of, 14t–15t
Semiempirical model, 163, 267
Shelf life effect on
  OH, 93–94
  PEF, 96–98
Silica particles, colloidal, 154
Sodium alginate gel–starch mixture, 246–252
  biopolymer properties in, 247
  cryoscopic point depression in, 250–251, 250f–251f
  effect of, 247, 248f
  emulsions responses in, 247
  Ostwald–de Waele parameters for, 249, 249t
  protector effect of, 249
  rheological and textural properties of, 247
Sodium tripolyphosphate ionic crosslinking technique, 71
Sonication cavities, 139
Source, encapsulation of antioxidants, 62–66
  natural antioxidants, 64–66, 64f
    benefits, 65
    free radicals, 65
    phenolic and polyphenolic compounds, 64, 64f
  synthetic antioxidants, 62–64, 63f
    phenolic compounds, 63, 63f
    toxicological behaviors, 64
Soy protein isolate (SPI) microsphere, 253
Spinneret orifice, 273
Spray-drying method, 69, 252
Squalene nanoemulsions, 155
Starch
  apparent viscosity, 249–250, 250f
  chemically modified, 251

Starch (*Continued*)
  granules, 245, 248–249, 251–252
  hydrolysis, 243–244, 243f, 247–251, 248f
  Ostwald–de Waele parameters for, 249, 249t
  sodium alginate on, 246–252
Statgraphics Plus 5.0, 137
Stefan–Boltzmann law, 5
Stilbene, 33–34, 35f
Surface tension
  beads, 265
  of polymer solution, 262, 273–275
Suspension crystallization method, 186
Synthetic antioxidants, 62–64
  phenolic compounds, 63, 63f
  toxicological behaviors, 64

T

TAA. *See* Total antioxidant activity (TAA)
TAC. *See* Total antioxidant capacity (TAC)
Taylor cone structure, 261–263, 271, 273
TCD. *See* Tip to collector distance (TCD)
Tea polyphenol, 71
Tertiary butylhydroquinone (TBHQ), 63, 63f
Thawing, IR, 15–16
Thermal technologies, OH, 44
Thermoplastic flour (TPF), 74
Thermoreversible polymer network, 246
Thiamine (vitamin B1), 39–40, 84–85
Tip to collector distance (TCD), 271, 273–274, 274f
Tocopherol (vitamin E), 39–40, 64, 64f, 66, 83, 85
Total antioxidant activity (TAA), 231, 233–234
Total antioxidant capacity (TAC), 133–134, 136–137, 140t
  bioactive compounds and, 143–144
  correlation between vitamin C and, 144f
  impact of USNs on, 142–143
Total phenolic compounds (TPC), 229, 230t, 234
  in orange juice skimmed milk beverage, 143t
Total phenolics, 136
  HPP effect on, 118–120
  impact of USNs on, 139
TPC. *See* Total phenolic compounds (TPC)
Trace elements and function, 42t
Transducer excitation signal, 132
Treatment time, PEF, 94–96
Trolox (6-hydroxy-2,5,7,8-tetramethylchroman-2-carboxylic acid), 135–137
Two-stage IR treatment of wheat, 16

U

Ultrafine fibers, 268, 280
Ultrahigh pressure (UHP) process. *See* High pressure process (HPP)

Ultrasonics treatment system, 135–136
Ultrasonification, 74
Ultrasound (USN) technology, 131
  cavitation, 132
  defined, 132
  effectiveness of, 132
  in food processing, 133
  materials and methods, 134–137
    chemicals and reagents, 135
    sample, 134–135
    statistical analysis, 137
    TAC, 136–137
    total phenolics, 136
    ultrasonics treatment system, 135–136
    vitamin C determination, 136
  to preserve
    fluid foods, 132
    nutritive fluid foods, 133
  processing equipment, 132, 133f
  results and discussion, impacts, 137–144
    on ascorbic acid content, 137–139
    bioactive compounds and TAC, 143–144
    on color, 139–142
    at different amplitude, 141f, 142f
    on TAC, 142–143
    on total phenolics, 139
  treatments, 135t, 136, 139–140
Ultrathin uniform PVP fibers, 270
Uptake/absorption of bioactive compounds, 168–169
USN. *See* Ultrasound (USN) technology

V

Vacuum impregnation (VI), 32, 52
  OH and, 46–50
Vegetable matrix
  as carriers of bioactive compounds, 51–54
    enrichment with, 51–53
    enzymatic inactivation, 53–54
  models, 222–224
Vegetables, dehydration of, 11t–13t, 18
Vibrating unit, dryer, 18, 20f
Vitamin A (retinol), 39, 68, 83, 96, 197
Vitamin B (folic acid), 39, 52, 279
Vitamin B1 (thiamine), 39–40, 84–85
Vitamin B2 (riboflavin), 39–40, 84–85, 97, 253–254
Vitamin B3 (niacin), 39–40, 84
Vitamin B5 (pantothenic acid), 39–40, 84
Vitamin B6 (pyridoxine/pyridoxal/pyridoxamine), 39–41, 84
Vitamin B12 (cobalamin), 39, 41, 85
Vitamin C (ascorbic acid), 39–40, 84–86, 196–197. *See also* Ascorbic acid (AA)
  air velocity, 207
  as antioxidant, 197

Index

degradation, 87–89, 95, 110, 126
determination method, 136
DHAA, 206–207
effect of drying temperature, 205–206
features, 205–206
human health, 197
oxidation of ascorbic acid, 206–207
Vitamin D (calciferol), 39, 83, 260
Vitamin E (tocopherol), 39–40, 64, 64f, 66, 83, 85
Vitamin K (phylloquinone), 39–40, 83–84
Vitamins, 38–41, 133–134
    B complex, 40–41
    biotin, 39, 41
    choline, 39, 41
    classification, 39f, 83f
    defined, 38, 82
    degradation mechanisms, 85–89
    electric field treatments
        degradation mechanisms, 85–89
        future trends, 98
        stability in fruits pulp, 89–98
    HPP-treated fruit purée, 111–114
    liposoluble (lipid-soluble) vitamins (A, D, E, and K), 39–40, 83
    stability in fruits pulp, 89–98
        ohmic heating, 90–94
        pulsed electric fields, 94–98
    structure and properties, 82–85
        fat-soluble, 83–84
        water-soluble, 84–85
    vitamin C, 39–40

Volatile aromas, 189
Voltage, electrospinning, 270–271, 272f

W

Water-soluble bioactive compounds, 157
Water-soluble chitosan, 154
Water-soluble vitamins, 84–85
Wheat, two-stage IR treatment of, 16
Whey protein isolate (WPI) protein, 278
Whipping instability, 261, 263
Wien's displacement law, 5
WPC protein, 278, 281

X

Xanthan gum, 247, 249
Xanthophylls, 197, 227
Xerophthalmia, 84, 197

Y

Yerba mate (*Ilex paraguariensis*), 65
    encapsulation of liquid and freeze-dried, 69–70, 70f
    infusion, 190
    polyphenols from, 70, 72

Z

Zein, 278–279
Zinc, 43